Developments in Soil Science 9

FORMATION AND PROPERTIES OF CLAY—POLYMER COMPLEXES

Further Titles in this Series

1. I. VALETON
BAUXITES

2. I.A.H.R.
FUNDAMENTALS OF TRANSPORT PHENOMENA IN POROUS MEDIA

3. F.E. ALLISON
SOIL ORGANIC MATTER AND ITS ROLE IN CROP PRODUCTION

4. R.W. SIMONSON (Editor)
NON-AGRICULTURAL APPLICATIONS OF SOIL SURVEYS

5. G.H. BOLT (Editor)
SOIL CHEMISTRY (two volumes)

6. H.E. DREGNE
SOILS OF ARID REGIONS

7. H. AUBERT and M. PINTA
TRACE ELEMENTS IN SOILS

8. M. SCHNITZER and S.U. KHAN (Editors)
SOIL ORGANIC MATTER

Developments in Soil Science 9

FORMATION AND PROPERTIES OF CLAY-POLYMER COMPLEXES

B.K.G. THENG

Soil Bureau, Department of Scientific and Industrial Research,
Lower Hutt, New Zealand

ELSEVIER SCIENTIFIC PUBLISHING COMPANY
Amsterdam Oxford New York 1979

ELSEVIER SCIENTIFIC PUBLISHING COMPANY
335 Jan van Galenstraat
P.O. Box 211, 1000 AE Amsterdam, The Netherlands

Distributors for the United States and Canada:

ELSEVIER/NORTH-HOLLAND INC.
52, Vanderbilt Avenue
New York, N.Y. 10017

Library of Congress Cataloging in Publication Data

Theng, B. K. G.
Formation and properties of clay-polymer complexes.

(Developments in soil science ; v. 9)
Bibliography: p.
Includes index.
1. Clay minerals. 2. Organic geochemistry.
3. Soil chemistry. I. Title. II. Series.
QE389.625.T48 631.4'17 78-13704

ISBN: 0-444-41706-0 (Vol. 9)
ISBN: 0-444-40882-7 (series)

Printed in The Netherlands

To the memory of
MORICE FIELDES
In appreciation of his
integrity and vision

CONTENTS

PREFACE

This book has been in planning since 1968 when I first began to search and abstract the literature on clay—organic systems. The intention then was to produce a critical and comprehensive account of the many and varied aspects of clay—organic chemistry, embracing those reactions in which organic polymers were involved. I still had this idea in mind when the opportunity for writing such a review arose in 1971 with the announcement (by the London publishing house of Adam Hilger Ltd) of a competition for the Adam Hilger Prize 1972. It would not have been possible, however, for me to enter the competition without the active support and enthusiastic encouragement from the late Dr. M. Fieldes who was then Director of the Soil Bureau. It seems fitting, therefore, that I should dedicate this book to his memory.

On sifting through the published material, it soon became evident that the reactions of clays with organic polymers should not be discussed concurrently with those involving small, well-defined organic species. To do so would not only lead to excessive length but would also result in a certain loss of perspective since, although both systems share a great many features, complex formation between clays and polymers is controlled by variables which are essentially different from those encountered in the adsorption of organic micromolecules.

Once the decision was taken to treat each system separately, it seemed only logical to begin with summarising the extensive data on the behaviour of small organic compounds at clay mineral surfaces. This work came out in the form of a monograph (Theng, 1974) in which I stated that the clay—polymer interaction would form the subject of a separate book. What I have done here, therefore, is simply to complete the task I set out to do and keep the promise I made to myself. In writing the present review, I have been much encouraged by the kind comments which my peers have expressed with regard to my previous effort. Also, the fact that a synthesis of the data on clay—polymer systems has not been previously attempted, has provided me with the necessary inspiration at moments when I would have preferred to be in the laboratory.

The systematic study of clay—organic reactions, as a whole, is a young science since the essential crystallinity of the main groups of clay minerals and the macromolecular concept have not become generally accepted until relatively recently. However, complex formation between clays and the organic constituents of soils, has been recognised and described for over a century (Schloesing, 1874). Indeed, the clay—organic interaction has been implicated in the very origin of life on earth; certainly, this process is ". . . as vital to the continuance of life as, and less understood than, photosynthesis . . ." (Jacks, 1973).

Despite its late beginning much progress has been made over the last two decades in understanding the behaviour of polymeric materials at the solid/solution interface, research being stimulated by the importance and potential of the solid—polymer interaction in agriculture, biology, and industry. It can even be said that theory has outrun practice in the case of uncharged, linear homopolymers. The picture is less clear as regards polyelectrolytes which, of course, are more relevant than their non-ionic counterparts to soil and biological systems. For this reason, much of the data on the clay—polymer interaction, in general, are not amenable to a quantitative interpretation. Nevertheless, it seems timely to survey the advances that have so far been made in this subject and to outline the current state of knowledge. In this task I have been much assisted by the availability of a number of excellent reports on some of the topics under discussion. I have attempted, as much as possible, to present a balanced and coherent account of each topic, pointing out areas of discordance and uncertainty, and suggesting ways in which such discrepancies may be resolved. In doing so, I hope to have given something more than just an annotated catalogue of references. How well or otherwise I have succeeded in this direction is for the reader to judge.

The growing literature on clay—polymer systems has made it necessary to be selective as to what aspects and how much, of a given topic, are to be included if the size of the book is to be kept within manageable bounds without sacrificing essential depth and detail. The practical applications of the clay—polymer interaction, for example, are not discussed at great length because much of the information on this topic is published in the form of patents which are difficult to assess and collate. The rapid growth of the subject has also meant that some parts of the book, at least, will become out of date in a very short time. One also had to draw a line as regards the selection of published material which, apart from the few last minute additions, does not go beyond the end of 1976.

For the sake of clarity and convenience, I have divided the contents of the book into three parts. Part I gives an outline of clay minerals structures, the properties of aqueous clay suspensions, and the behaviour of polymers at clay and mineral surfaces. Part II describes the reactions of clays with synthetic polymers and Part III deals with complex formation between clays and various classes of naturally occurring polymers. I have tried to write each chapter in Parts II and III in the form of a review paper ending with a rather comprehensive list of references to the original literature. This was done for the benefit of those readers who may wish only to get quickly acquainted with the state of the art in a topic of their own specialty.

Although this book is directed primarily to soil scientists and agronomists, it may also appeal and be of interest to a variety of chemists who, at some time in their respective professional careers, have to deal with "environmental" problems in which clays, polymers, and their reaction products feature. Besides serving as a work of reference, the book may be found useful for teaching at the graduate and higher levels of agricultural and soil science courses.

It goes without saying that the reliance of a book of this type on previous publications extends to illustrated and tabulated data. In this connection, I wish to acknowledge the many authors, councils of scientific societies, and publishers who have kindly permitted me to reproduce figures and tables. Thanks are also due to the draughting staff of the Science Information Division of D.S.I.R. for drawing up the numerous diagrams. I am indebted to the library and typing staff of the Soil Bureau for their unstinting assistance, respectively, in procuring journal articles from other libraries and in converting a long stream of untidily written drafts into neat typescripts. I am very grateful to Dr. R.B. Miller, Director of the Soil Bureau, for allowing me to devote myself entirely to bringing this book to completion over an extended period of time. Lastly, I would like to thank my family for their forbearance and understanding.

Lower Hutt
December 1977

B.K.G. THENG

REFERENCES

Jacks, G.V., 1973. The biological nature of soil productivity. Soils and Fertilizers, 26: 147—150.

Schloesing, Th., 1874. Détermination de l'argile dans la terre arable. Comptes Rendus Hebdomadaires des Séances de l'Académie des Sciences, 78: 1276—1279.

Theng, B.K.G., 1974. The Chemistry of Clay—Organic Reactions. Adam Hilger Ltd, London, 343 pp.

Part I

CLAY MINERALS AND POLYMER ADSORPTION

TABLE 1.1

Classification scheme for the crystalline clay minerals and related phyllosilicates (from Mackenzie and Mitchell, 1966)

Class	Type	Formula unit charge (x)	Group	Series
Phyllosilicates or layer silicates	1 : 1	$\sim$0	Kaolinite	Dioctahedral
			Serpentine	Trioctahedral
	2 : 1	$\sim$0	Pyrophyllite—talc	Dioctahedral Trioctahedral
		$\sim$0.25—0.6	Smectite or Montmorillonite	Dioctahedral
				Trioctahedral
		$\sim$0.6—0.9	Vermiculite	Dioctahedral Trioctahedral
		$\sim$0.9	Illite	Dioctahedral Trioctahedral
		$\sim$1.0	Mica	Dioctahedral Trioctahedral
		$\sim$2.0	Brittle mica	Dioctahedral Trioctahedral
	2 : 1 : 1	Variable	Chlorite	Dioctahedral Di-tri-octahedral Trioctahedral
Pseudo-layer silicates [b]	(2 : 1)	$\sim$0.1	Palygorskite—sepiolite [c]	Di-tri-octahedral Trioctahedral

[a] Only the more common are listed. The nature of isomorphous substitution indicated refers to the predominant cation types involved.
[b] This is the name recommended by the Nomenclature Committee of Association International pour l'Étude des Argiles (AIPEA), Pedro (1970), replacing the term "chain-lattice silicates".
[c] The name "hormite" for this subdivision has been rejected by the AIPEA Nomenclature Committee.
[d] See text.

Chapter 1

THE CLAY MINERALS

1.1. STRUCTURAL ASPECTS

Clays are hydrous silicates or alumino silicates and may broadly be defined as those minerals which dominantly make up the colloidal fraction of soils, sediments, rocks, and waters. The crystalline nature of the majority of clay minerals is now generally accepted. It should be recalled, however, that this is a relatively recent concept, dating from the late nineteen-thirties. By this time the crystallinity and structures of the main groups of layer silicates related to clay minerals had been established, largely through the pioneering studies of such workers as Ross (1927), Hendricks (1929), Hendricks and Fry (1930), and Pauling (1930a, b).

In the soils literature, the term clay or clay fraction denotes a textural class of minerals consisting of particles with an equivalent spherical diameter of less than 2 μm, not all of which may be crystalline. Indeed, the finely divided, highly disordered minerals in soil are probably the most active in terms of their behaviour towards plant nutrients and organic constituents.

For a detailed discussion of the structures and properties of layer silicate minerals together with their identification by X-ray diffraction, infrared spectroscopy, and differential thermal analysis, the reader should consult the series of monographs written or edited by Brown (1961), Marshall (1964), Carroll (1970), Mackenzie (1970), Farmer (1974), Gieseking (1975), and Thorez (1975). A number of books dealing with some particular aspect of clay chemistry are also available, notably that by Van Olphen (1963) on the colloid chemistry, by Weaver and Pollard (1973) on the mineral chemistry, and by Theng (1974) on the clay—organic interaction.

The majority of the clay minerals belong to the class of layer silicates or phyllosilicates because their structural framework is basically composed of layers comprising silica and alumina sheets, joined together in varying proportions and stacked on top of each other in a certain way. A silica sheet has two planes of oxygen/hydroxyl ions, one of which consists of the bases and the other of the tips or apices of linked $Si(O, OH)_4$ tetrahedra. Similarly, an alumina sheet has an upper and a lower plane both consisting of hydroxyl ions between which is a plane of Al^{3+} ions, octahedrally coordinated to the hydroxyl groups (Fig. 1.1).

Condensation in a 1 : 1 ratio of a silica sheet with an alumina sheet gives rise to a 1 : 1, dimorphic or two-sheet mineral as in kaolinite (cf. Fig. 1.2). Here, the tips of the silica tetrahedra project into an hydroxyl plane of the octahedral sheet, replacing two-thirds of the hydroxyl ions. In a 2 : 1, trimorphic or three-sheet mineral, two silica sheets condense with an alumina

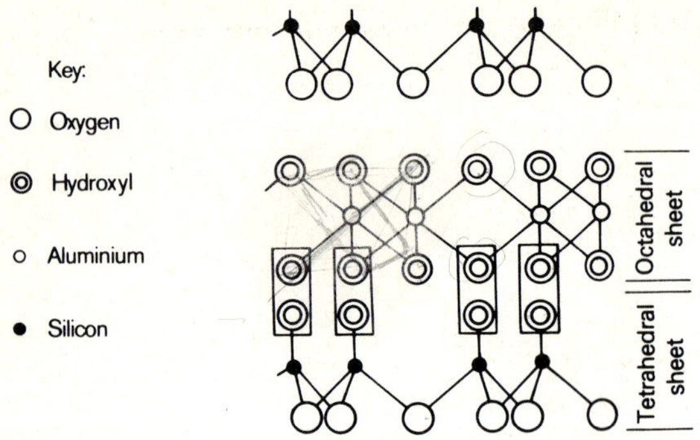

Fig. 1.1. Schematic representation of a silica tetrahedral sheet and an alumina octahedral sheet. The mode of condensation between the two sheets is indicated by the rectangles.

sheet. The resultant layer structure is such that the alumina sheet is sandwiched between two sheets of inward-pointing, linked silica tetrahedra, as exemplified by pyrophyllite (cf. Fig. 1.4). The possibility also exists that a trimorphic layer alternates regularly with an hydroxide sheet to form a 2 : 1 : 1, tetramorphic or four-sheet mineral, such as in chlorite. In the so-called pseudo-layer silicates ("chain-lattice silicates"), the trimorphic layers are arranged in chains or bands which are linked together through oxygen ions. Here, the silica sheets are continuous but the apices of the tetrahedra in adjacent chains point in opposite directions. The structures so formed have a corrugated surface with channels running parallel to the chain as in palygorskite (cf. Fig. 1.9). The clay fraction of soils, derived from volcanic ash, is mainly composed of highly disordered or non-crystalline hydrous alumino-silicates collectively termed allophane. In many instances, a gel-like material consisting of bundles of fine tubes, called imogolite, occurs in conjunction with allophane (Quantin, 1972).

Since the type of cation occupying tetrahedral and octahedral sites is limited more by ionic size and coordination than by valency, there is considerable scope for isomorphous replacement or substitution in these structures. The site and extent of this substitution are at the basis for distinguishing between groups, series, species, and varieties of the phyllosilicates. As indicated above, where Al^{3+} ions occupy octahedral positions, two out of three sites are filled, giving rise to dioctahedral minerals. In the trioctahedral minerals, 3 Mg^{2+} replace 2 Al^{3+} thereby filling all octahedral positions. The more common situation is the partial substitution of Si^{4+} and Al^{3+} in tetrahedral and octahedral sites, respectively, for cations of similar size and coordination but of different (usually lower) valency. Such a process gives rise to structures with a permanent net negative charge. Apart from the small compensating effect of internal substitution, this positive charge deficiency is

balanced by sorption of extraneous cations which may or may not be exchangeable.

The charge (x) per formula unit or half layer-unit-cell which ranges from ~0 to ~2 electron charges, is therefore an important parameter in the classification of the phyllosilicate and related minerals. Another criterion entering the scheme is the manner in which successive layers are stacked within a crystal. Thus, many of the groups listed in Table 1.1, have a number of polymorphs or structural varieties depending on the relative order or disorder of layer stacking. As might be expected, a crystal may also consist of different types of layers which are either randomly or regularly interstratified. Many pedogenic clay materials have this kind of structural arrangement.

In the structures just described, the symmetry is basically hexagonal. In practice, however, there is some departure from this ideal. Thus, the arrangement of the surface oxygens of the silica sheet, for example, is more ditrigonal than hexagonal (e.g. Radoslovich, 1975). This is ascribed to the opposed rotation of alternate silica tetrahedra which, in addition, may be slightly tilted with respect to the plane of surface oxygens. The extent of rotation and tilting is influenced by such factors as the geometric fit between the larger tetrahedral and the smaller octahedral sheets, the amount and site of isomorphous substitution, and the nature of the charge-balancing cation.

The classification scheme for the phyllosilicates proposed by the Nomenclature Committee of the Clay Minerals Society (Bailey et al., 1971) is now widely accepted. However, we shall adopt the earlier scheme of Mackenzie and Mitchell (1966) here because it seems more appropriate to our purposes. Thus, this scheme, set out in Table 1.1, includes the pseudo-layer silicates (Pedro, 1970) and also allots a place for illite or hydromica, a group of min erals which have been used as adsorbents of organic polymers. Of the oth minerals listed, those belonging to the kaolinite and smectite (or mor morillonite) groups are of particular relevance. Indeed, the bulk of the d on the clay—polymer interaction relates to montmorillonite.

Allophane and imogolite which are not included in the phyllosilicate c have also featured as adsorbents mainly of proteins (Milestone, 1971) humic substances (Wada and Inoue, 1967; Inoue and Wada, 1971). R ences to chlorite, mica, vermiculite and the palygorskite—sepiolite gro minerals are rare (Orlov et al., 1973; Kodama and Schnitzer, 1974). brief mention will therefore be made of these minerals here.

1.1.1. The kaolinite group

The layer structure of kaolinite is shown in Fig. 1.2. The indica thickness of ~0.72 nm also represents the basal or $d(001)$ spacing nite. When viewed under the electron microscope kaolinite appea or less well-defined, hexagonal platelets, ranging in thickness fr

Species	Structural varieties [a]	Chemical varieties [a]
Kaolinite	Kaolinite Nacrite Dickite *b*-Axis disordered kaolinite	
Halloysite		Metahalloysite (dehydrated) [d]
Antigorite		Amesite (Fe^{2+})
Chrysotile		Cronstedtite (Fe^{2+}, Fe^{3+})
Pyrophyllite		
Talc		Minnesotaite (Fe^{2+}, Fe^{3+}, Al^{3+}, H^{+})
Montmorillonite		Beidellite (Al^{3+} for Si^{4+}) Nontronite (Fe^{3+} for Al^{3+}) Volkonskoite (Cr^{3+} for Al^{3+})
Saponite		Sauconite (Zn^{2+} for Mg^{2+}) Hectorite (Li^{+} for Mg^{2+})
Vermiculite		
Vermiculite		
Illite		
Ledikite		
Muscovite	2M, 3H, 1M	Paragonite (Na^{+} for K^{+})
Phlogopite	1M, 2M, etc	Biotite (Fe^{3+} for Al^{3+}) Lepidolite (Li^{+} for Al^{3+}) Zinnwaldite (Li^{+}, Fe^{2+})
Margarite		
Clintonite		
Donbassite		
Sudoite		
Pennine		
Clinochlore		
Palygorskite		
Sepiolite		Xylotile (Fe^{3+} for Al^{3+})

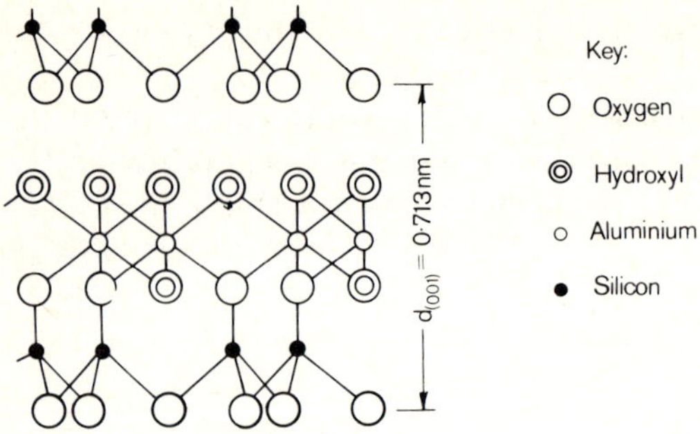

Fig. 1.2. The layer structure of kaolinite viewed along the *a*-axis. The basal or *d*(001) spacing of ~0.7 nm is equal to the layer thickness.

2 μm. A single crystal must therefore be made up of numerous such layers which are continuous in the *a*- and *b*-directions and stacked in a given order along the *c*-axis. The inability of kaolinite crystals to show interlayer (intra-crystalline) expansion or swelling is indicative of strong interlayer bonding. The forces holding the layers together have been attributed of O—H...O hydrogen bonding (involving 4 out of 6 hydroxyls per unit cell) through the superposition of oxygen and hydroxyl planes of successive layers, amplified by Van der Waals attraction. This view, however, has been challenged by Cruz et al. (1972) who considered that the interlayer attraction is largely electrostatic in origin. Whatever the underlying mechanism may be, kaolinite can only be induced to swell under special conditions. This occurs, for example, when the crystal is immersed in some highly polar, proton-accepting organic liquids and then only slowly and usually incompletely (see Theng, 1974 for references).

Summation of the anionic and cationic charges in the structure as depicted in Fig. 1.2, shows that the kaolinite layer is electrically neutral. In reality, however, most if not all kaolinites carry a net negative surface charge, the origin of which is still a matter of some controversy. The problem relates to the fact that any isomorphous replacement which might occur is limited in extent and so cannot be readily deduced from analytical chemical data. By the same token, the negative charge arising from this source may easily be masked or swamped by small amounts of surface-adsorbed impurities.

Ferris and Jepson (1975), for example, have rationalised their counterion adsorption results for a Cornish kaolinite in terms of the presence at the surface of an alumino-silicate gel coating. The weight of evidence, however, favours the concept that the pH-independent negative charge, which forms a substantial part of the total charge, is ascribable to isomorphous substitution within the structure (e.g. Bolland et al., 1976). In addition, kaolinite may carry a small amount of pH-dependent negative charge due to exposed silica as well as positive charges on the edge or cleavage face of the crystal ascribable to exposed alumina.

It is widely recognised that the edge surface of kaolinite and of layer silicates, in general, may acquire negative and positive charges depending on the pH of the medium (Van Olphen, 1963; Swartzen-Allen and Matijević, 1974). The former may arise through dissociation of silanol groups under alkaline conditions and the latter through protonation of edge aluminols at acid pH values. These points are illustrated in Fig. 1.3 which is based on the early work by Schofield and Samson (1953, 1954).

This view which is supported by subsequent measurements, is also borne out by the recent rheological data of Rand and Melton (1975) using aqueous suspensions of a sodium-saturated Cornish kaolinite. Thus, the plots of Bingham yield values against pH (in the presence of NaCl of varying concentrations) intersect at pH = 7.3, a value which they suggest represents the apparent isoelectric point of the edge surface. This value, however, would be expected to vary slightly among kaolinite samples depending on the mode of their preparation. The amphoteric nature of the crystal edge surface has an important bearing on the colloidal properties of clay suspensions as we shall see later on.

It is interesting to note that the net negative charge (at pH 7) of a number of kaolinites ranges from 1.4 to 3.6 meq/100 g corresponding to 1.1–1.5×10^{-7} meq cm^{-2} which is of the same order of magnitude as for montmorillonites (Peigneur et al., 1975; Bolland et al., 1976). Bolland et al. (1976) have pointed out that if much of this charge arises from isomorphous substitution, it must reside quite closely to the cleavage surface of the crystal. This accords with the results of other studies using widely different techniques (Hofmann et al., 1961; Weiss and Russow, 1963; Follett, 1965; McBride, 1976) in that the cation exchange sites appear to be confined to the tetrahedral basal surface of kaolinite rather than being uniformly distributed throughout the crystal.

Another noteworthy feature is that the kaolinite edge surface occupies an appreciable proportion (10–20%) of the total crystal area of some 15–40

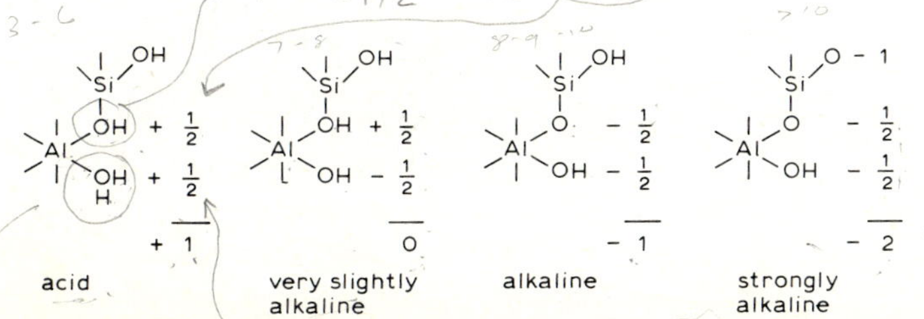

Fig. 1.3. Variations in charge characteristics of the edge surface of kaolinite-type crystals under acid and alkaline conditions (after Schofield and Samson, 1953).

$m^2\ g^{-1}$ whereas in the expanding 2 : 1 type phyllosilicates, such as montmorillonite, this value seldom exceeds 5%. The influence of the crystal edge on various surface chemical reactions, such as anion uptake (e.g. Quirk, 1960) as well as colour and polymerisation reactions (Solomon, 1968; Theng and Walker, 1970; Theng, 1971) is therefore more pronounced with kaolinite than with montmorillonite.

Halloysite is also made up of dimorphic layers essentially identical to those of kaolinite, but the individual layers are, or have been, separated by water. From a survey of the literature, Churchman and Carr (1975) have concluded that this intercalation of water in its natural state or during its past history is the single, most characteristic feature of halloysite and one which distinguishes it from kaolinite.

It is known that halloysites can occur with varying degrees of hydration, ranging from the fully hydrated variety with a unit formula of $Al_2Si_2O_5(OH)_4 \cdot 2\ H_2O$ and a basal spacing of ~1 nm to the fully dehydrated form having the formula $Al_2Si_2O_5(OH)_4$ and a spacing of ~0.72 nm. The latter which is the most stable form, thermodynamically, has been termed "metahalloysite" (MacEwan, 1947). Churchman and Carr (1975) have advocated the use of prefixes, such as hydrated, partially hydrated, and nonhydrated.

As compared with kaolinite, the stacking of successive layers in halloysites is disordered both in the *a*- and *b*-directions. In addition, many halloysites have a tubular morphology when viewed under the electron microscope. Here again, however, there is a continuous gradation both in degree of mutual layer displacement and in morphological properties ranging from the ordered, well-crystallised, platy kaolinites to the disordered, tubular halloysites (Grim, 1968; Weaver and Pollard, 1973; Churchman and Carr, 1975).

Since the layers in (hydrated) halloysite are already separated, a wide variety of organic compounds can be intercalated. Although this process essentially involves the displacement of the interlayer water by the organics, the underlying mechanisms are still poorly understood. There is also a difference between halloysite and montmorillonite in their respective ability to form interlayer complexes in that halloysite is not known to intercalate more than a single layer of organic micromolecules (MacEwan, 1948). Its behaviour towards polymeric compounds has not been examined in much detail but it seems unlikely that organic macromolecules can penetrate the interlayer space of halloysite; their adsorption will mostly be confined to external crystal surfaces.

1.1.2. The smectite group

In this group the dioctahedral montmorillonite as a mineral species is undoubtedly the most important and widely used adsorbent for organic compounds. Its reactions with small, well defined organics have been described in some detail by Theng (1974). As will be seen in Parts II and III of the present monograph, montmorillonite has also been used extensively as the adsorbing substrate for organic polymers.

Montmorillonite owes this attention to its ability to show extensive interlayer expansion or swelling which, in turn, is related to its peculiar structure. By this means, a large active surface area (700—800 $m^2\ g^{-1}$) is exposed, allowing an enormous range, both in number and variety, of guest molecules to be intercalated. Due to its high negative surface charge (cation exchange capacity) and its layer morphology, montmorillonite also serves as nature's own model for a platy, hydrophobic colloid of the "constant charge" type.

The layer structure of montmorillonite was deduced by Hofmann et al. (1933) on the basis of its similarity to that of pyrophyllite. As seen in Fig. 1.4 the pyrophyllite layer with a unit formula of $Al_2Si_4O_{10}(OH)_2$ is electrically neutral. The absence of isomorphous substitution means that there are no positively charged hydratable counterions in the interlayers and consequently, pyrophyllite does not show interlayer swelling (in water). Indeed, the energy required in order to separate the individual layers is quite considerable, being composed of Van der Waals (16.7 kcal mol^{-1}) and electrostatic (6.5 kcal mol^{-1}) attractive forces (Giese, 1975).

In terms of layer charge or x-value (Table 1.1) the brittle micas may be regarded as representing minerals at the other end of the scale. Here, there is extensive isomorphous substitution leading to a loss of swelling ability due to very strong electrostatic attraction between successive layers and the interlayer Ca^{2+} ions. It should be recalled that complete replacement of Al^{3+} by Mg^{2+} ions in the octahedral positions (i.e. 3 Mg^{2+} for 2 Al^{3+}) of pyrophyllite gives rise to the electrically neutral and hence, non-expanding talc structure.

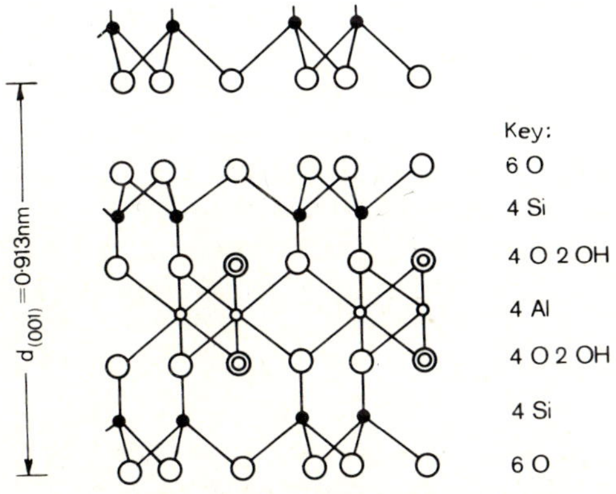

Fig. 1.4. The pyrophyllite layer structure viewed along the a-axis. The basal or $d(001)$ spacing and layer composition are also indicated.

For montmorillonite to have the high exchange capacity (70—150 me/100 g) and peculiar hydration/swelling properties that it has, requires a relatively low amount of isomorphous substitution ($x \sim 0.26$—0.60; Table 1.1) in the basic pyrophyllite layer structure. Another feature is that, unlike pyrophyllite, the stacking of successive layers in montmorillonite is more or less random (Marshall, 1935; Maegdefrau and Hofmann, 1937; Hendricks, 1942). The Hofmann—Endell—Wilm—Marshall—Maegdefrau—Hendricks structure, or Hofmann structure for short (as distinct from the Edelman—Favejee (1940) structure in which every alternate tetrahedron in the silica sheet is inverted) is now generally accepted.

The Hofmann structure is shown in Fig. 1.5 which indicates the partial substitution of Al^{3+} for other cations (e.g. Mg^{2+}, Fe^{2+}, Fe^{3+}) in octahedral sites and, less frequently, of Si^{4+} for Al^{3+} in the tetrahedral sheet of pyrophyllite (cf. Fig. 1.4). The average values of cation content, deduced from an analysis of a large number of montmorillonites—beidellites, are 1.49 and 0.19 for octahedral aluminium and iron, respectively, and 0.16 for tetrahedral aluminium, all being expressed on a per formula unit basis (Weaver and Pollard, 1973). The structural formulas of some common montmorillonite varieties are listed in Table 1.2. It should be noted that variations exist even between samples from a given source or locality. Thus, the net layer charge for the Otay-type samples resides almost entirely in the octahedral sheet whereas for the Chambers-type samples between 15 and 50% of this charge is in the tetrahedral sheet (Schultz, 1969).

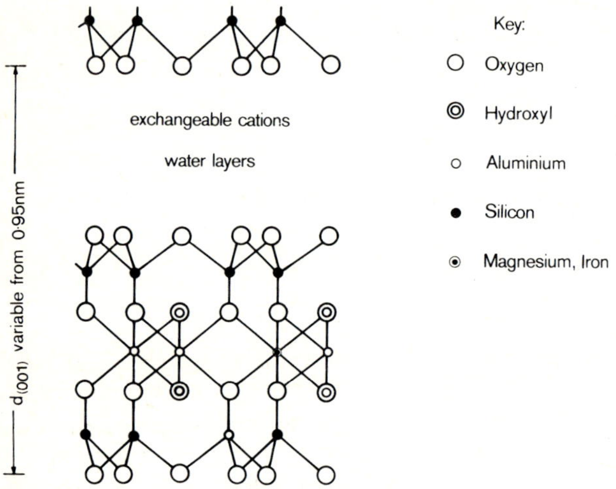

Fig. 1.5. The layer structure of montmorillonite according to Hofmann, Endell, Wilm, Marshall, Maegdefrau and Hendricks (see text), showing the occurrence of isomorphous replacement and the occupancy of the interlayer space by exchangeable cations and water.

TABLE 1.2

Structural formulas of some montmorillonite types from the United States of America, based on the analyses by different authors (from Weaver and Pollard, 1973, pp. 69 and 73)

Type	Locality	Formula
Clay Spur	Wyoming	$(Al_{1.54}Fe_{0.16}Mg_{0.33})(Si_{3.91}Al_{0.09})O_{10}(OH)_2$
Belle Fourche	South Dakota	$(Al_{1.61}Fe_{0.18}Mg_{0.23})(Si_{3.87}Al_{0.13})O_{10}(OH)_2$
Upton	Wyoming	$(Al_{1.55}Fe_{0.21}Mg_{0.26})(Si_{3.88}Al_{0.12})O_{10}(OH)_2$
Bentonite	Wyoming	$(Al_{1.56}Fe_{0.03}Mg_{0.19})(Si_{3.85}Al_{0.15})O_{10}(OH)_2$
Polkville	Mississippi	$(Al_{1.39}Fe_{0.10}Mg_{0.58})(Si_{3.94}Al_{0.06})O_{10}(OH)_2$
Chambers	Arizona	$(Al_{1.37}Fe_{0.19}Mg_{0.46})(Si_{3.84}Al_{0.16})O_{10}(OH)_2$
Otay	California	$(Al_{1.32}Fe_{0.06}Mg_{0.71})(Si_{3.95}Al_{0.05})O_{10}(OH)_2$

A further point of note is that the layer charge, indicated by a given structural formula, should only be regarded as an average over the whole crystal because this charge may vary (between certain limits) from layer to layer. Indeed, such intracrystal charge variations appear to be more the rule than the exception in that relatively few montmorillonite samples display a homogeneous charge distribution. The experimental evidence for this view and its implications for the colloid stability of montmorillonite suspensions have been described by Lagaly and co-workers in a series of papers (e.g. Lagaly et al., 1972; Lagaly and Weiss, 1975).

In the 2 : 1 type phyllosilicates, only a relatively small proportion of the extraneous (inorganic) cations balancing the negative layer charge, is located at external crystal surfaces, the majority being present in the interlayer space (cf. Fig. 1.5). In montmorillonites and vermiculites, all of these cations are normally exchangeable for other species in solution. This property is related to the expandability of these minerals in water and aqueous electrolyte solutions. On the other hand, the crystals of illites, micas, and brittle micas do not swell in aqueous media so that the interlayer cations are either not exchangeable or can only be replaced by special treatments (Weiss, 1958, 1969; Mackintosh et al., 1971).

Besides the charge-balancing cations, water is normally present in the interlayer space where it tends to associate with the cations, forming hydration shells around them. Water apparently enters the interlayer space in an integral number of molecular layers depending on the relative humidity and the nature of the interlayer cation. Thus, for the small alkali ions (Li^+, Na^+) up to four water layers may be intercalated giving rise to a series of basal spacings from ~0.95 nm for the fully collapsed state to ~2.25 nm, corresponding to an interlayer separation or distance of ~1.30 nm, for the fully expanded form. Under optimum conditions, Li^+- and Na^+-montmorillonites in aqueous suspensions may, in fact, swell beyond the $d(001) \sim 2.25$ nm "barrier" to give interlayer separations of some tens of nanometres (Norrish, 1954). The swelling and colloidal properties of montmorillonite minerals will be described more fully in the following section. On the other hand, the basal spacing of montmorillonites saturated with divalent ions does not exceed ~1.9 nm, even in distilled water.

Water coordinated to the exchangeable cations plays an important part in the interaction of montmorillonite with small, uncharged organic compounds (Mortland, 1970; Theng, 1974). This is so because the polar groups of such compounds compete with water for the same ligand sites around the cation. Cation—dipole interactions of this type might also be expected to operate in the attachment of uncharged, polyfunctional organic polymers to the clay surface. Indeed, the translational entropy gained in the system in replacing many water molecules from the mineral surface by a single macromolecular chain, provides the driving force for polymer adsorption (Greenland, 1972).

From water vapour adsorption and infrared data on hectorite, Prost (1975) has been able to deduce the probable states of adsorbed water in function of relative humidity and the nature of the saturating counterions. At a relative humidity (R.H.) of ~0%, some water is trapped between the layers of hectorite, containing Na^+, K^+, and Cs^+ ions, as the layers collapse. For the sample saturated with Li^+, Mg^{2+}, Ca^{2+}, Sr^{2+}, and Ba^{2+} ions, however, the interlayer water is largely coordinated to the cation, forming part of the ion's inner or primary hydration shell ("solvation water"). At R.H. < 50% a monolayer of water molecules is present at the clay surface, being partitioned between the inner and outer hydration ("filling water") shells, and the external crystal surface. The two hydroxyls of a water molecule in the inner shell are not equivalent, one being involved in hydrogen bonding to a neighbouring molecule ($O—H_b$) and the other lying perpendicular to the silicate layer ($O—H_a$). At R.H. > 50%, the clay contains multilayers of water molecules. The solvation water is no longer distinguishable by means of infrared spectroscopy. In this state of hydration, the properties of adsorbed water probably resemble those of bulk water. These points are summarised in Table 1.3.

1.1.3. The vermiculite group

The close association of water molecules with the interlayer cations is more clearly demonstrable in vermiculite, the structure of which has been fully elucidated by Mathieson and Walker (1954) from single crystal X-ray data. Fig. 1.6 gives a schematic representation of the structure of a normal, hydrated, macroscopic vermiculite in the magnesium form ($d(001) = 1.456$ nm).

The interlayer water molecules are arranged in two sheets held together

TABLE 1.3

The probable states of adsorbed water in hectorite as influenced by the relative humidity and the nature of the exchangeable cation (from Prost, 1975)

Relative humidity (%)	Saturating (exchangeable) cation	
	Na^+, K^+, Cs^+	Li^+, Mg^{2+}, Ca^{2+}, Sr^{2+}, Ba^{2+}
~0	Trapped water	$M^{n+}\cdots O(H_a)-H_b$ solvation water
<50	$M^{n+}\cdots O(H_a)-H_b\cdots O(H_a)-H_b\cdots O(H_a)-H_b$ + $(H_2O)_x$ solvation water, filling water, external water	
>50	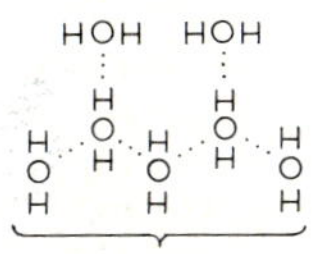 + $(H_2O)_x$ interlayer water, external water	

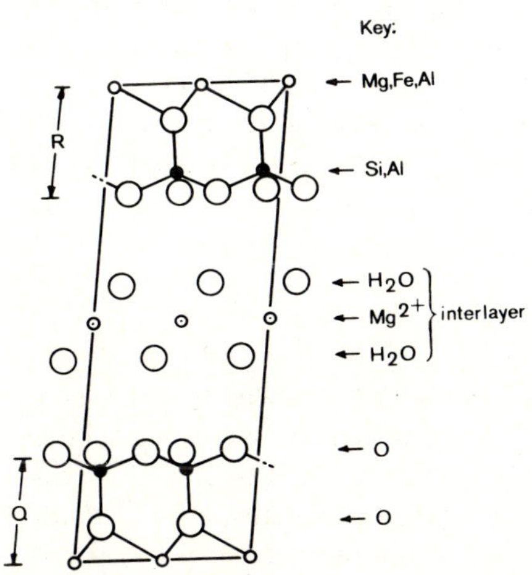

Fig. 1.6. The structure of hydrated magnesium-vermiculite according to Mathieson and Walker (1954). Q and R denote silicate half layers.

by Mg^{2+} ions which take up positions midway between two opposing silicate layers. The water sites are arranged in hexagonal array being determined by the configuration of the oxygen ions on the silicate surface. The mutual relationship of the water sheets is dictated by the requirement for octahedral coordination of the water molecules around the Mg^{2+} ion.

There is more isomorphous substitution in vermiculite than in montmorillonite, much of this occurring in the tetrahedral sheet. The structural formula of a sample from West Chester, Pennsylvania, given below is rather typical

$$(Mg_{1.92}Fe^{3+}_{0.46}Al_{0.22}Ti_{0.11}Fe^{2+}_{0.08})(Si_{2.72}Al_{1.28})O_{10}(OH)_2$$

Macroscopic vermiculites which are always trioctahedral do not occur in soils. However, fine-grained vermiculite-like minerals or clay vermiculites which may be either dioctahedral or trioctahedral, have been found in a number of soils (Walker, 1975).

1.1.4. The mica and chlorite groups

Like true vermiculites, macroscopic forms of mica and chlorite are, strictly speaking, not to be found in soils nor can they be regarded as clay minerals. However, fine-grained micaceous minerals occur in nearly all soils. Indeed, in terms of their abundance and distribution, these minerals are the most important soil constituents (Von Reichenbach and Rich, 1975). Clay chlorites have also been reported to occur in some soils.

The structure of an ideal muscovite may be written as $K\,Al_2(Si_3Al)O_{10}(OH)_2$, that is to say, the positive charge deficiency due to substitution of Al^{3+} for Si^{4+} in the tetrahedral sheet is balanced by interlayer K^+ ions, giving a layer thickness or basal spacing of 1 nm (Fig. 1.7). The K^+ ions are in twelve-fold

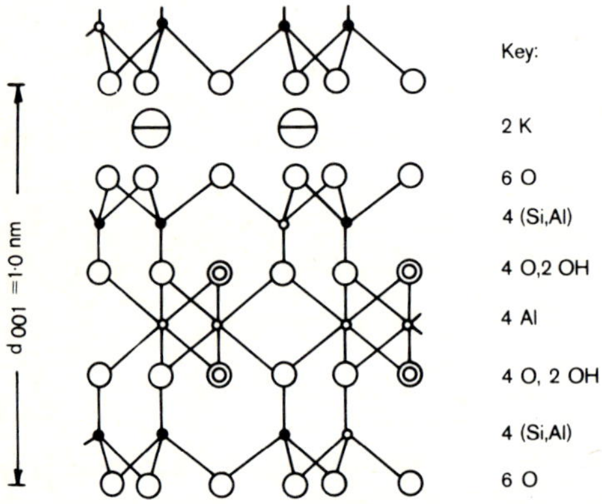

Fig. 1.7. The layer structure of muscovite viewed along the *a*-axis showing interlayer potassium ions balancing the negative layer charge.

coordination, situated centrally on the lines joining the centres of the ditrigonal network of oxygen ions of opposing silicate layers. Muscovite is dioctahedral whereas biotite ($K(MgFe)_3(Si_3Al)O_{10}(OH)_2$) and phlogopite ($KMg_3(Si_3Al)O_{10}(OH)_2$) are trioctahedral. In chlorite, the trioctahedral mica-like layers alternate regularly with a hydroxide layer, having the general composition of $(MgFe)_3(SiAl)_4O_{10}(OH)_2$ and $(MgAl)_3(OH)_6$, respectively, giving a repeat distance of ~1.42 nm. The charge deficiency in the mica-like layer is partly balanced by the positively charged hydroxide layer.

Many varieties and polymorphs exist in both the mica and chlorite groups (Bailey, 1975; Radoslovich, 1975).

Illites may be loosely defined as clay-sized micaceous minerals, occurring in soils and sediments, having a low degree of stacking order. They are characterised by a basal spacing of ~1 nm which does not increase on treating the mineral with ethylene glycol. Most illites are dioctahedral and have a lower potassium content than muscovite. Because of the higher water content as compared with the macroscopic micas, the terms "illite, "hydromica" or "hydrous mica" have been used interchangeably in the literature. The nature and origin of this "extra" water is not well understood; part of the surplus may be present as H_3O^+, replacing K^+ in interlayer positions, associated with chemically altered hydrous surface layers, or trapped in the interlayers (Wentworth, 1970).

The layer charge on illites is less than on micas. Much of the isomorphous substitution, like in micas, occurs in the tetrahedral sheet but some replacement of Al^{3+} for Mg^{2+}, Fe^{2+}, and Fe^{3+} takes place in octahedral positions.

As might be appreciated, many soil micas are admixtures or interstratified structures of illite, montmorillonite, and vermiculite. This fact would account for the marked variations in their composition and physico-chemical properties, such as exchange capacity and crystallinity. Similarly, some fine-grained or clay chlorites may be interstratified with montmorillonite-type layers, and swelling chlorites have been reported (Marshall, 1964).

These findings are hardly surprising since clay minerals in soils undergo a sequence of transformations and alterations induced by weathering. The rate and products of the weathering process are influenced by such factors as biological activity, climate, cultivation practice, parent material, and topography. The following scheme (Fig. 1.8) taken from Jackson (1959) indicates some of the possible sequences of layer silicate transformation in soil.

1.1.5. The palygorskite—sepiolite group

The name "attapulgite" is frequently used for palygorskite, and sepiolite is probably better known as meerschaum. The presence, in the structure, of channels or grooves which normally contain (zeolitic) water accounts for the

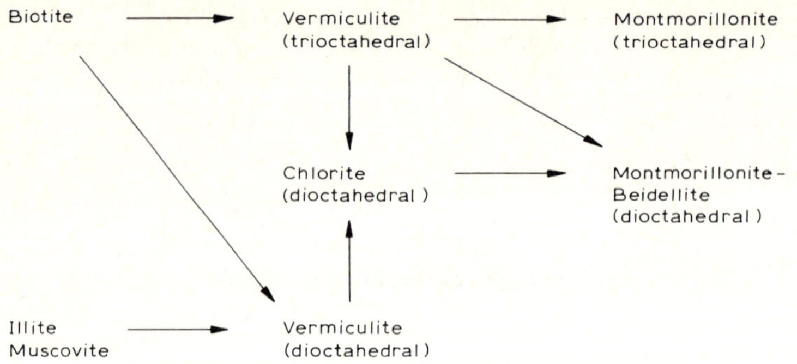

Fig. 1.8. Possible transformational or weathering sequences involving 2 : 1 type layer silicates in soil, as suggested by Jackson (1959).

high adsorptive capacity of the pseudo-layer silicates. For the same reason, palygorskite and sepiolite have found application in the pharmaceutical industry and a variety of industrial processes (Robertson, 1957; Theng, 1974).

Besides being characterised by the fibrous appearance under the electron microscope, the differential thermal analysis curves of these minerals are diagnostic for their occurrence in admixtures with layer silicates. However, palygorskite and sepiolite are found in few soils and then only under special conditions (Hénin and Caillère, 1975; Gorbunov and Shchurina, 1976; Yaalon and Wieder, 1976).

Some isomorphous substutition occurs in the structure, the resulting charge deficiency being normally balanced by sorption of Ca^{2+} ions. Where palygorskite is intermediate between dioctahedral and trioctahedral, sepiolite is apparently nearly purely trioctahedral. Another difference is that the unit cell dimension of palygorskite is smaller and its channels narrower as compared with sepiolite. A typical structural formula for these minerals (in the dehydrated form) is

$(Si_{7.99}Al_{0.01})(Al_{1.13}Fe^{3+}_{0.10}Mg_{3.02})O_{21}Ca_{0.10}$ for palygorskite and

$(Si_{11.8}Al_{0.16}Fe^{3+}_{0.04})(Fe^{3+}_{0.47}Fe^{2+}_{0.13}Mg_{7.14})O_{32}Ca_{0.11}$ for sepiolite.

The structure of sepiolite as proposed by Brauner and Preisinger (1956) is depicted in Fig. 1.9, indicating the sites and number of both bound and zeolitic water together with the approximate dimensions of the unit cell. On removal of the zeolitic water the bound or coordinated water molecules, located at the crystal edge, can enter into hydrogen bonding interactions with neighbouring oxygen or hydroxyl ions. This is indicated in Fig. 1.9 by the dotted lines (1) and (2). When one such water molecule at each site is lost on further heating, this type of intracrystal bridging is eliminated, allow-

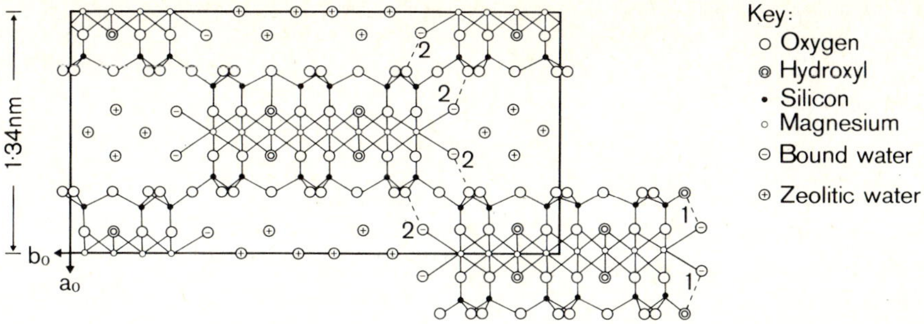

Fig. 1.9. The structure of sepiolite according to Brauner and Preisinger (1956). The unit cell dimension is indicated by the solid rectangle. At the edge of the mineral fibre, the bound water molecules can interact with either neightbouring silanol groups (1) or oxygen ions (2), as suggested by Serna et al. (1975).

ing the crystal to fold (Serna et al., 1975). This process which occurs at 154—175°C under vacuum, leads to a marked reduction in surface area and a collapse of the channel structure or micropores (Dandy and Nadiye-Tabbiruka, 1975). At ~500°C (under vacuum), a true anhydride is formed with little further structural change. This anhydride can rehydrate when exposed to water vapour at 60°C.

1.1.6. Allophane and imogolite

Allophane, either by itself or frequently together with imogolite, occurs in soils derived from volcanic ash where it is commonly the dominant material in the clay fraction (Mitchell et al., 1964; Quantin, 1972; Fieldes and Claridge, 1975). Because of their variable composition, allophanes, like illites should be regarded more as a series of minerals than as a mineral species. A tentative definition of allophanes is "members of a series of naturally occurring minerals which are hydrous aluminium silicates of widely varying chemical composition, characterised by short range order, by the presence of Si—O—Al bonds, and by a differential thermal analysis curve displaying a low temperature endotherm and a high temperature exotherm with no intermediate endotherm" (Van Olphen, 1971). Wada (1967) has suggested that the two end members of the series have the composition of 2 $SiO_2 \cdot Al_2O_3 \cdot$ 3 H_2O and $SiO_2 \cdot Al_2O_3 \cdot 2\ H_2O$. The structure of the material with a Si/Al ratio of 1/1 is thought to consist of a short silica tetrahedral chain joined to an alumina octahedral chain through oxygen ions. For the other end member with a Si/Al ratio of 1/2, a second alumina octahedral chain is added to the previous structure. Wada's structural scheme for allophane is shown in Fig. 1.10.

This concept, however, does not give due consideration to the presence

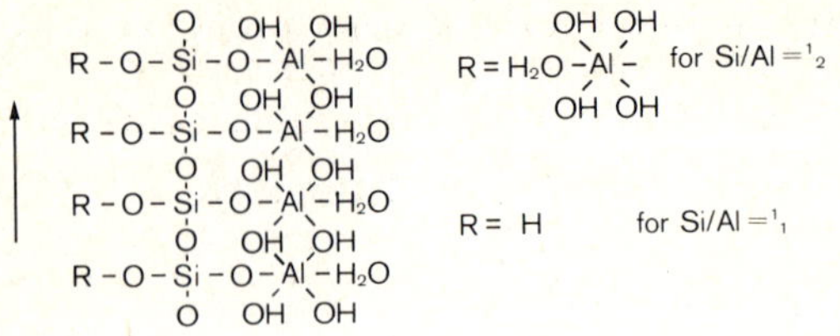

Fig. 1.10. The chain structure of allophane for two limiting Si/Al ratios, as proposed by Wada (1967). Arrow indicates the direction of chain propagation.

in allophane of aluminium in four-fold coordination for which there is a fair amount of experimental evidence (Fieldes and Claridge, 1975; Henmi and Wada, 1976). Nor does it take into account the fact that, under the electron microscope, most allophane samples appear as microaggregates made up of more or less spherical "unit particles" of about 5.5 nm in diameter (Kitagawa, 1971).

Several workers (e.g. Brindley and Fancher, 1969; Udagawa et al., 1969; Okada et al., 1975) have therefore proposed a sheet structure for allophane, somewhat akin to that of kaolinite. The observed curling to form a spherule may result from strains imposed by a misfit between the tetrahedral and octahedral sheets or by the uneven rate of their condensation. This would mean that the ratio of aluminium in four-fold (Al^{IV}) to that in six-fold (Al^{VI}) coordination may vary depending on the aluminium content, hydration state, and pH (Léonard et al., 1964).

Earlier, De Villiers and Jackson (1967) proposed a model consisting of a silica core in which some substitution of Si^{4+} for Al^{3+} occurred. The resulting excess negative charge is partly balanced by a positively charged hydroxy-Al^{VI} layer surrounding the core but a large portion of it is neutralised by sorption of exchangeable cations (Cloos et al., 1969).

A structural concept which perhaps comes closest to reconciling the chemical, morphological, and physical properties of allophane is that of a hollow spherule (Wada and Wada, 1977). As shown in Fig. 1.11, the outer diameter of such a unit particle of allophane is of the order of 5 nm. Its wall (~1 nm thick) is thought to consist of a defective kaolinite-like layer with more defects occurring in the Al-O, OH octahedral than in the Si, Al-O, OH tetrahedral sheet. The presence of such defects would account for the observed curling of the layer; where large clustered vacant sites occur, apertures are created through which small molecules (e.g. water) may enter, as indicated by the ease by which OH groups within the layer are deuterated. Substitution of aluminium for silicon in the tetrahedral sheet gives rise to a negative layer charge which may be partially compensated by sorption of protons. This together with changes in the coordination state of aluminium, induced by variations in the hydration and pH status of the system, would

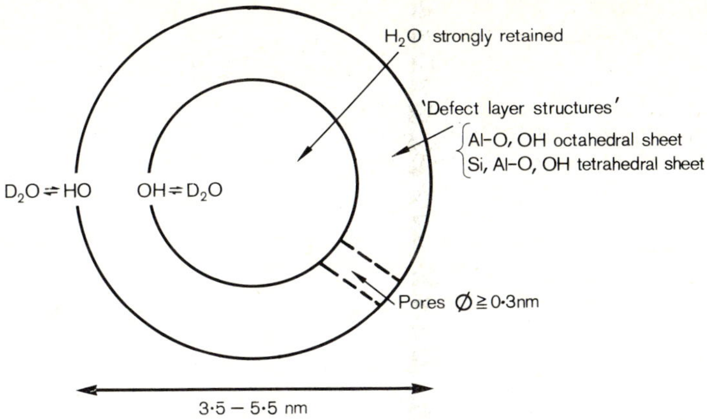

Fig. 1.11. Diagram showing the structure of a unit particle of allophane, as proposed by Wada and Wada (1977).

explain the cation exchange capacity data of allophane samples (Henmi and Wada, 1976).

Imogolite is paracrystalline and the essential features of its structure have been worked out by Cradwick et al. (1972) on the basis of chemical, infrared, and X-ray/electron diffraction analyses.

An imogolite bundle consists of many fine tubes each of which has a diameter of ~2 nm. The repeat distance along and perpendicular to the tube axis is 0.84 and 2.2—2.3 nm, respectively, the latter being the centre-to-centre tube separation in the desiccated material. Imogolite is basically made up of a gibbsite sheet of which an orthosilicate anion forms an integral part. This anion occupies the vacant octahedral site in the sheet, thereby displacing hydrogen from the three hydroxyl groups which surround such a site. The fourth Si—O bond of the anion is directed away from the sheet and picks up a proton to form SiOH. On this basis, the structural formula of imogolite is given by $(OH)_3Al_2O_3Si\,OH$.

The insertion of an orthosilicate anion in the gibbsite sheet, as proposed, leads to a contraction of the O—O distances around the octahedral site from 0.32 nm (in gibbsite) to less than 0.3 nm (in imogolite) and to a shortening in repeat distance from 0.86 to 0.84 nm. These changes could conceivably cause the gibbsite sheet to curl (Fig. 1.12) giving rise to the observed tubular morphology of imogolite.

The high adsorptive capacity of allophane and imogolite for humic and fulvic acids in volcanic ash soils is related to the porous structure of these minerals. Soil polymers may even be capable of penetrating the imogolite tubes (cf. Chapter 12).

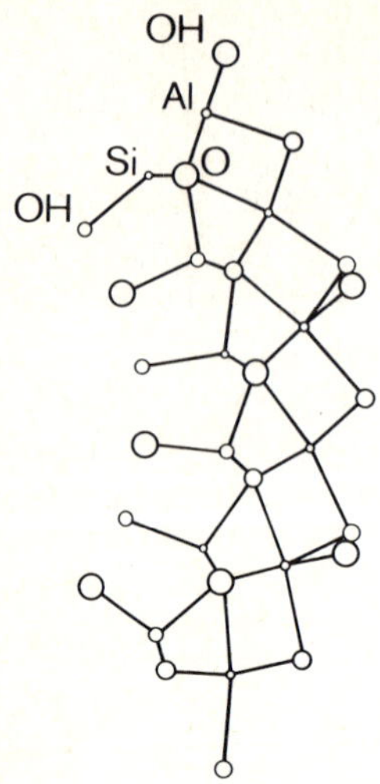

Fig. 1.12. Diagram showing the insertion into the gibbsite structure of an orthosilicate anion (SiO_3OH). The attendant surface contraction and sheet curling may account for the tubular morphology of imogolite (after Cradwick et al., 1972).

1.2. SWELLING PROPERTIES

In order to appreciate the clay—polymer interaction in aqueous media, it seems useful to append a description of the swelling behaviour of clay minerals. Only the outlines of clay colloid chemistry will be given here since this topic has been treated in some detail by various authors (Marshall, 1949; Van Olphen, 1963; Norrish, 1954, 1972; Low, 1968; Swartzen-Allen and Matijević, 1974) who have also listed the relevant references to the extensive literature. Further, our discussion will largely concern montmorillonite which has served as a model for many years because of its high swelling capacity and ability to form stable, aqueous dispersions.

We have already seen that montmorillonite is capable of intercalating a number of water layers through which the basal spacing increases stepwise from ~1 to ~2.2 nm, corresponding to interlayer separations of ~0 to ~1.2 nm. This type of expansion is referred to as intracrystalline, or simply, crystalline swelling, signifying that the gross crystal morphology is preserved. Other terms used for this process are "region 1" and "stage 1" swelling as distinct from those events which may take place on subsequent uptake of water (Norrish, 1972). Swelling in this region is primarily due to hydration of the interlayer cations (cf. Table 1.3) which probably take up positions midway between two opposing silicate layers (cf. Fig. 1.6). Montmorillonite saturated with polyvalent cations do not normally expand beyond an interlayer separation of ~1 nm, presumably because the repulsive effect of ion hydration is more than offset by electrostatic attraction between cation and silicate layer.

On the other hand, montmorillonite containing small, monovalent cations, such as Li^+ and Na^+, can take up more water under suitable conditions. The interlayer spacing then increases abruptly to 3—4 nm and continues to rise more or less linearly to some tens of nanometres with water content. This is termed region or stage 2 swelling and is ascribable to the formation of diffuse electrical double layers on interlayer surfaces. The interaction of these layers generates a repulsive, osmotic force which is opposed only to a limited extent by Van der Waals forces; for the most part, it is restrained by frictional forces arising from edge-to-face particle (layer) association. At this stage, the consistency of the clay—water systems is that of a paste or thick gel.

In the near absence of edge-to-face bonding or when this type of particle association has been disrupted by gentle shaking, the system passes into region or state 3 swelling. The individual silicate layers are separated by large distances, the extent of separation being limited only by the volume of water in the system. When this is large, the system is a dilute sol in which the particles are kept apart mainly through thermal motion. In relatively concentrated suspensions or gels, diffuse double layer interaction is the stabilising factor together with some residual edge-to-face and edge-to-edge bonding.

The changes occurring with montmorillonite systems which give all three regions of swelling, are summarised in Table 1.4. Besides being dependent on the nature of the saturating cation, the ability of montmorillonite to show extensive swelling is influenced by the amount and site of isomorphous substitution as well as by particle size. Thus, beidellites in which this substitution occurs mainly in the tetrahedral sheet, and consisting of particles with an equivalent spherical diameter of >2 μm, do not tend to swell beyond region 1. However, they may be made to swell extensively by mechanical treatment (e.g. ultrasonification). It should also be noted that non-expanding 2 : 1 type phyllosilicates, such as the illites and most vermiculites, do not form diffuse double layers on their interlayer surfaces but may do so on external crystal faces (Quirk, 1968). Vermiculites containing Li^+, small alkylammonium ions, and some amino acids, however, can swell extensively (Barshad, 1952; Walker and Garrett, 1961; Garrett and Walker, 1962; Norrish and Rausell-Colom, 1963; Rausell-Colom and Salvador, 1971). Interestingly, montmorillonite containing about equal amounts of Ca^{2+} and short-chain alkylammonium ions swells almost as extensively as the sodium clay, although the samples fully saturated with either Ca^{2+} or alkylammonium ions show only crystalline (region 1) expansion (Theng et al., 1968).

It is clear that when extensive swelling can occur, the extent of surface exposed or accessible to organic polymers will be greatly increased. This would account for the observation that Na^+-montmorillonite generally ad-

TABLE 1.4

The swelling behaviour of montmorillonite in water (from Norrish, 1972)

Region or stage	State of clay—water system	X-ray diffraction pattern	Water content ($g\ g^{-1}$)	Swelling pressure ($dyne\ cm^{-2}$)	Nature of interlayer forces	
					Repulsive	Attractive
1	Solid, crystalline	Sharp [a]	0—~0.7	$\sim 4 \times 10^9$	Cation hydration	Electrostatic
2	Paste—gel	Very diffuse maxima [b]	0.7—~20	10^7—10^5	Diffuse double layer interaction	Frictional, edge-to-face bonds
3	Gel—sol	Only central scatter	>20	$<10^5$	Diffuse double layer interaction and thermal motion	Edge-to-face and edge-to-edge bonds

[a] Basal spacings in this region are limited to ~2.2 nm.
[b] Basal spacings in excess of 4 nm.

sorbs much greater amounts of water-soluble, uncharged polymers as compared with samples containing polyvalent cations (cf. Chapter 3). On the other hand, the flocculation of dilute clay suspensions by polymers, which involves both adsorption of and interparticle bridging by the macromolecule, will be facilitated when the clay particles have previously been brought close together as, for example, by electrolyte addition.

As already remarked on, clay particles or unit silicate layers in suspension

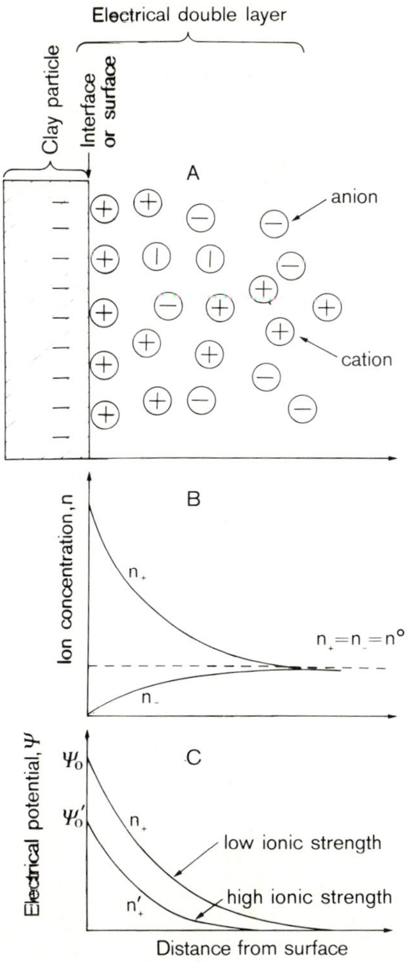

Fig. 1.13. A, the structure of a diffuse electrical double layer at the surface of a clay particle or silicate layer; B, variations of the concentration of cations (n_+) and anions (n_-) with distance from the surface in the case of a symmetrical electrolyte with a bulk concentration of n^0; C, variations of the electrical potential and cation concentration with distance from the surface at low and high ionic strengths.

owe their stability to mutual repulsion when their interacting diffuse electrical double layers interact on approach. In clay—water systems, a double layer is made up of the negative surface charge and the balancing cation charge (Fig. 1.13A). Since the surface charge arises from isomorphous substitution within the structural framework, clay colloids are of the "constant charge" type as distinct from the "constant potential" kind, such as the AgI sol. The cation charge is said to be diffuse near the interface because the cation concentration at a given point can be described by the Boltzmann equation

$$n_+ = n_+^0 \exp(-ze^* \psi /kT) \tag{1.1}$$

where n_+ and n_+^0 are the local and bulk cation concentration (number of ions per unit volume), respectively; z is the valency; e^* the electronic charge; ψ the local electric potential; k the Boltzmann constant; and T the absolute temperature. Frequently, the system also contains an indifferent electrolyte which contributes both cations and anions to the total ionic strength (I). In this case, the respective anion concentrations (n_- and n_-^0) will obey a similar but converse equation to (1.1), that is

$$n_- = n_-^0 \exp(ze^* \psi /kT) \tag{1.2}$$

Fig. 1.13B shows the curves relating n_+ and n_- to the distance (x) from the charged surface. The difference between n_+ and n_- then gives the charge density at that point (ρ). When ρ is now inserted in the Poisson equation

$$\mathrm{d}^2\psi/\mathrm{d}x^2 = -(4\pi/\epsilon)\rho \tag{1.3}$$

where ϵ is the dielectric constant of the medium, we arrive at the basic equation for a flat double layer and symmetrical electrolytes

$$\mathrm{d}^2 y/\mathrm{d}X^2 = \sinh y \tag{1.4}$$

Here $y = ze^*\psi/kT$ and $X = \kappa x$ where

$$\kappa^2 = 8\pi n(e^*)^2 z^2/\epsilon kT\,(\mathrm{cm}^{-2}) \tag{1.5}$$

Equation (1.4) can be integrated for two boundary conditions, (i) $\mathrm{d}y/\mathrm{d}X = 0$ at $X = \infty$; and (ii) $y = y_0 = z\,e^*\psi_0/kT$ at $X = 0$.

The above treatment is known as the Gouy—Chapman theory (Gouy, 1910, 1917; Chapman, 1913) which predicts an approximately exponential decay of ψ with x (Fig. 1.13C). Only for small potentials ($\psi < 25$ mV, $y_0 < 1$) in this decay purely exponential. Besides the now classic book by Verwey and Overbeek (1948), the reviews inter alia by Van Olphen (1963), Napper (1967), Lyklema (1968), Gregory (1975) and Overbeek (1977) should be consulted for more details.

Of particular note is the fact that for $X = 1$, $x = 1/\kappa$ at which distance the potential has fallen to the $1/e$ th (~0.37) of its surface value, where e is the base of the natural logarithm. $1/\kappa$ is therefore a measure of the thickness or extension into the medium of the electrical double layer which is sensitive to the presence of electrolytes in the system. It is therefore useful to express equation (1.5) in terms of the ionic strength of the solution defined as $I = (\Sigma\ c\ z^2)/2$ where c is the molar concentration. For aqueous systems at room temperature, the double layer thickness may then simply be written as

$$1/\kappa \sim 0.3/\sqrt{I}\ \text{(nm)} \tag{1.6}$$

indicating that as the ionic strength of the system increases, the double layer becomes more compressed or thinner. The effect of electrolyte addition on $1/\kappa$, estimated by means of equation (1.6), is summarised in Table 1.5 and shown schematically in Fig. 1.13C.

The Gouy—Chapman theory assumes the surface charge to be smeared out and the counterions to be point charges, leading to unrealistically large ion concentrations at high potentials (equation 1.1). To correct this deficiency, Stern (1924) postulated the existence of an adsorbed layer of finite-sized counterions adjacent to the surface (Fig. 1.14). The potential in the Stern layer falls off linearly from its surface value (ψ_0) to the so called Stern potential (ψ_δ). Beyond the Stern layer (of thickness $\delta \sim$ 0.5 nm for clay systems; cf. Table 1.6) the normal Gouy—Chapman or diffuse double layer extends into the bulk solution and the potential decay follows the pattern described previously.

A further refinement was proposed by Grahame (1947) who recognised an inner and outer Helmholtz layer or plane. The latter is thought to lie just outside the Stern layer, marking the boundary between the fixed and mobile part of the diffuse double layer ("plane of shear" or "slipping plane"). At this plane the potential is zeta (ζ), being the only potential which is experimentally accessible. In practice, however, it is frequently assumed that $\zeta \sim \psi_\delta$. These concepts are illustrated in Fig. 1.14, and Table 1.6 gives values of ψ_0, ζ, and distance from the surface of the plane of shear for some clays, derived by Cremers (1968) from surface conductivity and diffuse double layer theory.

TABLE 1.5

Electrical double layer thickness ($1/\kappa$) as a function of concentration of different symmetrical electrolytes in the system

Electrolyte molarity	Ionic strength for symmetrical electrolytes			Approximate values of $1/\kappa$ (nm) for different ion valencies		
	1 : 1	2 : 2	3 : 3	z = 1	z = 2	z = 3
10^{-4}	10^{-4}	4×10^{-4}	9×10^{-4}	30	15	10
10^{-3}	10^{-3}	4×10^{-3}	9×10^{-3}	9.6	4.8	3.2
10^{-2}	10^{-2}	4×10^{-3}	9×10^{-2}	3	1.5	1

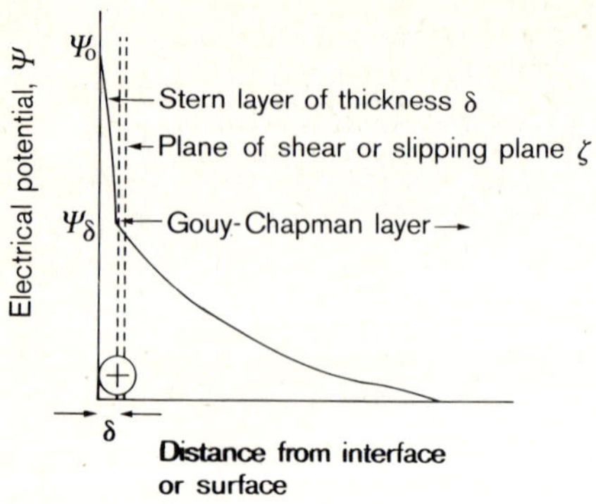

Fig. 1.14. Variations of the electrical potential with distance from the surface according to Stern (1924) and Grahame (1947). The broken line nearest to the surface marks the extension of the Stern layer of thickness δ, coinciding approximately with the inner Helmholtz plane (I.H.P.) of potential ψ_δ. The other broken line indicates the outer Helmholtz plane (O.H.P.), identified with the "plane of shear" or "slipping plane" of potential ζ.

A theoretical analysis of the interaction between colloidal particles has been developed by Derjaguin and Landau (1941) and, independently, by Verwey and Overbeek (1948). The fundamental feature of what is now known as the DLVO theory is that this interaction and hence, colloid stability, is determined by a combination or superposition of the interparticle double layer repulsion energy (V_R) and the Van der Waals attractive energy (V_A).

V_R is a complex function of surface potential (ψ_0) and interparticle separation (d). A number of expressions for V_R can be derived depending on

TABLE 1.6

Surface potential (ψ_0), zeta potential (ζ), and distance from the surface of the plane of shear (l_s) for kaolinite, illite, and montmorillonite at 298°C (from Cremers, 1968)

Clay species	NaCl concentration (M)	Surface charge (meq cm^{-2}) × 10^7	ψ_0 (mV)	ζ (mV)	l_s [a] (nm)
Kaolinite 1	0.0045	2.15	197	130	0.42
	0.00025		270	197	0.58
Kaolinite 2	0.0065	2.85	203	145	0.30
	0.0003		281	193	0.67
Illite	0.02	1.50	143	88	0.40
Montmorillonite	0.02	1.10	128	78	0.50

[a] Assumed to equal the Stern layer thickness (δ).

whether the particles are of the constant potential or constant charge type, whether they are platy or spherical, and whether the surface potentials for a pair of interacting particles are equal or different (e.g. Gregory, 1975). The simplest form, applicable to large plate separations and moderate-to-high, equal surface potentials, is

$$V_R = \frac{64\,nkT}{\kappa}\,\gamma^2 \exp(-\kappa d) \tag{1.7}$$

where $\gamma = [\exp(ze^*\psi_0)/2\,kT - 1]/[\exp(ze^*\psi_0)/2\,kT + 1]$
For the attractive energy between two flat plates, we may write

$$V_A = -A_{1(2)}/12\pi d^2 \tag{1.8}$$

where $A_{1(2)}$ is the so-called Hamaker (1937) constant.

The net or total interaction energy (V_T) is then the sum of V_R and V_A. A schematic representation of V_R, V_A, and V_T in function of interplate separation for low electrolyte concentrations ($<10^{-2}$ M) and high surface potentials (>50 mV), is shown in Fig. 1.15. Of prime importance is the occurrence of a maximum (V_m) at intermediate distances. This maximum may be considered as an energy barrier which the particles must overcome if they are to "fall" into the deep primary minimum (at close distances) and so come together. The height of V_m therefore determines the relative stability of the system. Depending on the suspension strength, a value of 15—25 kT is normally required for long-term stability (Van Olphen, 1963). At large interparticle separations, a secondary minimum (not shown in Fig. 1.15) may occur since V_R falls off more rapidly with increasing distances than does V_A.

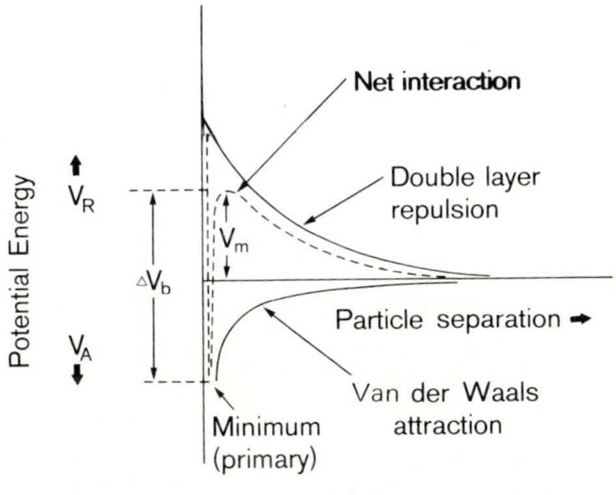

Fig. 1.15. Diagram showing the energy of double layer repulsion (V_R) and of Van der Waals attraction (V_A), together with the net or total interaction energy (V_T) for parallel flat plates, as a function of particle (plate) separation. The interaction curve shows a deep (primary) minimum at close separations. V_m is the energy barrier to particle coagulation and ΔV_b is the barrier to redispersion.

This minimum, however, is usually quite shallow, being of the order of the thermal energy so that particle coagulation taking place here is fairly readily reversible. On the other hand, the barrier to redispersion or peptisation for particles coagulated at the primary minimum (ΔV_b) is very high.

Another noteworthy feature is the sensitivity of V_R to ionic strength or electrolyte addition which influences V_R through κ (equation 1.7). By compressing the double layer thickness (cf. Fig. 1.13C), the addition of electrolytes to the system will reduce V_m and so promote coagulation. It can be shown that for high surface potentials, the concentration of electrolyte required to reduce V_m to zero is inversely proportional to the sixth power of the counterion valency, in line with the empirical Schulze—Hardy rule. The effect of adsorbed polymers on the structure of the diffuse double layer and hence, on V_R and V_A will be discussed in the following chapter.

To conclude this section, we wish to remark on the applicability of the DLVO theory to clay systems. Of the many attempts, using different approaches, which have been made in this connection, only relatively few have met with some measure of success (see Quirk (1968) and Swartzen-Allen and Matijević (1974) for references). In many instances, modifications have to be introduced into both the repulsion and attraction terms before experimental data can be satisfactorily interpreted.

Barclay et al. (1972) for example, have found that for dilute Na^+-montmorillonite dispersions, the repulsion at short distances is smaller than predicted by the DLVO theory. Similarly, the surface areas of montmorillonites estimated from anion exclusion data are generally less than theory would suggest. However, by allowing for such things as specific counterion adsorption, ion—surface affinity, and ion hydration, a better agreement between experiment and theory can be achieved. Other factors which affect interparticle repulsion are the discreteness of the negative surface charge allowing for ion-pairing (Edwards et al., 1965), degree of face-to-face association (Banin and Lahav, 1968), and the unequal surface potentials of two interacting silicate layers arising from non-uniform isomorphous substitution (Lagaly et al., 1972).

By comparison, the attractive forces between clay particles are more difficult to rationalise since their nature is incompletely understood (Norrish, 1972). We have already indicated that the Van der Waals forces can account for but a very small part of the interparticle attraction in region 2 (Table 1.4). Apparently, frictional forces arising from edge-to-face bonding are far more significant here. Van Olphen (1963) has raised the possibility that the edge surfaces carry a positive diffuse double layer in neutral and acidic suspensions although the overall electrophoretic mobility of the particles is negative.

Besides the natural face-to-face association between unit layers in a crystal (sometimes referred to as "tactoid" formation), clay particles may also associate through edge-to-face and edge-to-edge bonding. The evidence for this has come from light scattering, viscosity, and rheological measurements on clay suspensions (see Van Olphen, 1963; Swartzen-Allen and Matijević, 1974). Edge-to-face association gives rise to a "card-house" structure in more concentrated suspensions. It may be disrupted by raising the pH of, or by adding polyanions to, the suspension through which treatments the positive edge charge is neutralised and then reversed. In this connection, it is interesting to note that the addition of small amounts of electrolytes to a relatively dilute (~0.2% w/v) Na^+-montmorillonite suspension enhances particle dispersion or dissociation as indicated by a reduction in viscosity of the system (Van Olphen, 1963). Although the diffuse double layers are thereby slightly compressed, the resulting decrease in edge-to-face interaction apparently exceeds that in face-to-face interaction, leading to a breakdown of the card-house structure.

The swelling behaviour in water and electrolyte solutions of condensed, natural clay systems, such as do occur in soils, is vastly more complex. For such systems the DLVO theory cannot be used to predict adsorbed water film thicknesses (Quirk, 1968). This is so because wetting results in strain relaxation and the formation of a gel structure within which much of the water is enmeshed.

By analogy with the swelling characteristics of divalent ion clays, Aylmore and Quirk (1960) have proposed the concept of a domain structure for condensed clay systems. They envisage a domain to be composed of a number of clay crystals in near parallel arrangement. These units, each about 1 μm in thickness, are arranged in turbostratic array. The swelling between, say, Ca^{2+}-illite crystals within a domain may be compared with that between layers of a single montmorillonite crystal saturated with divalent cations. In both situations, much of the swelling occurs in molar electrolyte solutions.

In a rather similar vein, Norrish (1972) has suggested that the failure of some sedimentary and soil clays to show extensive swelling — even in the apparent absence of any cementing agents in the interlayers — is due to crystal growth and overlap. Joints established at a few summits of surface irregularities would suffice to prevent much particle rearrangement and hence large swelling from occurring.

The presence of an uncharged polymer around a domain or in the interlayer space of montmorillonite, does not necessarily inhibit intradomain or interlayer swelling (Theng et al., 1967; Parfitt and Greenland, 1970). Indeed, Schott (1968) has reported that the intercalation of polyoxyethylated nonionic detergents by montmorillonite promotes, rather than restricts, dispersion. This effect is analogous to that of the addition of small amounts of electrolytes, described above, and may be explained in similar terms.

REFERENCES

Aylmore, L.A.G. and Quirk, J.P., 1960. Domain or turbostratic structure of clays. Nature (London), 187: 1046—1048.

Bailey, S.W., 1975. Chlorites. In: J.E. Gieseking (Editor), Soil Components, Vol. 2, Inorganic Components. Springer Verlag, Berlin, pp. 191—263.

Bailey, S.W., Brindley, G.W., Johns, W.D., Martin, R.T. and Ross, M., 1971. Summary of national and international recommendations on clay mineral nomenclature. 1969—1970 CMS Nomenclature Committee. Clays and Clay Minerals, 19: 129—132.

Banin, A. and Lahav, N., 1968. Particle size and optical properties of montmorillonite in suspension. Israel Journal of Chemistry, 6: 235—250.

Barclay, L., Harrington, A. and Ottewill, R.H., 1972. The measurement of forces between particles in disperse systems. Kolloid Zeitschrift und Zeitschrift für Polymere, 250: 655—666.

Barshad, I., 1952. Factors affecting the interlayer expansion of vermiculite and montmorillonite with organic substances. Soil Science Society of America Proceedings, 16: 176—182.

Bolland, M.D.A., Posner, A.M. and Quirk, J.P., 1976. Surface charge on kaolinites in aqueous suspension. Australian Journal of Soil Research, 14: 197—216.

Brauner, K. and Preisinger, A., 1956. Struktur und Enstehung des Sepiolits. Tschermaks Mineralogische und Petrographische Mitteilungen, 6: 120—140.

Brindley, G.W. and Fancher, D., 1969. Kaolinite defect structures; possible relation to allophanes. Proceedings of the International Clay Conference, Tokyo, 2: 29—34.

Brown, G. (Editor), 1961. The X-ray Identification and Crystal Structures of Clay Minerals, 2nd Edition. Mineralogical Society, London, 544 pp.

Carroll, D., 1970. Clay Minerals: A Guide to Their X-ray Identification. The Geological Society of America, Special Paper 126, 80 pp.

Chapman, D.L., 1913. A contribution to the theory of electrocapillarity. Philosophical Magazine, 25: 475—481.

Churchman, G.J. and Carr, R.M., 1975. The definition and nomenclature of halloysites. Clays and Clay Minerals, 23: 382—388.

Cloos, P., Léonard, A.J., Moreau, J.P., Herbillon, A. and Fripiat, J.J., 1969. Structural organization in amorphous silicoaluminas. Clays and Clay Minerals, 17: 270—287.

Cradwick, P.D.G., Farmer, V.C., Russell, J.D., Masson, C.R., Wada, K. and Yoshinaga, N., 1972. Imogolite, a hydrated aluminium silicate of tubular structure. Nature Physical Science (London), 240: 187—189.

Cremers, A., 1968. Surface conductivity in sodium clays. Israel Journal of Chemistry, 6: 195—202.

Cruz, M., Jacobs, H. and Fripiat, J.J., 1972. The nature of the interlayer bonding in kaolin minerals. Proceedings of the International Clay Conference, Madrid, pp. 35—44.

Dandy, A.J. and Nadiye-Tabbiruka, M.S., 1975. The effect of heating in vacuo on the microporosity of sepiolite. Clays and Clay Minerals, 23: 428—430.

Derjaguin, B.V. and Landau, L.D., 1941. Theory of the stability of strongly charged lyphobic sols and of the adhesion of strongly charged particles in solutions of electrolytes. Acta Physicochimica of the United Soviet Socialist Republics, 14: 633—662.

De Villiers, J.M and Jackson, M.L., 1967. Cation exchange capacity variations with pH in soil clays. Soil Science Society of America Proceedings, 31: 473—476.

Edelman, C.H. and Favejee, J.C.L., 1940. On the crystal structure of montmorillonite and halloysite. Zeitschrift für Kristallographie, Kristallgeometrie, Kristallphysik, Kristallchemie, A, 102: 417—431.

Edwards, D.G., Posner, A.M. and Quirk, J.P., 1965. Discreteness of charge on clay surfaces. Nature (London), 206: 168—170.

Farmer, V.C., 1974. The layer silicates. In: V.C. Farmer (Editor), The Infrared Spectra of Minerals. Mineralogical Society, London, pp. 331—363.

Ferris, A.P. and Jepson, W.B., 1975. The exchange capacities of kaolinite and the preparation of homoionic clays. Journal of Colloid and Interface Science, 51: 245—259.

Fieldes, M. and Claridge, G.G.C., 1975. Allophane. In: J.E. Gieseking (Editor), Soil Components, Vol. 2, Inorganic Components. Springer Verlag, Berlin, pp. 351—393.

Follett, E.A.C., 1965. Retention of amorphous colloidal "ferric hydroxide" by kaolinites. The Journal of Soil Science, 16: 542—548.

Garrett, W.G. and Walker, G.F., 1962. Swelling of some vermiculite—organic complexes in water. Clays and Clay Minerals, 9: 557—567.

Giese, R.F., Jr, 1975. Interlayer bonding in talc and pyrophyllite. Clays and Clay Minerals, 23: 165—166.

Gieseking, J.E. (Editor), 1975. Soil Components, Vol. 2, Inorganic Components. Springer Verlag, Berlin, 684 pp.

Gorbunov, N.I. and Shchurina, G.N., 1976. Occurrence and formation of palygorskite and sepiolite in Sierozems and Takyos. Pochvovedenie, No. 4: 133—139.

Gouy, G., 1910. Sur la constitution de la charge électrique a la surface d'un électrolyte. Journal de Physique, 9: 457—468.

Gouy, G., 1917. Sur la fonction électrocapillaire. Annales de Physique, 7: 129—184.

Grahame, D.C., 1947. The electrical double layer and the theory of electrocapillarity. Chemical Reviews, 41: 441—501.

Greenland, D.J., 1972. Interactions between organic polymers and inorganic soil particles. Mededelingen Fakulteit Landbouwwetenschappen Rijksuniversiteit Gent, 37: 897—914.

Gregory, J., 1975. Interfacial phenomena. In: K.J. Ives (Editor), The Scientific Basis of Filtration. Noordhoff, Leyden, pp. 53—90.

Grim, R.E., 1968. Clay Mineralogy, 2nd Edition. McGraw-Hill, New York, N.Y., 596 pp.

Hamaker, H.C., 1937. The London—van der Waals attraction between spherical particles. Physica, 4: 1058—1072.

Hendricks, S.B., 1929. Diffraction of X-ray radiation from crystalline aggregates. Zeitschrift für Kristallographie, Kristallgeometrie, Kristallphysik, Kristallchemie, 71: 269—273.

Hendricks, S.B., 1942. Lattice structure of clay minerals and some properties of clays. Journal of Geology, 50: 276—290.

Hendricks, S.B. and Fry, W.H., 1930. The results of X-ray and microscopal examinations of soil colloids. Soil Science, 29: 457—478.

Hénin, S. and Caillère, S., 1975. Fibrous minerals. In: J.E. Gieseking (Editor), Soil Components, Vol. 2, Inorganic Components. Springer Verlag, Berlin, pp. 335—349.

Henmi, T. and Wada, K., 1976. Morphology and composition of allophane. The American Mineralogist, 61: 379—390.

Hofmann, U., Boehm, H.P. and Gromes, W., 1961. Die Abmessungen der Kristalle der Tonminerale. Zeitschrift für Anorganische und Allgemeine Chemie, 308: 143—154.

Hofmann, U., Endell, K. and Wilm, D., 1933. Kristallstruktur und Quellung von Montmorillonit. Zeitschrift für Kristallographie, Kristallgeometrie, Kristallphysik, Kristallchemie, 86: 340—348.

Inoue, T. and Wada, K., 1971. Reactions between humified clover extract and imogolite as a model of humus—clay interaction. Clay Science, 4: 61—80.

Jackson, M.L., 1959. Frequency distribution of clay minerals in major great soil groups as related to the factors of soil formation. Clays and Clay Minerals, 6: 133—143.

Kitagawa, Y., 1971. The "unit particle" of allophane. The American Mineralogist, 56: 456—475.

Kodama, H. and Schnitzer, M., 1974. Adsorption of fulvic acid by non-expanding clay minerals. Transactions of the 10th International Congress of Soil Science, Moscow, 2: 51—56.

Lagaly, G. and Weiss, A., 1975. The layer charge of smectite layer silicates. Proceedings of the International Clay Conference, Mexico City, pp. 157—172.

Lagaly, G., Schön, G. and Weiss, A., 1972. Über den Einfluss einer unsymmetrischen Ladungsverteilung auf die Wechselwirkung zwischen plättchenformigen Kolloidteilchen. Kolloid Zeitschrift und Zeitschrift für Polymere, 250: 667—674.

Léonard, A.J., Suzuki, Sho, Fripiat, J.J. and De Kimpe, C., 1964. Structure and properties of amorphous silicoaluminas. The Journal of Physical Chemistry, 68: 2608—2617.

Low, P.F., 1968. Mineralogical data requirements in soil physical investigations. In: Mineralogy in Soil Science and Engineering. Soil Science Society of America Special Publication, No. 3: 1—34.

Lyklema, J., 1968. Principles of the stability of lyophobic colloidal dispersions in non-aqueous media. Advances in Colloid and Interface Science, 2: 65—114.

MacEwan, D.M.C., 1947. The nomenclature of the halloysite minerals. Mineralogical Magazine, 28: 36—44.

MacEwan, D.M.C., 1948. Complexes of clays with organic compounds. I. Complex formation between montmorillonite and halloysite and certain organic liquids. Transactions of the Faraday Society, 44: 349—367.

Mackenzie, R.C., 1970. Simple phyllosilicates based on gibbsite- and brucite-like sheets. In: R.C. Mackenzie (Editor), Differential Thermal Analysis, Vol. 1. Academic Press, London, pp. 497—537.

Mackenzie, R.C. and Mitchell, B.D., 1966. Clay mineralogy. Earth Science Reviews, 2: 47—91.

Mackintosh, E.E., Lewis, D.G. and Greenland, D.J., 1971. Dodecylammonium mica complexes. I. Factors affecting the exchange reaction. Clays and Clay Minerals, 19: 209—218.

Maegdefrau, E. and Hofmann, U., 1937. Die Kristallstruktur des Montmorillonits. Zeitschrift für Kristallographie, Kritstallgeometrie, Kristallphysik, Kristallchemie, 98: 299—323.

Marshall, C.E., 1935. Layer lattices and base-exchange clays. Zeitschrift für Kristallographie, Kristallgeometrie, Kristallphysik, Kristallchemie, 91: 433—449.

Marshall, C.E., 1949. The Colloid Chemistry of the Silicate Minerals. Academic Press, New York, N.Y., 195 pp.

Marshall, C.E., 1964. The Physical Chemistry and Mineralogy of Soils, Vol. 1, Soil Materials. John Wiley, New York, N.Y., 388 pp.

Mathieson, A.McL. and Walker, G.F., 1954. Crystal structure of magnesium-vermiculite. The American Mineralogist, 39: 231—255.

McBride, M.B., 1976. Origin and position of exchange sites in kaolinite: an ESR study. Clays and Clay Minerals, 24: 88—92.

Milestone, N.B., 1971. Allophane, its structure and possible uses. Journal of the New Zealand Institute of Chemistry, 35: 191—197.

Mitchell, B.D., Farmer, V.C. and McHardy, W.J., 1964. Amorphous inorganic materials in soils. Advances in Agronomy, 16: 327—383.

Mortland, M.M., 1970. Clay—organic complexes and interactions. Advances in Agronomy, 22: 75—117.

Napper, D.H., 1967. Modern theories of colloid stability. Science Progress (Oxford), 55: 91—109.

Norrish, K., 1954. Swelling of montmorillonite. Discussions of the Faraday Society, No. 18: 120—134.

Norrish, K., 1972. Forces between clay particles. Proceedings of the International Clay Conference, Madrid, pp. 375—383.

Norrish, K. and Rausell-Colom, J.A., 1963. Low-angle X-ray diffraction studies of the swelling of montmorillonite and vermiculite. Clays and Clay Minerals, 10: 123—149.

Okada, K., Morikawa, S., Iwai, S., Ohira, Y. and Ossaka, J., 1975. A structure model of allophane. Clay Science, 8: 87—99.

Orlov, D.S., Pivovarova, I.A. and Gorbunov, N.I., 1973. Interaction of humic substances with minerals and the nature of their bond. A review. Agrokhimiya, 9: 140—153.

Overbeek, J.Th.G., 1977. Recent developments in the understanding of colloid stability. Journal of Colloid and Interface Science, 58: 408—422.

Parfitt, R.L. and Greenland, D.J., 1970. Adsorption of water by montmorillonite—poly-(ethylene glycol) adsorption products. Clay Minerals, 8: 317—324.

Pauling, L., 1930a. Structure of the micas and related minerals. Proceedings of the National Academy of Sciences of the United States of America, 16: 123—129.

Pauling, L., 1930b. Structure of the chlorites. Proceedings of the National Academy of Sciences of the United States of America, 16: 578—582.

Pedro, G., 1970. Report of the AIPEA Nomenclature Committee. AIPEA Newsletter, No. 4: 3—4.

Peigneur, P., Maes, A. and Cremers, A., 1975. Heterogeneity of charge density distribution in montmorillonite as inferred from cobalt adsorption. Clays and Clay Minerals, 23: 71—75.

Prost, R., 1975. Interactions between adsorbed water molecules and the structure of clay minerals: hydration mechanism of smectites. Proceedings of the International Clay Conference, Mexico City, pp. 351—359.

Quantin, P., 1972. Les andosols. Revue bibliographique des connaissances actuelles. Cahiers de l'Office de la Recherche Scientifique et Technique Outre-Mer (O.R.S.T.O.M.), Série Pédologie, 10: 273—301.

Quirk, J.P., 1960. Negative and positive adsorption of chloride by kaolinite. Nature (London), 188: 253—254.

Quirk, J.P., 1968. Particle interaction and soil swelling. Israel Journal of Chemistry, 6: 213—234.

Radoslovich, E.W., 1975. Micas in macroscopic forms. In: J.E. Gieseking (Editor), Soil Components, Vol. 2, Inorganic Components. Springer Verlag, Berling, pp. 27—57.

Rand, B. and Melton, I.E., 1975. Isoelectric point of the edge surface of kaolinite. Nature (London), 257: 214—216.

Rausell-Colom, J.A. and Salvador, P.S., 1971. Complexes vermiculite—amino acides. Clay Minerals, 9: 139—149.

Robertson, R.H.S., 1957. Sepiolite: a versatile raw material. Chemistry and Industry, 1492—1495.

Ross, C.S., 1927. The mineralogy of clays. Transactions of the 1st International Congress of Soil Science, Washington, D.C., 4: 555—561.

Schofield, R.K. and Samson, H.R., 1953. The deflocculation of kaolinite suspensions and the accompanying change-over from positive to negative chloride adsorption. Clay Minerals Bulletin, 2: 45—51.

Schofield, R.K. and Samson, H.R., 1954. Flocculation of kaolinite due to the attraction of oppositely charged crystal faces. Discussions of the Faraday Society, No. 18: 138—145.

Schott, H., 1968. Deflocculation of swelling clays by nonionic and anionic detergents. Journal of Colloid and Interface Science, 26: 133—139.

Schultz, L.G., 1969. Lithium and potassium absorption, dehydroxylation temperature, and structural water content of aluminous smectites. Clays and Clay Minerals, 17: 115—149.

Serna, C., Ahlrichs, J.L. and Serratosa, J.M., 1975. Folding in sepiolite crystals. Clays and Clay Minerals, 23: 452—457.

Solomon, D.H., 1968. Clay minerals as electron acceptors and/or electron donors in organic reactions. Clays and Clay Minerals, 16: 31—39.

Stern, O., 1924. Zur Theorie der elektrolytischen Doppelschicht. Zeitschrift für Elektrochemie, 30: 508—516.

Swartzen-Allen, S.L. and Matijević, E., 1974. Surface and colloid chemistry of clays. Chemical Reviews, 74: 385—400.

Theng, B.K.G., 1971. Mechanisms of formation of colored clay—organic complexes. A review. Clays and Clay Minerals, 19: 383—390.

Theng, B.K.G., 1974. The Chemistry of Clay—Organic Reactions. Adam Hilger Ltd, London, 343 pp.

Theng, B.K.G. and Walker, G.F., 1970. Interactions of clay minerals with organic monomers. Israel Journal of Chemistry, 8: 417—424.

Theng, B.K.G., Greenland, D.J. and Quirk, J.P., 1967. The swelling in water of complexes of montmorillonite with polyvinyl alcohol. Australian Journal of Soil Research, 5: 69—75.

Theng, B.K.G., Greenland, D.J. and Quirk, J.P., 1968. The effect of exchangeable alkylammonium ions on the swelling of montmorillonite in water. Clay Minerals, 7: 271—293.

Thorez, J., 1975. Phyllosilicates and Clay Minerals: A Laboratory Handbook for Their X-ray Diffraction Analysis. G. Lelotte, Dison, Belgium, 579 pp.

Udagawa, S., Nakada, T. and Nakahira, M., 1969. Molecular structure of allophane as revealed by its thermal transformation. Proceedings of the International Clay Conference, Tokyo, 1: 151—159.

Van Olphen, H., 1963. An Introduction to Clay Colloid Chemistry. Interscience, New York, N.Y., 301 pp.

Van Olphen, H., 1971. Amorphous clay minerals. Science, 171: 91—92.

Verwey, E.J.W. and Overbeek, J.Th.G., 1948. Theory of the Stability of Lyophobic Colloids. Elsevier, Amsterdam, 216 pp.

Von Reichenbach, H. Graf and Rich, C.I., 1975. Fine-grained micas in soils. In: J.E. Gieseking (Editor), Soil Components, Vol. 2, Inorganic Components. Springer Verlag, Berlin, pp. 59—95.

Wada, K., 1967. A structural scheme of soil allophane. The American Mineralogist, 52: 690—708.

Wada, K. and Inoue, T., 1967. Retention of humic substances derived from rotted clover leaves on soils containing montmorillonite and allophane. Soil Science and Plant Nutrition (Tokyo), 13: 9—16.

Wada, S. and Wada, K., 1977. Density and structure of allophane. Clay Minerals, 12: 289—298.

Walker, G.F., 1975. Vermiculites. In: J.E. Gieseking (Editor), Soil Components, Vol. 2. Inorganic Components. Springer Verlag, Berlin, pp. 155—189.

Walker, G.F. and Garrett, W.G., 1961. Complexes of vermiculite with amino acids. Nature (London), 191: 1389.

Weaver, C.E. and Pollard, L.D., 1973. The Chemistry of Clay Minerals. Elsevier, Amsterdam, 213 pp.

Weiss, A., 1958. Der Kationenaustausch bei den Mineralen der Glimmer-, Vermikulit- und Montmorillonitgruppe. Zeitschrift für Anorganische und Allgemeine Chemie, 297: 257—286.

Weiss, A., 1969. Organic derivatives of clay minerals, zeolites, and related minerals. In: G. Eglinton and M.T.J. Murphy (Editors), Organic Geochemistry. Springer Verlag, Berlin, pp. 737—781.

Weiss, A. and Russow, J., 1963. Über die Lage der austauschbaren Kationen bei Kaolinit. Proceedings of the International Clay Conference, Stockholm, 1: 203—213.

Wentworth, S.A., 1970. Illite. Clay Science, 3: 140—155.

Yaalon, D.H. and Wieder, M., 1976. Pedogenic palygorskite in some arid brown (calciorthid) soils of Israel. Clay Minerals, 11: 73—80.

Chapter 2

POLYMER BEHAVIOUR AT CLAY AND SOLID SURFACES

2.1. INTRODUCTION

Since polymer adsorption is involved in, and holds the key to an understanding of, the formation and properties of clay—polymer complexes, it seems appropriate to sketch out the outlines of this process. Many of the underlying mechanisms have come to light as a result of experimental and theoretical studies carried out over the past two decades in response to the importance and potential of the solid—polymer interaction in agricultural and industrial applications. Although difficulties still exist in matching experiment with theory, a self-consistent picture of polymer behaviour at solid/solution interfaces is slowly emerging.

It should be added here that the majority of the experimental data refers to polymer adsorption from non-aqueous solvents. However, processes in clay and soil systems with which we are chiefly concerned essentially occur in an aqueous environment. The role of water must therefore be borne in mind if we are to extrapolate such data to aqueous systems. Thus, water will compete strongly with the polymer for adsorption sites on the solid surface, particularly when hydrogen bonding can occur. Being bifunctional, however, water can act as both a hydrogen bond donor and acceptor so that its presence at the adsorbent surface may either promote or inhibit polymer adsorption. In this connection it is important to recall that water molecules at clay mineral surfaces are associated more with the exchangeable cations than with the silicate surface as such. This is so because the basal oxygens of 2 : 1 type layer silicates (siloxane surfaces) are weak electron donors; hydrogen bonding interactions between water and surface are thus, at best, not very strong. The experimental evidence for this conclusion has been summarised by Theng (1974). Such interactions will occur more readily with the aluminol surfaces of 1 : 1 type phyllosilicates.

A feature common to both aqueous and non-aqueous systems is that a large number of solvent molecules must be desorbed in order to accommodate a single polymer molecule. The translational entropy so gained by the system provides the driving force for polymer adsorption (Fig. 2.1). Such entropy effects are also partly responsible for the very strong attachment of polymers despite the very small or virtually zero enthalpy change of the adsorption process.

On the theoretical side, treatments of polymer adsorption have mostly been developed for uncharged linear flexible homopolymers at a plane surface. Indeed, progress in our understanding of this system is at present limited more by experimental than by theoretical inadequacies (Silberberg,

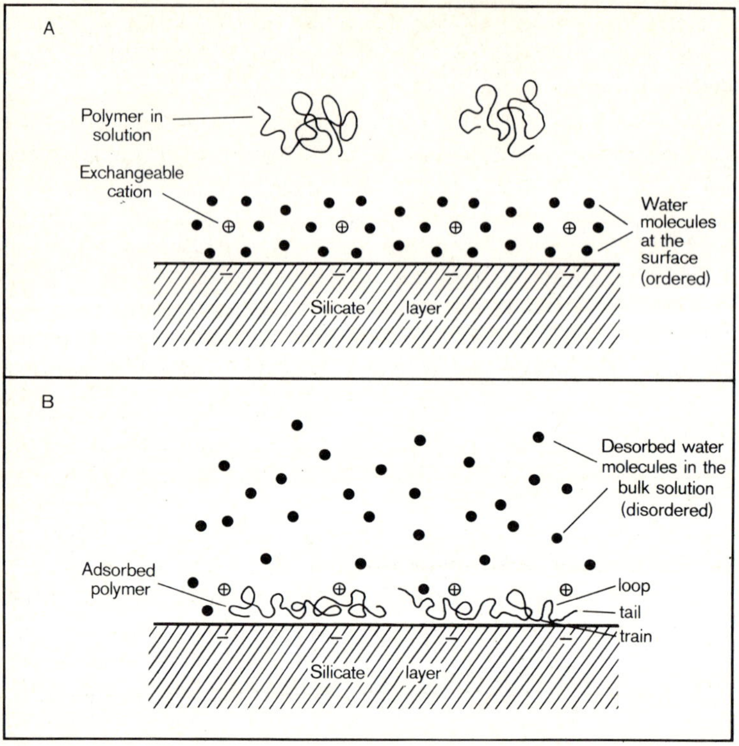

Fig. 2.1. Diagram illustrating the desorption of numerous water molecules from the clay surface during the adsorption of a single uncharged polymer molecule. The change in polymer conformation on adsorption is also indicated.

1971). By comparison, the theoretical basis for the adsorption of compact macromolecules, copolymers, and polyelectrolytes, has not been adequately established (Hesselink, 1971, 1972; Silberberg, 1973). This seeming neglect is a reflection of the inherent complexity of such systems rather than of a lack of interest. By the same token, relatively little attention has been paid to adsorption on other types of surfaces although some attempt has been made to analyse the behaviour of chain molecules between two plates (Di Marzio and Rubin, 1971; Hesselink, 1971) and in a pore (Pouchly, 1970).

The lack of an adequate theory of polyelectrolyte adsorption is quite a serious drawback in view of the importance of charged macromolecules in biological systems. Similarly, humic substances which consist largely of negatively charged polymers, are the most abundant and ubiquitous in soils and waters. The interactions of naturally occurring polyelectrolytes with clays and soil colloids, dealt with in Part III, are therefore only amenable to a qualitative or, at best, a semi-quantitative description.

A number of excellent general reviews on the adsorption of polymers from solution onto solid adsorbents have been written previously (Hughes and Von Frankenberg, 1963; Patat et al., 1964; Kipling, 1963; Stromberg, 1967; Fontana, 1971; Vincent, 1974). It is not intended here to give yet another comprehensive account of this topic. Rather, we shall attempt to bring together the main conclusions, pointing out those features which may be applicable and relevant to the clay-polymer interaction.

The mechanisms underlying the reactions of simple organic compounds with clays are now fairly well understood (e.g. Theng, 1974) and this basic information is of much value to the interpretation of clay-polymer complex formation. Thus, positively charged organic micromolecules are adsorbed by an exchange reaction with the charge-balancing inorganic cations at the clay surface, and uncharged polar compounds principally by interaction between the functional group and the exchangeable cation. Adsorption is normally exothermic and increases with molecular weight of the compound, reflecting the increased contribution of Van der Waals forces to the overall adsorption energy.

Although these principles would apply to the adsorption of organic macromolecules, their interactions with clays differ in some important respects from those of simple non-polymeric species. These differences essentially arise from the fact that besides being long, a polymer chain is flexible and often polyfunctional. We have already remarked on entropy effects in relation to chain length. Chain flexibility and polyfunctionality further allow the polymer to adopt various shapes or conformational states at the surface and to be attached to the solid by numerous segment—surface bonds. A complete quantitative description of polymer behaviour at the solid/solution interface would therefore require a knowledge of both the solution and surface properties of the polymer in question, a condition which is not often met by clay—polymer systems.

2.2. ADSORPTION ISOTHERMS

The determination of adsorption isotherms is perhaps the most useful, single technique or approach for characterising the properties of polymer solutions in the presence of an adsorbing surface. Hughes and Von Frankenberg (1963) have likened the solid or adsorbent surface to a probe which, on insertion into a polymer solution, is capable of "scanning" the thermodynamic condition of the system. However, the determination of adsorption isotherms can only yield a limited amount of information about the state of the adsorbed layer. Supplementary measurements need to be carried out if we wish to gain some insight into the conformation of the adsorbed polymer.

A feature of polymer adsorption is that equilibrium conditions may take a considerably long time ($>10^2$ s) to become established. This is particularly true when porous adsorbents, such as clays, are used. Equilibrium is normally assumed to have been attained when the amount of polymer in solution remains sensibly constant with time. Only in a few instances (e.g. Koral et al., 1958) has the reversibility of the process

$$(\text{Polymer})_b \rightleftharpoons (\text{Polymer})_s \tag{2.1}$$

been examined, where the quantities in brackets refer to concentration, and the subscripts b and s to the bulk (solution) and surface (adsorbed) phases, respectively. In general, equilibrium (2.1) is shifted far to the right and this is reflected in the small temperature effect of the process and in the shape of the isotherm, as described below. By the same token the rate of desorption is very small in systems where solvent—surface interactions are weak; so much so that adsorption is often regarded as being "irreversible". This is perhaps hardly surprising because of the small probability that all of the adsorbed segments can be simultaneously detached from the surface and remain so sufficiently long for the polymer to move away from the interface. However, the difficulty in desorbing an adsorbed polymer (as compared with a small molecule) is only one of degree and the process is normally amenable to being treated by equilibrium thermodynamics (Silberberg, 1962b). The overall rate of adsorption may, in fact, consist of two separate rates, one of arrival or initial attachment with very few segment—surface contacts followed by one of spreading which leads to the formation of many contacts (Peterson and Kwei, 1961; Grant et al., 1975).

In the case of micromolecules, the shape of the isotherm is diagnostic of the mechanism underlying the adsorption process. Thus, on the basis of the initial slope, Giles et al. (1960) have been able to distinguish four classes or types of solid—solution adsorption (SSA) isotherms, referred to as S, L (for "Langmuir"), H (for "high affinity") and C (for "constant partition") (Fig. 2.2). Subsequently, a theoretical basis for each type together with the corresponding experimental verification were developed using a thermodynamic approach (Giles et al., 1974a, b).

S-type isotherms are rarely encountered with polymer systems. They occur when the activation energy for removal of the adsorbed species (E_A) is concentration-dependent and/or is reduced by a large negative contribution from that of the solvent (E_B), or from a second solute (E_C) if present. L-curves are found in situations where none of these energy terms is dependent on concentration. When $E_A >> E_B + E_C$, H-curves obtain since here the affinity of the solute for the substrate is high even at very low concentrations. They occur, for example, when the solute molecules which individually may be small, are adsorbed as a relatively large unit or aggregate. Since this is the case for a single macromolecule, the SSA isotherms of most polymers are of the H-type. As the name suggests, C-curves are found when there is a constant partition of solute between the bulk solution and the surface phase during adsorption.

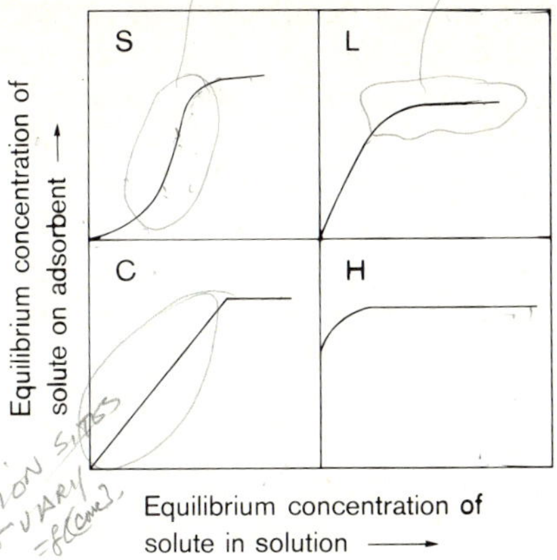

Fig. 2.2. The different classes or types of solid—solution adsorption isotherms according to Giles et al. (1960).

The initial slope may be regarded as a measure of the ease with which solute molecules can find vacant sites on the surface. The increase in slope with concentration for S-type isotherms indicates that there is a corresponding rise in the number of adsorbing sites. On the same basis, the decrease in slope for L-curves may be ascribed to the greater difficulty the solute has in finding surface vacancies as its solution concentration is increased. Experimentally, the SSA isotherms for a number of polymers are of the L-type particularly when they are measured at low concentration and surface coverage. However, the apparent fit of the data to the Langmuir equation (cf. equation 2.4) is probably fortuitous since linear flexible macromolecules can assume different conformations at the interface as well as show varying degrees of attachment to the surface, as adsorption progresses. The constancy in slope shown by the C-curves, would indicate that the number of adsorbing sites does not vary with concentration. This implies that new sites are being created as adsorption progresses. Only porous adsorbents seem capable of giving rise to linear isotherms. As the solute penetrates the solid structure through a micropore, more sites become exposed and accessible to the following molecules. Interestingly, the negatively charged humic and fulvic acid polymers give rise to C-curves when they adsorb onto montmorillonite (cf. Chapter 12).

If the determination of adsorption isotherms is experimentally straightforward, their interpretation is often difficult because the adsorption process, as mentioned earlier, is strongly influenced by polymer conformation. Uncharged linear flexible homopolymers in solution tend to adopt the conformation of a random coil rather than that of an extended, stretched out chain. At the solid/solution interface, however, the macromolecule prefers a conformation which allows for maximum segment—surface contact. Since the attachment of one segment would increase the probability of neighbouring segments to be adsorbed and the number of segments and/or functional groups per molecule is large, multiple bonding between polymer and surface is favoured. The net result is normally an interfacial conformation in which contiguous sequences of adsorbed segments or "trains" alternate with free three-dimensional "loops" extending away from the surface, the chain terminating at either end in two free-dangling "tails" (Fig. 2.1). Such a conformation may be regarded as a compromise between a random coil in solution and a randomly twisted or buckled two-dimensional surface structure of a fully collapsed chain.

The considerable amount of distortion and alteration in polymer shape occurring on adsorption leads to a large decrease in conformational energy which will oppose uptake. This effect, however, is more than offset by the greater increase in the interaction energy due to the establishment of numerous segment—surface contacts, even though the free energy increase per contact may be quite small. As we shall discuss more fully in the following section, all that is required for adsorption to occur is that the interaction energy per segment—surface attachment is greater than a certain critical value. Once this barrier is overcome, however, the excess energy per polymer chain rises rapidly and a large increase in adsorption occurs. In this regard, entropy effects play an important part.

One consequence of the strong polymer—surface bonding is that a large amount of material can be removed even from very dilute solutions. H-type isotherms therefore typify the adsorption from solution of polymeric compounds by solids, including clay minerals. Indeed, the most informative region of the isotherm corresponding to the initial rise where interactions between adsorbed chains and segments within a chain are minimal, is often experimentally inaccessible because the slope here is so steep.

Of the other factors which influence polymer adsorption, we may just briefly mention molecular weight, temperature, and the nature of the adsorbent and of the solvent.

There are situations where the plateau value or maximum extent of adsorption is virtually independent of polymer size. For reasons already stated, however, a more general observation is that adsorption increases with chain length up to a limiting value of molecular weight. This is at least true for non-porous adsorbents where adsorption is restricted only by the extent of the accessible surface. The molecular weight effect extends to polydisperse systems where the high molecular weight components are usually preferentially taken up. This behaviour might be expected on the ground that the frequency of segment—surface bonding will increase with chain length and with the number of functional groups per molecule. Clay and soil systems, however, are porous and an apparent reversal of the molecular weight effect is by no means uncommon. This is so because entry into the intercrystalline (intradomain) and/or intracrystalline (interlayer) spaces of the clay system or into the pores of a soil aggregate, may be physically impossible for polymers of a certain size range.

Similarly, the effect of temperature on adsorption is not as straightforward as it might at first sight appear. Thus, we might expect to find a negative temperature coefficient, or adsorption to decrease with a rise in temperature, as is normally the case with simple, monofunctional solutes. However, a zero or small positive temperature coefficient is more generally observed with polymeric compounds and this relates to entropy effects referred to above. Since the position of equilibrium (2.1) is far to the right, this observation may also be explained in terms of the way in which temperature affects the number of attached segments or the area required per attached segment (Fontana, 1971). Thus, theoretical considerations (Silberberg, 1962a, b) predict that train size decreases whilst the number of segments contained in loops increases with rising temperature.

On the other hand, solvent effects are qualitatively easier to predict in terms of polymer—solvent and surface—solvent interactions. When the latter are weak, adsorption increases with a decrease in polymer solubility. This is, of course, a reflection of the way in which a given solvent influences the coil dimensions of the polymer in solution. Since the macromolecular chain will be more tightly coiled in a "poor" than in a "good" solvent, the surface requirement in the former situation is less and the polymer can be more densely packed at the surface. The nature of the solvent, however, can be left out of the picture when only one type of solvent, e.g. water, is being used. This is the case with most clay and soil systems where the role of water has already been emphasised. Essentially, the polymer here must compete with water for sites around the exchangeable cation and, to a lesser extent, with water which is directly associated with the silicate surface. By analogy with the behaviour of simple organic molecules, the water in the primary hydration shell of the cation may be retained and act as a bridge between a polar group of the polymer and the cation.

Another important parameter in the solid—polymer interaction is the thickness of the adsorbed polymer layer which again is related to the coil dimension in solution. Thicknesses between 10 and 100 nm have been reported for a variety of polymer—solvent—mineral combinations, depending on the method of measurement and solute concentration (e.g. Stromberg, 1967; Fontana, 1971). Whereas a polymer may adopt a relatively flat conformation when adsorbed from dilute solutions, the size of its loops will increase with rising concentration until some maximum loop size is reached.

Likewise, the fraction of adsorbed segments is concentration dependent in that train size or the number of segment—surface contacts will decrease as loop size increases. However, for polymers possessing numerous functional groups capable of interacting with surface sites, no large conformational changes may occur as adsorption progresses. Using infrared spectroscopy, Fontana and Thomas (1961), for example, were able to demonstrate that the adsorption of polylauryl methacrylate (from organic solvents) onto silica occurred through hydrogen bonding between the ester carbonyl groups of the polymer and the surface silanols. This interaction gave rise to a shift in the carbonyl absorption frequency from that of the unattached segments, enabling the fraction of adsorbed segments to be determined without ambiguity. For this system about 36% of the polymer segments was in contact with the surface, a value which was virtually invariant with molecular weight and surface coverage. Using similar techniques, Thies et al. (1967) found a value of ~25% for polymethyl methacrylate. For polystyrene and poly(4-vinyl pyridine), however, the fraction of attached segments decreased with increasing surface coverage.

Due to interference by the solvent, such a direct approach is not well suited for aqueous clay—polymer systems. However, where intercalation can occur as it does with montmorillonite, X-ray diffraction analysis of the complex can provide a means of estimating the average depth and hence, the probable conformation of the polymer on interlayer surfaces (Greenland, 1963; Francis, 1973). Adsorbed layer thicknesses deduced in this way, however, may not apply to that portion of the polymer which adsorbs onto external crystal surfaces. Even if the polymer system were monodispersed, there is more scope for the molecules to form longer loops or thicker layers on these than on interlayer surfaces. Furthermore, different effective layer thicknesses or loop sizes may be possible for a given proportion of segments in trains.

2.3. THEORETICAL ASPECTS

Any theory of polymer adsorption must clearly take into account the conformational properties of the adsorbed chain and so requires a formidable amount of statistical mechanics. In addition, theoretical treatments must be capable of explaining the essential features of polymer behaviour, that is, of predicting the effect of the various parameters on adsorption. Existing theories on the adsorption of uncharged, flexible homopolymers at the solid/solution interface can adequately reproduce most of the observed facts. However, as already remarked on, quantitative experimental verification of theoretical predictions is not without difficulty because of the limited breadth of measurements and the uncertainty regarding the nature of the adsorbent. This situation is particularly true of clay—polymer systems. In addition, the different approaches used in developing these theories have also led to somewhat divergent conclusions since no single treatment gives equal weight to all aspects of polymer behaviour.

Most of the available theories consider the adsorbed polymer in isolation so that interactions between chains can be neglected. They further assume dilute solutions and an accessible surface whose area is much larger than that occupied by the polymer. Only the salient features and important conclusions of the various treatments will be given here.

The first serious attempt to develop a theory of polymer adsorption was made by Frisch, Simha, and Eirich in the nineteen-fifties (Frisch et al., 1953; Simha et al., 1953; Frisch and Simha, 1954, 1957; Frisch, 1955). Adsorption is formulated in terms of a diffusion equation, regarding the surface as a reflecting wall and the polymer as a Gaussian coil whose segments are attached on single surface sites.

The average number of adsorbed segments (ν) in a chain consisting of n segments is given by

$$\nu \sim f^{-0.5} n^{0.5} \tag{2.2}$$

where f is an inverse function of chain flexibility. For a situation where the attached sequence comprises only one segment and solvent adsorption is absent, the Frisch—Simha—Eirich (FSE) isotherm equation is

$$\frac{\theta}{1-\theta} \exp(2K_1\theta) = (Kc)^{1/\nu} \tag{2.3}$$

where K_1 and K are constants; K_1 is a measure of the interaction energy between adsorbed segments (in excess of their interactions in the bulk phase); K is the equilibrium constant; θ is the fraction of total surface covered by the polymer and c is the polymer concentration in solution. For $K_1 = 0$ and $\nu = 1$ equation (2.3) reduces to

$$\frac{\theta}{1-\theta} = Kc \tag{2.4}$$

which is the well-known Langmuir expression for the adsorption of simple (non-polymeric) monofunctional solutes.

In terms of isotherm shape, the FSE theory predicts a steeper initial rise, followed by a slower approach to a plateau value than does the Langmuir equation. More importantly, the polymer is attached to the surface by relatively few segments, the bulk of it being contained in loops. Further, adsorption may either increase or decrease with rising temperature; uptake, however, would be favoured from a poor solvent since here f would be small.

Using a thermodynamic approach, Gilliland and Gutoff (1960) as well as Forsman and Hughes (1963) have developed theories which lead to essentially similar conclusions. The latter predict that adsorption will increase with molecular weight (M) as reflected in the more rapid rise to a plateau of the isotherm with increasing n. Only at low solute concentration and for weak polymer—surface interactions would the isotherm resemble the Langmuir form and θ become proportional to $M^{0.5}$.

The FSE theory and related treatments of polymer adsorption tend to overcount the number of distinguishable chain conformations at the surface (Di Marzio, 1965). In order to avoid this, Silberberg (1962a, b) approached the problem by considering the possible equilibrium conformations of an isolated (linear, flexible) homopolymer at an infinite plane surface. He then derived equilibrium values for all variables from the partition function of the whole system.

Unlike the FSE theory, his treatment predicts an essentially extended (flat) surface conformation. In other words, the polymer is adsorbed with small loops and a high fraction (p) of segments in trains even when the segment—surface interaction energy (ϵ) is low. Thus, for $\epsilon \sim kT$ where k is the Boltzmann constant and T the absolute temperature, p exceeds 0.7. As indicated in the preceding section, the experimentally derived values of p are generally somewhat smaller (0.25—0.4), reflecting the influence of steric factors and chain interference in the surface layer on adsorption. Only when there is a good geometrical "fit" between polymer functional groups and surface sites, as between polyesters and silica, does p conform closely to the theoretical prediction (Joppien, 1974).

Nonetheless, uncharged linear homopolymers are apparently adsorbed with an appreciable proportion of their segments in direct contact with the surface. If the magnitude of p is relatively insensitive to variations in molecular weight and solute concentration, the amount adsorbed normally increases with molecular weight at moderately low concentrations. This dependence of adsorption on molecular weight, however, decreases for $n \sim 10^3$ becoming quite small for very large molecules. Hoeve's (1965) analysis indicates that large loops and small p values are possible when the polymer chain is sufficiently flexible and ϵ is small, a finding which is more in line with the FSE than the Silberberg treatment.

As a corollary, loop size will decrease and most segments will tend to be adsorbed as the polymer chain becomes less flexible. More recently, Skvortsov and Birshtein (1976) have come to broadly similar conclusions regarding the dependence of surface conformation on ϵ and chain flexibility.

The magnitude of ϵ is clearly of prime importance to polymer adsorption and a number of theoretical analyses have been carried out in order to evaluate this value. Of note here is the attempt of Di Marzio and McCrackin (1965) who, by using a lattice-walk model, have been able to define a critical or transitional point (ϵ_0) separating regions of "low" and "high" segment—surface adsorption energies. Thus in a plot of ϵ against p, this point occurs at $\epsilon_0 = kT \ln 2 \sim 0.7\ kT$ for $n = \infty$. It is at this point that the number of attached segments (ν) varies as the square root of n (cf. equation 2.2). For $\epsilon > \epsilon_0$, however, ν is proportional to n whilst for $\epsilon < \epsilon_0$, ν is constant and small. In the case of a real polymer, that is, for finite values of n and for $\epsilon > \epsilon_0$, the molecule lies essentially flat on the surface, a conformation which is more or less independent of molecular weight. Similar conclusions were reached by Rubin (1965) and by Motomura and Matuura (1969) with slight variations in the assigned value of ϵ_0.

Another approach to the problem is the statistical mechanical formulations used, inter alia, by Roe (1965), Hoeve et al. (1965), and Hesselink (1969). A feature of Roe's treatment is that for $\epsilon \leqslant \epsilon_0$ most of the unattached segments of the polymer are contained in tails rather than in loops, a situation which would favour polymer desorption.

More recently, Chan et al. (1975) have used a continuum theory which they claimed has the advantage of being able to quantitatively assess a whole range of conformational properties of the adsorbed polymer in function of the adsorption energy parameter. The density distribution of adsorbed segments with distance away from the surface, for example, is dependent on the relative magnitude of this parameter. Earlier, Di Marzio and McCrackin (1965) found that for $-\epsilon/kT = 1$, this density falls off exponentially with an increase in distance, most of the polymer segments being present close to the surface. Hoeve (1965) has stated similarly for a situation where the distance from the interface exceeds the effective thickness of the adsorbed layer, whilst Hesselink (1971) has derived density—distance relationships for three different loop size distributions.

More elaborate models have been described by Silberberg (1967, 1968) who takes into account the effects of self-exclusion, solute concentration, degree of polymerisation, as well as chain flexibility, polymer—solvent, and polymer—surface interactions, on adsorption behaviour. He concluded that whereas the concentration of trains is primarily dependent on polymer—surface interactions, that of loops and loop size are determined by solute concentration. The existence of relatively thick adsorbed layers at the surface should therefore not be identified with "true" multilayer formation (Silberberg, 1972). However, the development of a multilayer in the sense of polymer layers which have no single contact with the surface is possible in theory and, by no means, uncommon in practice. One criterion is that the number of molecules in these layers is less than that present in the "monolayer".

The question arises as to whether compact macromolecules would behave similarly to open-chain, flexible polymers discussed above. Silberberg's (1973) analysis would suggest that this is so. It is perhaps surprising that the adsorption of compact polymers onto solid surfaces has not received the attention it deserves in view of the fact that many proteins, enzymes, and soil humic substances occur as more or less tight, globular or spherical units. One reason suggested by Silberberg, is that besides being compact these molecules are normally charged. Their adsorption is therefore influenced by electrochemical factors in addition to those which apply to uncharged polymers in general. However, in situations where the number and the density of charges on the molecule are not too large and where adsorption is carried out at high ionic strength, such factors may be less important than size and shape effects in determining their behaviour at solid surfaces.

Polyelectrolyte chains can also undergo a stretching—coiling transformation together with the associated phenomenon of swelling—shrinking, in response to changes in solution pH and added electrolyte. Thus, as early as 1941 Staudinger noted that the addition of sodium hydroxide to polyacrylic acid solutions had a dramatic effect on their viscosity (Flaig, 1976). This is shown in Fig. 2.3 where the initial rise in viscosity is ascribable to the gradual uncoiling of the polyanionic chain through an increase in the number of dissociated carboxyl groups within the chain, leading to intramolecular charge repulsion. The maximum in the curve may be identified with the situation where the polymer chain is fully extended or stretched out. Beyond this point, the addition of NaOH results in chain recoiling (shrinking), presumably through charge screening at high ionic strength. Another classic example is provided by Oth and Doty (1952) who found that polymethacrylic acid in dilute solutions showed a gradual expansion with increasing degree of ionisation (α). Thus, for the fraction with a molecular weight of 93 000 and a contour length of 266 nm, the root-mean-square end-to-end distance of the polymer coil varied from 17 nm for the unionised state (α = 0.001) to a maximum of 118 nm at $\alpha = 0.71$.

These effects are of utmost importance and relevance to the use of polyelectrolytes as aggregating agents of clay and colloidal dispersions/suspensions (e.g. Michaels, 1954). The relationship between coil dimension and flocculating effectiveness will be further discussed later.

Because of the extreme complexity of the system, an adequate theoretical analysis of the polyelectrolyte—solid interaction is yet to be developed. It might be expected that the net segment—surface interaction energy for polyelectrolyte adsorption onto oppositely charged surfaces will be greatly in excess of 1 kT. On this basis, the proportion of attached segments would

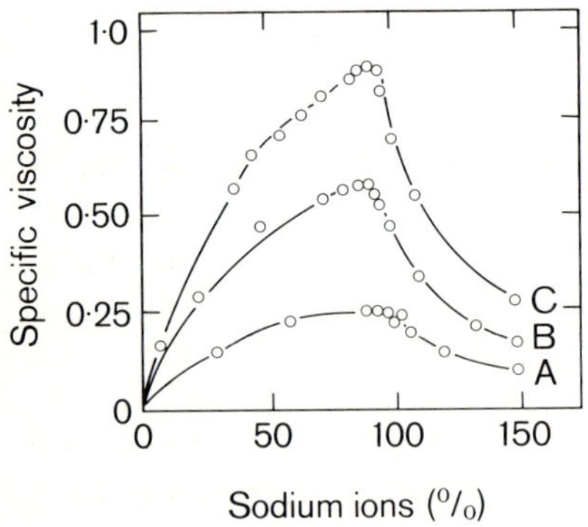

Fig. 2.3. Changes in the specific viscosity of polyacrylic acid solutions which occur on addition of increasing amounts of sodium hydroxide, as a function of the degree of polymerisation (n). Curve A, n = 640; curve B, n = 790; curve C, n = 1000 (after Staudinger, 1941 as quoted by Flaig, 1976).

approach unity, irrespective of chain length (Higuchi, 1961). This implies a very rapid collapse of the polymer chain onto the solid surface with hardly any looping, at least at low surface coverage. Some short loops may form when the clay surface becomes fully occupied by the polymer (Ueda and Harada, 1968). On the other hand, polyelectrolytes may be negatively adsorbed by surfaces of like charge due to mutual repulsion, as Parfitt and Greenland (1970) have observed for the interaction of polyuronides with Na^+-montmorillonite. This effect, however, can be nullified and reversed by saturating the exchange sites on the clay with polyvalent cations which are capable of acting as a bridge between the negatively charged functional groups of the polymer and the mineral surface. Indeed, the presence of polyvalent cations is generally necessary if anionic macromolecules are to adsorb onto negatively charged surfaces (Sommerauer et al., 1968). Polyanions may also be adsorbed onto the edges of clay crystals by hydrogen-bonding interactions (Emerson, 1956; Kohl and Taylor, 1961) or when the edge surface acquires positive charges at acid pH values (Warkentin and Miller, 1958; Mortensen, 1962).

More recently, Hesselink (1972) has derived equations for the adsorption of polyelectrolytes onto surfaces of varying charge in function of some solution variables. The expression for the driving force of adsorption (E) consists of electrical and non-electrical free energy gain terms. The effect of added electrolyte and pH are reflected in the dependence of E on the electrical double layer thickness (section 2.4) and the degree of ionisation of the polyelectrolyte. Thus, for $\alpha = 0$ the electrical term vanishes and the amount adsorbed becomes proportional to the solute concentration, giving rise to linear or C-type isotherms. For $\alpha = 0.1$, L-type isotherms are obtained when $n = 20$, and H-curves when $n = 200$. The theory further predicts that under certain conditions polyelectrolytes would not be adsorbed on a surface bearing charges of the same sign as the polymer.

2.4. COLLOID CHEMICAL ASPECTS

Most of the data on clay—polymer interactions relate to situations where solutions of the macromolecule are added to aggregated clay systems because the process has been studied in terms of its agronomic/pedological significance rather than in the colloid chemical sense.

However, an increasing amount of attention is being paid to the reactions of polymers with dilute aqueous suspensions of clays in which the particles are dispersed, that is, separated by relatively large distances. Indeed, the aggregating action of polymers, notably of water-soluble polyelectrolytes, on particulate colloidal dispersions is perhaps the most important aspect of the solid—polymer interaction (La Mer and Healy, 1963; Vincent, 1974). This is so because in aiding solid—liquid separation, this action forms the basis of a wide variety of industrial and technological processes, such as mineral processing, potable water and sewage treatment, paper making, sugar refining, and brewing (Dixon, 1967; Akers, 1975).

Of particular relevance to clay/soil systems is the use of synthetic polymers as "soil conditioners" in order to improve the physical conditions of soils which, in turn, influence soil fertility and crop production (De Boodt and Gabriels, 1976). It should be pointed out, however, that the polymer here is again added to an already aggregated system in which the soil/clay particles are in close proximity. Soil conditioners thus act more as an aggregate stabilising or strengthening agent than as a flocculant (cf. Chapter 6).

Besides acting as an aggregant, polymers can have a dispersive or protective effect on suspended solids when added in sufficiently large quantities. This action is ascribable to the formation of a coat or layer of the macromolecule around the solid particles, with loops and tails extending into the external solution. By this means, the particles are sterically prevented from entering each other's attraction sphere. The process is therefore generally known as steric stabilisation, the other type being charge stabilisation due to mutual repulsion of electrical double layers around the particles (cf. section 1.2). Theories of steric stabilisation have been proposed by a number of authors (Lyklema, 1968; Evans and Napper, 1973a, b; Vincent, 1974; Osmond et al., 1975). On the other hand, when very small amounts of a polymer are added to a charge-stabilised dispersion, no aggregation may occur but the particles become sensitised in that they are more susceptible to being coagulated by electrolytes.

It seems helpful at this point to define the various other terms used. In the colloid chemical literature the terms aggregation, coagulation, and flocculation are often used interchangeably. We prefer to use aggregation in the generic sense, reserving coagulation for particle aggregation induced by electrolyte addition, and flocculation for aggregation resulting from the linking or bridging of several particles by a polymer chain (La Mer, 1966). These concepts are schematically illustrated in Fig. 2.4.

A system in which hydrophobic colloidal particles usually of <1 μm equivalent spherical diameter (e.s.d.) are homogeneously distributed in the solvent phase, is referred to as a dispersion or sol. For larger particles which settle out relatively rapidly (under gravity), the term suspension is commonly used. However, as Van Olphen (1963) has pointed out, the distinction between these two states of particle division is purely arbitrary. This is particularly true for clay systems since, conventionally, clay consists of particles with an e.s.d. of <2 μm. For our purposes, "dispersion" and "suspension" will be used synonymously.

The effect of organic polymers on the stability of dispersed solids has been discussed in a series of reviews which have appeared over the last decade or so (Kuzkin and Nebera, 1966; Kini, 1970; Kitchener, 1972; Vincent, 1974; Akers, 1975). Here, we shall only touch on the salient fea-

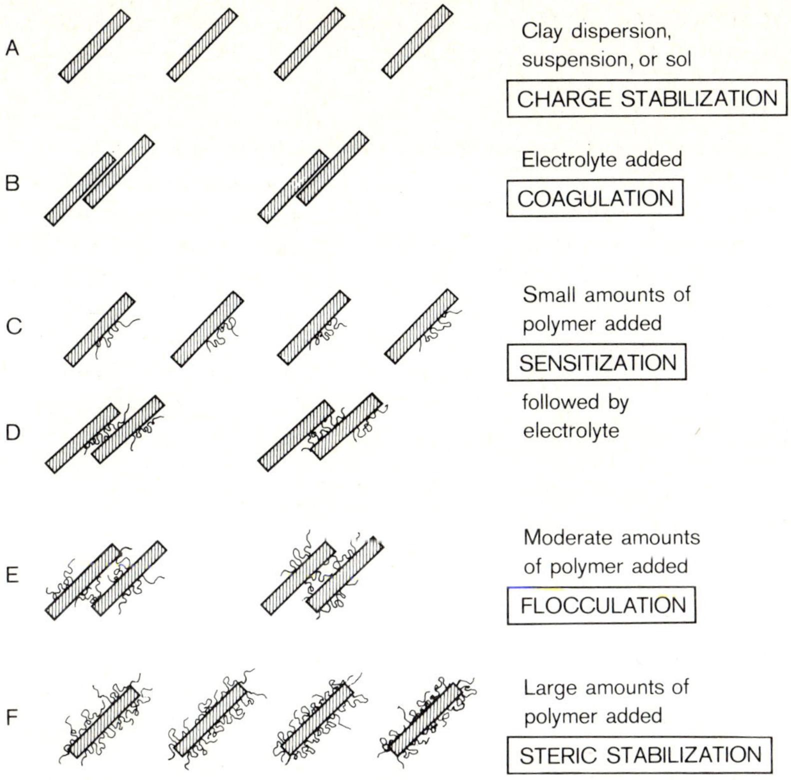

Fig. 2.4. Diagram showing the effect of electrolyte addition and polymer adsorption on the stability of aqueous clay dispersions.

tures, in particular on those points which seem relevant to clay mineral systems.

It is now generally accepted that the aggregation of particulate dispersions by polymers results from interparticle bridging by the macromolecule as it adsorbs onto more than one solid particle (Ruehrwein and Ward, 1952; Michaels, 1954; Linke and Booth, 1960). A direct, elegant demonstration of interparticle bridge formation in different clay—polymer systems has been provided by several workers using electron microscopy (Beutelspracher, 1955; Jordine, 1963; Audsley and Fursey, 1965). It should be mentioned in this connection that uncharged polymers are generally poor flocculants of clay suspensions since the molecules exist as randomly coiled units rather than as extended chains in solution. Non-ionic macromolecules, however, are very effective as soil conditioners because they can slowly spread out like a "coat of paint" over adjacent clay/soil surfaces (Greenland, 1963).

As already remarked on, the presence of polyvalent counterions is essential for, or at least promotes, the flocculation of particles by polyelectrolytes of like charge. Optimum flocculation normally occurs when the polymer chains achieve their maximum extension which depends on the pH and ionic strength of the system (Michaels, 1954; Slater et al., 1969). Both particle bridging by the polymer and charge reduction by the counterion are involved in the process. On the other hand, polyelectrolytes bearing charges opposite in sign to that of the particles, can cause aggregation in the absence of an inert electrolyte. Here, the neutralisation of the particle charge by the adsorbed polymer is of prime importance so that by our definition, both coagulation and flocculation are involved. Indeed, charge reversal has been reported for montmorillonite on adsorption of polycations beyond the exchange capacity of the mineral (Ueda and Harada, 1968). However, because polycations tend to collapse very rapidly onto the negatively charged clay surface, the extent of interparticle bridging formed is limited. By the same token, polycations are effective coagulants of clays and negatively charged colloids, in general, even at low concentrations (Black et al., 1966; Rebhun et al., 1969).

Experimentally, this effect may be assessed in a variety of ways, based on following the changes in a given property of either the total system (e.g. its rheology), the supernatant liquid (e.g. its optical transmission), or the aggregate formed (e.g. its filtration behaviour). Only in very few instances have the various methods been compared for a given system (Kuzkin et al., 1964; Slater and Kitchener, 1966).

La Mer and co-workers (see La Mer and Healy, 1963) favour the refiltration rate technique in which the collected filtrate (supernatant) is repassed through the aggregate or sediment bed, acting as a filter cake. Besides being simple, the method is claimed to be reproducible and amenable to theoretical analysis. Thus, in its simplest form (La Mer and Smellie, 1962), the initial polymer concentration (P_0) is related to the filtration rate through the sediment bed (Q) and the stable dispersion (Q_0), by the expression

$$\frac{P_0^{1/2}}{(Q - Q_0)^{1/8}} = A + BP_0 \tag{2.5}$$

where A and B are constants such that the ratio A/B is equal to the optimum concentration of flocculant (P_m).

Subsequently, Healy and La Mer (1962) have modified equation (2.5) taking into account the fact that only a fraction (p) of the total number of polymer segments (n) is attached to the solid surface. P_m is then given by

$$P_m = \left(1 + \frac{bk}{p}\right)^2 /b \tag{2.6}$$

where b and k are constants. For $P_0 = P_m$, the rate of filtration reaches a maximum (Q_m) which is considered to correspond to the condition of maximum flocculation or interparticle bridging. This is also the point at which the fraction of the surface covered by the polymer (θ) is ~0.5 (cf. equation 2.4) and where floc stability (at a constant shear rate) is maximal.

When tested on the calcium phosphate slime—polyacrylamide system, there was good agreement between experiment and theory. Equation (2.5) could also account satisfactorily for the behaviour of kaolinite suspensions flocculated with an anionic polyacrylamide but the theory broke down with the cationic and uncharged forms of the polymer (Dollimore and Horridge, 1972). The theory has also been criticised by Shyluk and Smith (1969) and more so by Slater and Kitchener (1966) who questioned the underlying assumptions.

Thus, in comparing several methods of analysing the dispersion state of a fluorite sample flocculated with a non-ionic polyacrylamide, the latter workers noted that only the refiltration rate curve showed a well-defined maximum (Fig. 2.5) whereas the maxima for the other parameters were broad, if at all present. This maximum, however, did not appear to correspond to maximum flocculation nor to optimum floc size and strength which continued to increase at polymer dosages in excess of P_m. The observed discrepancies might in part be due to the fact that the conditions of agitation and mixing, used by Slater and Kitchener, were less than optimal so that polymer adsorption/usage was "patchy"/inefficient (Healy, 1971). These factors,

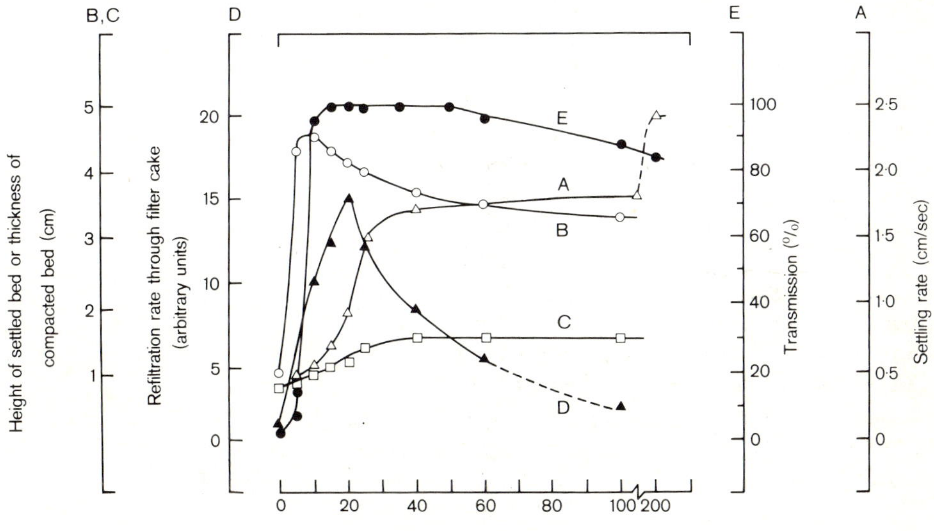

Fig. 2.5. Comparison of various methods of assessing the state of a fluorite dispersion in the presence of increasing amounts of polyacrylamide (after Slater and Kitchener, 1966).

as also the method of bringing the reagents together, can have a profound influence on the degree of flocculation (La Mer et al., 1957; Fleer and Lyklema, 1969).

The mixing effect is also reflected in the rate of flocculation which generally follows second-order kinetics. Thus, were bare and polymer-covered particles are brought together, the rate of disappearance of primary particles is given by (Fleer, 1971)

$$dN_1/dt = -kfN_1^2 \qquad (2.6)$$

N_1 is the number of particles per unit volume, t the time, k the rate constant, and f a function of the volume fraction of sol containing the covered particles. On the other hand, where the polymer is added in a single step to a stirred suspension of solids, f is related to the surface coverage by the polymer, provided that adsorption rapidly reaches a steady state with respect to the rate of particle collision (La Mer and Healy, 1963).

The adsorption of an uncharged homopolymer at a charged interface will clearly alter the structure of the electrical double layer around the particle. To a first approximation, we may conclude with Lyklema (1976) that any modification of the inner (Stern) layer is due to the presence of trains in this layer, whereas that of the outer (Gouy) layer is to be ascribed to loop and tail segments of the adsorbed macromolecule. This concept is illustrated in Fig. 2.6, where the extension of the Stern layer is assumed to coincide with the outer Helmholtz plane (O.H.P.) and the plane of shear (Fig. 1.14). As a consequence, both the long-range Van der Waals attraction and the repulsion

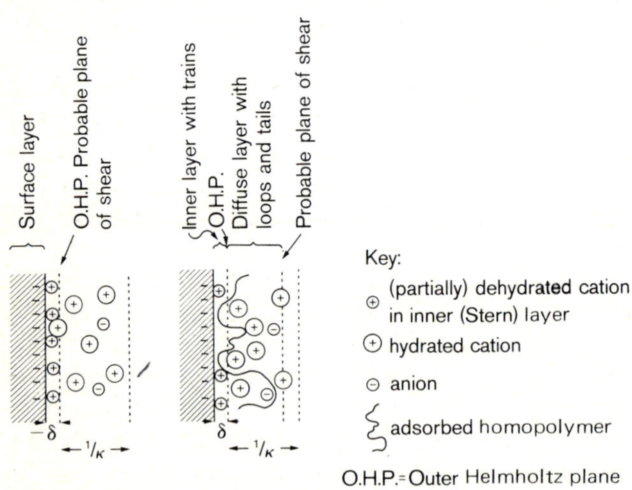

Fig. 2.6. The effect of adsorbed uncharged homopolymers on the structure of the electrical double layer at a charged surface (after Vincent, 1974).

terms describing the total interaction between particles, as formulated by the DLVO theory (cf. Fig. 1.15), have to be modified.

The presence of a homopolymer at the surface may either decrease or increase the interparticle attraction (V_A). However, if the distance between the particle centres is kept constant, the net effect is to enhance V_A (Vincent, 1974). The influence on V_A of adsorbed layer composition and segment density in this layer has been discussed by Vincent (1973).

With reference to Fig. 2.6, the repulsion term (V_R) in the DLVO theory is modified by the following factors (Vincent, 1974): (a) changes in surface charge, (b) displacement of specifically adsorbed counterions by trains, (c) displacement of oriented water dipoles, (d) changes in dielectric constant and thickness of the inner layer by train segments, and (e) changes in the outer layer by loops and tails.

Vincent has given an equation relating the effect of (a) to (d) on the potential ψ_δ at the outer Helmholtz plane. The effect under (e) should also be reflected in ψ_δ if the segment density in the outer layer is small as would be the case at low adsorption. For two spherical particles an expression for V_R in terms of ψ_δ, particle surface—surface separation, and Stern layer thickness, can be derived. However, the readily determined electrokinetic or zeta potential (ζ) will be reduced by the presence of an adsorbed polymer, provided that changes in potential and charge distribution due to adsorption are not significant. The decrease in ζ is ascribable to the displacement of the probable plane of shear away from the O.H.P. (Fig. 2.6). Since the approximation $\psi_\delta \sim \zeta$ is no longer valid, some relationship between ψ_δ and ζ must be established or assumed before V_R can be evaluated. A different kind of approach is necessary when the segment density in the outer diffuse layer is large (e.g. Lyklema, 1976) whilst for polyelectrolyte adsorption a new model would be required.

Theoretical analyses have also been carried out for the situation where an uncharged homopolymer forms a bridge between two parallel plates. Thus, Di Marzio and Rubin (1971) have calculated the changes in tension (τ) between plates as a function of segment—surface interaction energy (ϵ), plate separation (s), and segment length (l). The results are shown in Fig. 2.7 where the curves for $s = 4l$, $6l$ and $8l$ are seen to intersect at $\tau = 0$ and $\epsilon = 0.693\ kT$. It is interesting to note that for $\epsilon > 0.693\ kT$ (i.e. there is a net attraction between plates), the tension does not continue to rise with increasing ϵ. Rather, τ rises to a maximum and then declines. This behaviour is explained in terms of two opposing effects in that with increasing ϵ the number of segments in a bridge decreases giving rise to a tauter bridge. Concomitantly, the number of bridges decreases leading to a fall in tension. The latter effect predominates when ϵ becomes greater than a given maximum value. For $\epsilon < 0.693\ kT$, the tension is negative and there is a net repulsion which also increases as the separation between plates becomes smaller.

The other approach is due to Fleer (1971) who introduced three forces

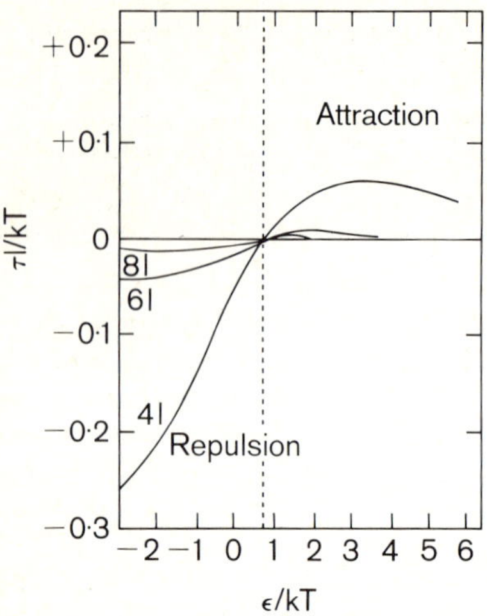

Fig. 2.7. Variations in tension (τ) between two parallel plates bridged by an uncharged polymer chain as a function of net segment—surface interaction energy (ϵ) and plate separation (s). The three curves shown were derived on theoretical grounds for $s = 4l$, $6l$, and $8l$ by Di Marzio and Rubin (1971), where l is the length of a segment.

additional to V_A and V_R. These are an adsorption attraction term, a steric repulsion term, and a mixing term, arising from changes in segment distribution and concentration among bridges, loops and trains between two adsorbing surfaces.

Experimentally, the action of polymeric flocculating agents on particulate colloidal dispersions is governed by a number of principles which Kitchener (1972) has summarised as follows: (i) the polymer must be soluble in the medium and yet capable of competing effectively with solvent molecules for surface sites, (ii) although there is an optimum dosage, the concentration for effective flocculation is much less than would be needed for surface saturation (θ = 0.5), (iii) at concentrations beyond the optimum, flocculation efficiency decreases until eventually the system becomes sterically stabilised, (iv) the effectiveness/dosage ratio is optimised when the polymer solution is added to the dispersion using a "flash mixing" technique since this leads to a rapid, uniform distribution of polymer over the accessible surfaces, (v) flocculation is initiated by the relatively rapid aggregation of sub-micron particles through diffusional collisions (perikinetic flocculation), followed by a slower growth of supra-micron size flocs through hydrodynamic collisions (orthokinetic flocculation), (vi) in order to promote the latter process and keep the flocs in suspension until all the particles have aggregated, a "conditioning" period with gentle stirring is recommended, (vii) although flocs are generally stronger than coagula, the former can be disrupted by applying a sufficiently large shearing force; the flocs so broken cannot reform to the original size on relaxing or removing the force, (viii) under steady state conditions (low rate of shear for a sufficiently long period), the size of flocs is related to the extent and strength of interparticle bridging which, in turn, depends on the polymer chain length and the segment—surface interaction energy parameter, (ix) differences in effectiveness of a flocculant between suspensions in a similar medium reflects differences in adsorption strength whereas differences between media for the same suspension may be ascribed to changes in either polymer—surface interaction or effective chain length of the polymer.

REFERENCES

Akers, R.J., 1975. Flocculation. The Institution of Chemical Engineers, London, 32 pp.

Audsley, A. and Fursey, A., 1965. Examination of a polysaccharide flocculant and flocculated kaolinite by electron microscopy. Nature (London), 208: 753—754.

Beutelspracher, H., 1955. Wechselwirkung zwischen anorganischen und organischen Kolloiden des Bodens. Zeitschrift für Pflanzenernährung, Düngung, Bodenkunde, 69: 108—115.

Black, A.P., Birkner, F.B. and Morgan, J.J., 1966. The effect of polymer adsorption on the electrokinetic stability of dilute clay suspensions. Journal of Colloid and Interface Science, 21: 626—648.

Chan, D., Mitchell, D.J. and White, L., 1975. Phase transitions in adsorbed polymer systems. Faraday Discussions of the Chemical Society, No. 59: 181—188.

De Boodt, M. and Gabriels, D. (Editors), 1976. Third International Symposium on Soil Conditioning. Mededelingen Fakulteit Landbouwwetenschappen Rijksuniversiteit Gent, 41: 464 pp.

Di Marzio, E.A., 1965. Proper accounting of conformations of a polymer near a surface. The Journal of Chemical Physics, 42: 2101—2106.

Di Marzio, E.A. and McCrackin, F.L., 1965. One-dimensional model of polymer adsorption. The Journal of Chemical Physics, 43: 539—547.

Di Marzio, E.A. and Rubin, R.J., 1971. Adsorption of a chain polymer between two plates. The Journal of Chemical Physics, 55: 4318—4336.

Dixon, J.K., 1967. Floccculation. Encyclopedia for Polymer Science and Technology, 7: 64—78.

Dollimore, D. and Horridge, T.A., 1972. The filtration of beds of china clay flocculated by polyacrylamide. Powder Technology, 5: 111—114.

Emerson, W.W., 1956. Synthetic soil conditioners. Journal of Agricultural Science (Cambridge), 47: 117—121.

Evans, R. and Napper, D.H., 1973a. Steric stabilization. I. Comparison of theories with experiment. Kolloid Zeitschrift und Zeitschrift für Polymere, 251: 409—414.

Evans, R. and Napper, D.H., 1973b. Steric stabilization. II. Generalization of Fischer's solvency theory. Kolloid Zeitschrift und Zeitschrift für Polymere, 251: 329—336.

Flaig, W., 1976. Contributions of biochemistry of soil organic matter to soil conditioning. Mededelingen Fakulteit Landbouwwetenschappen Rijksuniversiteit Gent, 41: 23—40.

Fleer, G.J., 1971. Polymer adsorption and its effect on colloidal stability. Theoretical and experimental study on the poly(vinyl alcohol)—silver iodide system. Mededelingen Landbouwhogeschool Wageningen, No. 71-20, 143 pp.

Fleer, G.J. and Lyklema, J., 1969. Stability of silver iodide sols in the presence of small amounts of poly(vinyl alcohol). Proceedings of the International Congress on Surface Activity, Barcelona, 2: 247—256.

Fontana, B.J., 1971. Adsorption of biological analog molecules on nonbiological surfaces. Polymers. In: M.L. Hair (Editor), The Chemistry of Biosurfaces. Marcel Dekker, New York, N.Y., pp. 83—142.

Fontana, B.J. and Thomas, J.R., 1961. The configuration of adsorbed alkyl methacrylate polymers by infrared and sedimentation studies. The Journal of Physical Chemistry, 65: 480—487.

Forsman, W.C. and Hughes, R.E., 1963. Adsorption theory for flexible linear polymer molecules. The Journal of Chemical Physics, 38: 2130—2135.

Francis, C.W., 1973. Adsorption of polyvinyl pyrrolidone on reference clay minerals. Soil Science, 115: 40—54.

Frisch, H.L., 1955. Polymer chain configuration near a boundary exerting forces. The Journal of Physical Chemistry, 59: 633—636.

Frisch, H.L. and Simha, R., 1954. The adsorption of flexible macromolecules. II. The Journal of Physical Chemistry, 58: 507—512.

Frisch, H.L. and Simha, R., 1957. Statistical mechanics of flexible high polymers at surfaces. The Journal of Chemical Physics, 27: 702—706.

Frisch, H.L., Simha, R. and Eirich, F.R., 1953. Statistical mechanics of polymer adsorption. The Journal of Chemical Physics, 21: 365—366.

Giles, C.H., MacEwan, T.H., Nakhwa, S.N. and Smith, D., 1960. Studies in adsorption. XI. A system of classification of solution adsorption isotherms and its use in diagnosis of adsorption mechanisms and in measurement of specific surface areas of solids. Journal of the Chemical Society, 3973—3993.

Giles, C.H., Smith, D. and Huitson, A., 1974a. A general treatment and classification of the solute adsorption isotherm. I. Theoretical. Journal of Colloid and Interface Science, 47: 755—765.

Giles, C.H., D'Silva, A.P. and Easton, I.A., 1974b. A general treatment and classification of the solute adsorption isotherm. II. Experimental interpretation. Journal of Colloid and Interface Science, 47: 766—778.

Gilliland, E.R. and Gutoff, E.B., 1960. The equilibrium adsorption of heterogeneous polymers. The Journal of Physical Chemistry, 64: 407—410.

Grant, W.H., Morrissey, B.W. and Stromberg, R.R., 1975. The conformation of adsorbed polystyrene. In: L.H. Lee (Editor), Adhesion Science and Technology. Plenum Press, New York, N.Y., pp. 43—54.

Greenland, D.J., 1963. Adsorption of polyvinyl alcohols by montmorillonite. Journal of Colloid Science, 18: 647—664.

Healy, T.W., 1971. Selective adsorption of organics on inorganic surfaces. In: S.D. Faust and J.V. Hunter (Editors), Organic Compounds in Aquatic Environments. Marcel Dekker, New York, N.Y., pp. 187—212.

Healy, T.W. and La Mer, V.K., 1962. The adsorption—flocculation reactions of a polymer with an aqueous colloidal dispersion. The Journal of Physical Chemistry, 66: 1835—1838.

Hesselink, F.Th., 1969. Density distribution of a terminally adsorbed macromolecule. The Journal of Physical Chemistry, 73: 3488—3490.

Hesselink, F.Th., 1971. On the theory of the stabilization of dispersions by adsorbed macromolecules. I. Statistics of the change of some configurational properties of adsorbed macromolecules on the approach of an impenetrable interface. The Journal of Physical Chemistry, 75: 65—71.

Hesselink, F.Th., 1972. On the adsorption of polyelectrolyte macromolecules on a flat interface. An approximate theory for low potentials. Journal of Electroanalytical Chemistry, 37: 317—325.

Higuchi, W.I., 1961. Effects of short range surface—segment forces on the configuration of an adsorbed flexible chain polymer. The Journal of Physical Chemistry, 65: 487—491.

Hoeve, C.A.J., 1965. Density distribution of polymer segments in the vicinity of an adsorbing interface. The Journal of Chemical Physics, 43: 3007—3008.

Hoeve, C.A.J., Di Marzio, E.A. and Peyser, P., 1965. Adsorption of polymer molecules at low surface coverage. The Journal of Chemical Physics, 42: 2558—2565.

Hughes, R.E. and Von Frankenberg, C.A., 1963. High polymers. Annual Review of Physical Chemistry, 14: 291—312.

Joppien, G.R., 1974. Strukturuntersuchungen an Adsorptionsschichten makromolekularer Stoffe. 1. Haftstellenzahlen von linearen Polyestern und Polystyrol an Aerosiloberflächen. Die Makromolekulare Chemie, 175: 1931—1954.

Jordine, E.St. A., 1963. Fine structure of polyelectrolytes and clay—organic complexes. Journal of Electron Microscopy, 12: 236—239.

Kini, K.A., 1970. Recent developments in the flocculation of suspensions of solids with organic polymers. Mineral Processing Information Note 6, Warren Spring Laboratory, Stevenage, 55 pp.

Kipling, J.J., 1963. Adsorption from Solutions of Non-electrolytes. Academic Press, New York, N.Y., pp. 134—159.

Kitchener, J.A., 1972. Principles of action of polymeric flocculants. British Polymer Journal, 4: 217—229.

Kohl, R.A. and Taylor, S.A., 1961. Hydrogen bonding between the carbonyl group and Wyoming bentonite. Soil Science, 91: 223—227.

Koral, J., Ullman, R. and Eirich, F.R., 1958. The adsorption of polyvinyl acetate. The Journal of Physical Chemistry, 62: 541—550.

Kuzkin, S.F. and Nebera, V.P., 1966. Synthetic Flocculants in Dewatering Processes. National Lending Library, Boston Spa, 344 pp.

Kuzkin, S.F., Nebera, V.P. and Zolin, S.N., 1964. Aspects of the theory of suspensions flocculation by polyacrylamides. Proceedings of the 7th International Mineral Processing Congress, New York, N.Y., Part 7: 347—357.

La Mer, V.K., 1966. Filtration of colloidal dispersions flocculated by anionic and cationic polyelectrolytes. Discussions of the Faraday Society, No. 42: 248—254.

La Mer, V.K. and Healy, T.W., 1963. Adsorption—flocculation reactions of macromolecules at the solid—liquid interface. Reviews of Pure and Applied Chemistry, 13: 112—133.

La Mer, V.K. and Smellie, R.H., 1962. Theory of flocculation, subsidence and refiltration rates of colloidal dispersions flocculated by polyelectrolytes. Clays and Clay Minerals, 9: 295—314.

La Mer, V.K., Smellie, R.H. and Lee, P.K., 1957. Flocculation, subsidence and filtration of phosphate slimes. Flocculation by gums and polyelectrolytes and their influence on filtration rate. Journal of Colloid Science, 12: 230—239.

Linke, W.F. and Booth, R.B., 1960. Physical chemical aspects of flocculation by polymers. Transactions of the American Institute of Mining, Metallurgical and Petroleum Engineers, 217: 364—371.

Lyklema, J., 1968. Principles of the stability of lyophobic colloidal dispersions in non-aqueous media. Advances in Colloid and Interface Science, 2: 65—114.

Lyklema, J., 1976. Inference of polymer adsorption from electrical double layer measurements. Pure and Applied Chemistry, 46: 149—156.

Michaels, A.S., 1954. Aggregation of suspensions by polyelectrolytes. Industrial and Engineering Chemistry, 46: 1485—1490.

Mortensen, J.L., 1962. Adsorption of hydrolyzed polyacrylonitrile on kaolinite. Clays and Clay Minerals, 9: 530—545.

Motomura, K. and Matuura, R., 1969. Conformation of adsorbed polymeric chain. II. The Journal of Chemical Physics, 50: 1281—1287.

Osmond, D.W.J., Vincent, B. and Waite, F.A., 1975. Steric stabilisation: A reappraisal of current theory. Colloid and Polymer Science, 253: 676—682.

Oth, A. and Doty, P., 1952. Macro-ions. II. Polymethacrylic acid. The Journal of Physical Chemistry, 56: 43—50.

Parfitt, R.L. and Greenland, D.J., 1970. Adsorption of polysaccharides by montmorillonite. Soil Science Society of America Proceedings, 34: 862—866.

Patat, F., Killmann, E. and Schliebener, C., 1964. Die Adsorption von Makromolekulen aus Lösung. Fortschritte der Hochpolymeren Forschung, 3: 332—393.

Peterson, C. and Kwei, T.K., 1961. The kinetics of polymer adsorption onto solid surfaces. The Journal of Physical Chemistry, 65: 1330—1333.

Pouchly, J., 1970. Behaviour of macromolecules on the phase boundary. II. Conformational statistics of the linear chain in a pore. The Journal of Chemical Physics, 52: 2567—2575.

Rebhun, M., Narkis, N. and Wachs, A.M., 1969. Effect of polyelectrolytes in conjunction with bentonitic clay on contaminants removal from secondary effluents. Water Research, 3: 345—355.

Roe, R.J., 1965. Conformation of an isolated polymer molecule at an interface. II. Dependence on molecular weight. The Journal of Chemical Physics, 43: 1591—1598.

Rubin, R.J., 1965. Random-walk model of chain-polymer adsorption at a surface. The Journal of Chemical Physics, 43: 2392—2407.

Ruehrwein, R.A. and Ward, D.W., 1952. Mechanism of clay aggregation by polyelectrolytes. Soil Science, 73: 485—492.

Shyluk, W.P. and Smith, R.W., 1969. Poly(1,2-dimethyl-5-vinylpyridinium methyl sulfate). IV. Flocculation of crystalline silica. Journal of Polymer Science, Part A, 2: 27—36.

Silberberg, A., 1962a. The adsorption of flexible macromolecules. Part I. The isolated macromolecule at a plane interface. The Journal of Physical Chemistry, 66: 1872—1883.

Silberberg, A., 1962b. The adsorption of flexible macromolecules. Part II. The shape of the adsorbed molecule; the adsorption isotherm, surface tension, and pressure. The Journal of Physical Chemistry, 66: 1884—1907.

Silberberg, A., 1967. Adsorption of flexible macromolecules. III. Generalized treatment of the isolated macromolecule; the effect of self-exclusion. The Journal of Chemical Physics, 46: 1105—1114.

Silberberg, A., 1968. Adsorption of flexible macromolecules. IV. Effect of solvent—solute interactions, solute concentration, and molecular weight. The Journal of Chemical Physics, 48: 2835—2851.

Silberberg, A., 1971. The behaviour of macromolecules at phase boundaries. Pure and Applied Chemistry, 26: 583—591.

Silberberg, A., 1972. Multilayer adsorption of macromolecules. Journal of Colloid and Interface Science, 38: 217—226.

Silberberg, A., 1973. The effect of conformation on the interaction of macromolecules with surfaces. The adsorption of compact macromolecules. Israel Journal of Chemistry, 11: 153—160.

Simha, R., Frisch, H.L. and Eirich, F.R., 1953. Adsorption of flexible macromolecules. The Journal of Physical Chemistry, 57: 584—589.

Skvortsov, A.M. and Birshstein, T.M., 1976. Study of the structure of adsorbed chains by the Monte-Carlo method. Vysokomolekulyarnye Soedineniya, Seriya A, 18: 2466—2471.

Slater, R.W. and Kitchener, J.A., 1966. Characteristics of flocculation of mineral suspensions by polymers. Discussions of the Faraday Society, No. 42: 267—284.

Slater, R.W., Clark, J.P. and Kitchener, J.A., 1969. Chemical factors in the flocculation of mineral slurries with polymeric flocculants. Proceedings of the British Ceramic Society, No. 13: 1—12.

Sommerauer, A., Sussman, D.L. and Stumm, W., 1968. The role of complex formation in the flocculation of negatively charged sols with anionic polyelectrolytes. Kolloid Zeitschrift und Zeitschrift für Polymere, 225: 147—154.

Staudinger, H., 1941. Organische Kolloidchemie, 2nd Edition. F. Vieweg Verlag, Braunschweig, 243 pp.

Stromberg, R.R., 1967. Adsorption of polymers. In: R.L. Patrick (Editor), Treatise on Adhesion and Adhesives, Vol. 1. Marcel Dekker, New York, N.Y., pp. 69—118.

Theng, B.K.G., 1974. The Chemistry of Clay—Organic Reactions. Adam Hilger Ltd., London, 343 pp.

Thies, C., Peyser, P. and Ullman, R., 1967. The determination of polymer structure at a liquid—solid interface by infrared analysis. Proceedings of the 4th International Congress on Surface Active Substances, Brussels, 2: 1041—1051.

Ueda, T. and Harada, S., 1968. Adsorption of cationic polysulfone on bentonite. Journal of Applied Polymer Science, 12: 2395—2401.

Van Olphen, H., 1963. An Introduction to Clay Colloid Chemistry. Interscience Publishers, New York, N.Y., 301 pp.

Vincent, B., 1973. The van der Waals attraction between colloid particles having adsorbed layers. II. Calculation of interaction curves. Journal of Colloid and Interface Science, 42: 270—285.

Vincent, B., 1974. The effect of adsorbed polymers on dispersion stability. Advances in Colloid and Interface Science, 4: 193—277.

Warkentin, B.P. and Miller, R.D., 1958. Conditions affecting the formation of the montmorillonite—polyacrylic acid bond. Soil Science, 85: 14—18.

Part II

INTERACTIONS OF CLAY MINERALS WITH SYNTHETIC POLYMERS

Chapter 3

UNCHARGED OR NON-IONIC POLYMERS

3.1. INTRODUCTION

Among the wide variety of uncharged, linear polymers which are commercially available or can be synthesised in the laboratory, only the water-soluble varieties have, for obvious reasons, been extensively studied as complex formers with clays. Such polymers invariably possess numerous polar groups along their chain, capable of being solvated by water. It is the functional groups which are also involved in interactions with clay mineral surfaces.

Unlike polyelectrolytes whose chains are more or less stretched out or extended due to intramolecular charge repulsion, non-ionic polymers tend to adopt a random coil conformation in solution. For this reason, the latter are relatively poor flocculants of dilute particulate dispersions unless the coil dimension is of the order of the mean interparticle separation. Accordingly, the flocculating efficiency of non-ionics increases with molecular weight. On contact with an adsorbing substrate, however, the molecules tend to collapse and spread out over the surface so that uncharged polymers are well suited for stabilising an existing structure, such as natural soil aggregates. Emerson (1956), for example, noted that soil crumbs containing 0.5% (w/w) of polyvinyl alcohol did not slake or disperse when immersed in water whereas the untreated aggregates did so. The use of synthetic polymers as soil conditioners, that is, agents capable of improving the physical conditions of soil, will be described more fully in Chapter 6.

Because polyvinyl alcohol is so effective in stabilising natural soil aggregates, its behaviour towards clays and soil minerals has received a great deal of attention. The other water-soluble, non-ionic polymers whose adsorption by clays has been studied in some detail, are polyacrylamide. polyethylene glycol (or polyethylene oxide), and polyvinyl pyrrolidone. Unlike most synthetic macromolecules, polyvinyl alcohol (PVA) is not made from its corresponding monomer but is obtained by the controlled hydrolysis of polyvinyl acetate. The degree of hydrolysis is usually expressed in terms of the number of residual acetyl groups per 100 segments, x. The chemical structures of these polymers are shown in the Appendix; the interactions of each type with clays and related minerals are described separately below.

3.2. FORMATION AND PROPERTIES OF COMPLEXES

Quite early on, Emerson (1960) reported that the (crystalline) swelling in water of Ca^{2+}-montmorillonite flakes was inhibited after being treated with

PVA. Since this resistance to slaking was maintained on washing the flakes with dilute solutions of either sodium hydroxide or pyrophosphate, he inferred that the polymer was linked to the basal surfaces of montmorillonite through a series of hydrogen bonds. On the other hand, carboxylated polymers, (e.g. polyacrylic acid) were likely to be attached to the edge surfaces of the clay crystals, forming "peripheral complexes". Hydrogen bonding between the hydroxyl groups of PVA and the basal oxygens of montmorillonite surfaces, was also proposed by Emerson and Raupach (1964) who further observed that adsorption of PVA did not affect the exchangeability of the inorganic cations initially present at the clay surface.

The factors influencing the montmorillonite—PVA interaction have been systematically examined by Greenland (1963). To this end, he used polymers containing 12 residual acetyl groups per 100 segments (the "G" series) with varying degrees of polymerisation, and montmorillonite saturated with Ca^{2+}, Cs^{+}, and Na^{+} ions. In some instances, the adsorption of the nearly completely hydrolysed polyvinyl acetates ("N" series) by the montmorillonite samples was also measured for comparative purposes. The properties of the polymers used are set out in Table 3.1.

The adsorption of PVA by montmorillonite is very sensitive to the (final) suspension strength of the clay—polymer system as well as to the nature of the exchangeable cation. As illustrated in Fig. 3.1, there was a marked increase in the maximum extent of adsorption or plateau value with decreasing suspension concentrations up to a certain level. For Na^{+}-montmorillonite this level corresponded to about 0.8 g PVA per g clay reached at a suspension concentration of 0.5% w/v. Kuznetsova and Serb-Serbina (1968) have reported similarly but the levelling out point for their clay sample occurred at a lower clay concentration.

TABLE 3.1

Some properties of polyvinyl alcohols (Goshenol products) supplied by the Nippon Synthetic Chemical Corporation, Japan (from Greenland, 1963)

Series	Designation	Degree of hydrolysis (%)	Residual acetyl groups per 100 segments (x)	Degree of poly-merisation	Molecular weight ($\times 10^3$)		
					(a)	(b)	(c)
"G"	GL-05	88	12	500	24	25	25
	GM-14	88	12	1400	67.2	70	—
	GH-20	88	12	2000	96	100	80
"N"	NL-05	>98.5	<1.5	500	24	22	—
	NH-22	>98.5	<1.5	2200	105.6	100	75

From: (a) degree of polymerisation; (b) viscosity data, and (c) ultracentrifugation.

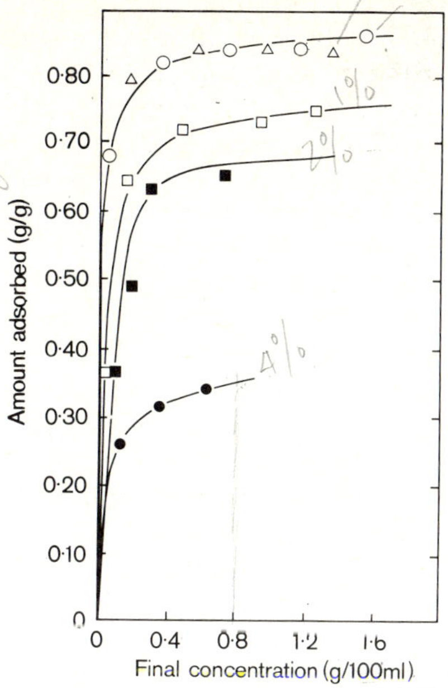

Fig. 3.1. Isotherms for the adsorption of polyvinyl alcohol (GH-20) from aqueous solutions by Na^+-montmorillonite. The polymer solutions were added to air-dry clay to form suspensions of different strength: (○) 0.25% w/v; (△) 0.5% w/v; (□) 1% w/v; (■) 2% w/v; and (●) 4% w/v (after Greenland, 1963).

The same trend was observed with the calcium and cesium clay systems except that uptake (of GL-05) reached a maximum at ~0.30 and 0.15 g g^{-1}, respectively (Fig. 3.2). The amount adsorbed at a given polymer concentration was also decreased in the presence of $NaNO_3$ solutions of increasing strength, as shown in Fig. 3.3 for the adsorption of GM-14 by Na^+-montmorillonite; the points at which the basal spacing of the parent (untreated) sodium clay reached a certain value are also indicated. Thus, at electrolyte concentrations for which basal spacings of ~1.9 and 1.5 nm were observed, the amount adsorbed by Na^+-montmorillonite was comparable to that given by the calcium and cesium clays, respectively (cf. Fig. 3.2).

These and other observations, described below, strongly suggest that surface accessibility is the most important single factor determining the interaction process. Accessibility, in turn is dependent on the distance separating individual silicate layers within a single clay crystal in relation to the dimension of the polymer coils in solution. In dilute (<1% w/v) aqueous suspensions of Na^+-montmorillonite, the average interlayer distance is greater than 13 nm whereas for the calcium and cesium forms, this distance does not exceed ~1 and ~0.4 nm, respectively (Norrish, 1954). These values should

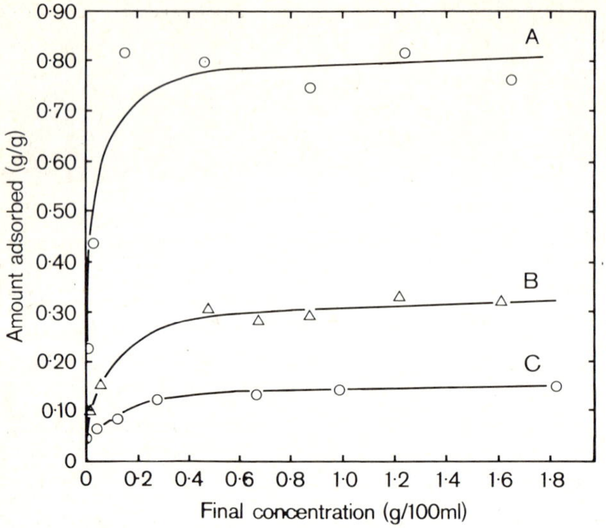

Fig. 3.2. Influence of saturating cation type on the adsorption of polyvinyl alcohol (GL-05) by montmorillonite. The polymer solutions were added to air-dry clay to form 0.5% w/v suspensions. A, Na^+-montmorillonite; B, Ca^{2+}-montmorillonite; C, Cs^+-montmorillonite (after Greenland, 1963).

be compared with the coil dimensions of PVA in solution for which Garvey et al. (1976) have reported values of 4.4 to 6.75 nm and 5.3 to 11.7 nm for the hydrodynamic and root mean square radii, respectively, of PVA fractions with a molecular weight range between 17 000 and 67 000.

On this basis, the polymer can gain unrestricted access to nearly all of the crystal (external and interlayer) surfaces of Na^+-montmorillonite in dilute suspension in which the individual silicate layers are more or less fully dissociated. On the other hand, little, if any, interlayer penetration by PVA would be expected to occur with the calcium- and cesium-saturated samples; adsorption here is confined mainly to external crystal surfaces. Although this expectation is not entirely realised as the X-ray data will show, the above analysis can qualitatively account for the relatively large uptake by the sodium clay as well as for the reduction in adsorption in more concentrated suspensions (Fig. 3.1) and at increasing ionic strength (Fig. 3.3).

As the suspension strength increases, there is less scope for interlayer swelling to occur (with Na^+-montmorillonite) or for the clay crystals to move apart and hence the extent of surface accessible to the polymer is reduced. By compressing the thickness of the diffuse double layers around interlayer or external clay surfaces, the presence of electrolytes produces a similar effect. In addition, an increase in suspension and/or ionic strength enhances the probability of a single polymer molecule spreading over adjacent clay

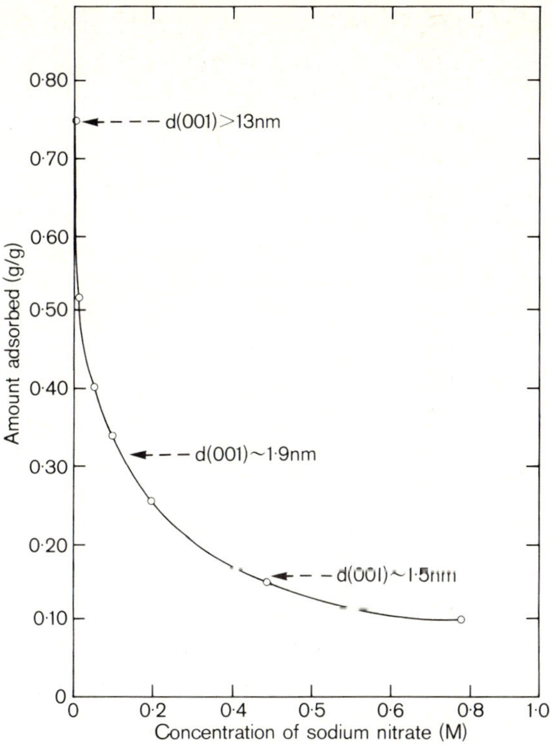

Fig. 3.3. Adsorption of polyvinyl alcohol (GM-14) by Na^+-montmorillonite in the presence of $NaNO_3$ solutions of different concentration. The polymer solutions were added to 6% w/v aqueous suspensions of clay to form 1% w/v final suspensions. The basal spacing of the complexes is also indicated (after Greenland, 1963).

surfaces and this will physically and sterically hinder other molecules from gaining access to the same surface. In this connection, it is relevant to note Emerson and Raupach's (1964) finding that the extent of interparticle bridging (in Na^+-montmorillonite) is influenced by the order in which PVA and salt are added to the clay suspension. More extensive bridging by the polymer and hence stronger flocs are apparently produced when the addition of PVA precedes that of NaCl than when this order is reversed. Indeed, the effectiveness of PVA and uncharged linear polymers, in general, as flocculating agents of aqueous clay suspensions may be greatly increased by adding electrolytes to the clay—polymer system. This observation is, as they pointed out, in contrast to the behaviour of anionic polymers, a point to which we shall return later.

The higher uptake of PVA by Ca^{2+}-montmorillonite as compared with the cesium clay (cf. Fig. 3.2) may likewise be ascribed to the greater accessibility of the interlayer surface in the former instance. In addition, it may reflect a difference in bonding mode between the two systems. Thus, simple n-alcohols are attached to montmorillonite surfaces primarily through cation—dipole interactions which may take one of two forms depending on the hydration status of the clay and the nature of the exchangeable cation (Mortland, 1970; Theng, 1974).

In dehydrated clay systems or when the cation is non-hydrated (e.g. Rb^+, Cs^+), the alcohol is directly coordinated to the exchangeable cation through the oxygen atom, as in (I). On the other hand, in an aqueous environment or with strongly hydrated cations, the cation—alcohol link is indirect, being mediated through a water molecule in the primary hydration shell of the cation ("water bridge" formation), as depicted in (II).

$$\underset{\text{(I)}}{M^{n+} \leftarrow \overset{\displaystyle \text{H} \atop |}{\text{O}}\text{—R}} \quad \text{and} \quad \underset{\text{(II)}}{M^{n+} \leftarrow \overset{\displaystyle \text{H} \atop |}{\text{O}}\text{—H}\cdots\overset{\displaystyle \text{H} \atop |}{\text{O}}\text{—R}}$$

where M^{n+} is an exchangeable cation of valency n and ROH is an alkanol. Ion—dipole interactions of this type are of general occurrence in systems involving clays and polar organic micromolecules, being more important than direct molecule—surface hydrogen bonding.

This is not to say, however, that O—H···O—Si bonding does not occur. Indeed, the considerable adsorption of PVA by freshly ignited silica which exposes a siloxane surface (Table 3.2) must largely take place through this type of interaction. The sheer number of hydroxyl groups per molecule of PVA would make such bonding probable, even in aqueous systems. By the same token, the contribution of O—H···O—Si bonding to the overall adsorption energy may be significant, the more so when the clay—PVA system is dehydrated since this would bring the polymer chain into close juxtaposition with the silicate surface. However, apart from the recent infrared spectroscopic studies by Varfolomeeva et al. (1977), there is little experimental evidence that PVA forms O—H···O—Si hydrogen-bonding with montmorillonite surfaces.

On the other hand, direct segment—surface interaction is commonly observed for silica—polymer systems, the active sites on the mineral being the surface silanol groups (Fontana, 1963; Griot and Kitchener, 1965a; Howard and McConnell, 1967; Iler, 1975). Similarly, Kavanagh et al. (1976) have suggested that polyvinyl alcohol was attached to gibbsite and goethite by hydrogen-bonding and Van der Waals interactions on the basis that adsorption was little affected by the magnitude and sign of the surface charge.

However, by analogy with the behaviour of simple n-alcohols, we may presume that in the case of montmorillonite and other layer silicate minerals containing exchangeable cations, ion—dipole interactions are probably at least as important as O—H···O—Si hydrogen-bonding suggested by some workers (Emerson, 1960; Greenland, 1963; Emerson and Raupach, 1964). In this connection it is relevant to mention Griot and Kitchener's (1965a) work on the adsorption of polyacrylamide by freshly prepared silica. They inferred from infrared studies that the polymer was attached by means of hydrogen-bonding to isolated surface silanol groups. Further, many simple hydrogen-bonding substances could inhibit flocculation of the mineral by the polymer or cause the flocculated system to redisperse. On the other hand, Emerson and Raupach (1964) failed to prevent PVA from reacting with montmorillonite by adding such an effective hydrogen-bonding competitor as dimethyl tetraethylene glycol to the system, irrespective of the order of addition both in the presence and absence of salt.

Although ion—dipole and segment—surface hydrogen-bonding interactions are presumably involved in the adsorption of PVA by clays in aqueous systems, the principal driving force of the process is provided by entropy effects. These arise from the displacement of water molecules, in the secondary and successively higher hydration shells around the exchangeable cations and those adsorbed onto the silicate surface, by a single polymer molecule (cf. Fig. 2.1). The importance of entropy effects in clay—polymer systems will be further discussed in connection with the adsorption of polyethylene oxide (glycol) by montmorillonite (Parfitt and Greenland, 1970a). Van der Waals (dispersion) forces between uncharged polymer segments and clay surface would also play a part in the overall adsorption process. As is the case with O—H···O—Si bonding, these secondary interactions will become significant when the clay—polymer complex is allowed to dry out, bringing the solute into close contact with the surface.

There is good evidence to indicate that surface accessibility and entropy effects are more important in determining the adsorption of uncharged polymers by montmorillonite than ion—dipole interactions and segment—surface hydrogen bonding. The influence of the exchangeable cation on adsorption is apparently connected with the extent to which the parent clays can show interlayer expansion in water and with the amount of water which is contained in the interlayer space.

Thus, a feature of ion—dipole interactions is that adsorption increases with the polarising power or ionic potential of the exchangeable cation (e.g. Theng, 1974). On this basis, the uptake of PVA should increase in the order of $Cs^+ < Na^+ < Ca^{2+}$ and not $Cs^+ < Ca^{2+} < Na^+$ as is found experimentally. Similarly, Dowdy (1972) observed the sequence of $Al^{3+} < Ca^{2+} < Na^+$ for polyvinyl alcohol, Parfitt and Greenland (1970a, b) that of $Al^{3+} < Ca^{2+} < Na^+$ for polyethylene glycols and $Ca^{2+} < Al^{3+} << Na^+$ for a dextran, and Francis (1973) that of $Ca^{2+} < Na^+$ for polyvinyl pyrrolidone.

The position of Na^+-montmorillonite in all these instances is undoubtedly ascribable to its high surface accessibility which overrules any effect the cation may have on the adsorption process. Comparisons among clay samples showing limited intracrystalline (interlayer) swelling in water are only valid if the extent of interlayer penetration and the amount of water displaced by the polymer in the process, can be assessed. The higher adsorption of PVA by Ca^{2+}-montmorillonite as compared with the cesium system, for example, is due to the ability of the polymer to enter the interlayer space of the former clay sample. By contrast, hardly any PVA is intercalated by Cs^{+}-montmorillonite as the X-ray diffraction data below will indicate.

The basal spacing of montmorillonite—PVA complexes generally increases with the amount of polymer adsorbed, as Greenland (1963) and Toyoshima (1973) have shown for Na^{+}-montmorillonite (Fig. 3.4). This behaviour is commonly observed during the intercalation of polymeric materials by montmorillonite (cf. Figs. 7.3 and 12.6) and may be explained in terms of layer segregation or the random interstratification of polymer-rich and inorganic-rich interlayers within a single crystal. The entry of a PVA molecule into a given interlayer presumably facilitates and promotes the intercalation of the following molecules into the same space. Interestingly, Emerson and Raupach (1964) reported two strong reflections at 1.85 and 1.26 nm together with a weak line at 0.96 nm, for their Na^{+}-montmorillonite complex containing 0.2—0.3 g PVA per g clay. This would indicate that segregation occurred between, rather than within, crystals. The basal spacing values, however, refer to complexes prepared in the presence of 3×10^{-3} M NaCl and this may well influence the intercalation process.

On the basis that each PVA segment occupies an area of 0.152 nm^2 and that the polymer is adsorbed in a fully collapsed form, Greenland (1963) has calculated that the maximum adsorption of 0.80 g per g of Na^{+}-montmorillonite would correspond to $\sim$1500 $m^2\ g^{-1}$ which is twice the total accessible surface of the clay. It would therefore appear that only half of the segments are adsorbed in trains, the rest being present in the form of loops and tails with a mean thickness of about two molecular layers ($\sim$1 nm). For the cesium clay where adsorption was mainly confined to external crystal surfaces of 120 $m^2\ g^{-1}$, a 1 nm deep layer would correspond to

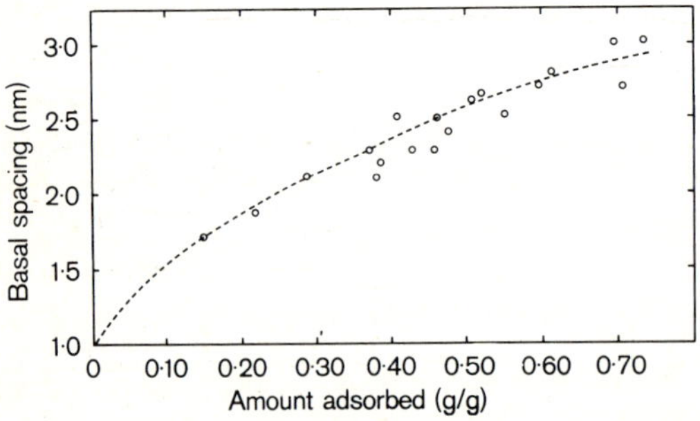

Fig. 3.4. Basal spacings of Na^{+}-montmorillonite complexes containing different amounts of polyvinyl alcohols (GL-05, GM-14, and GH-20) (after Greenland, 1963).

~0.13 g/g which agreed well with the amount adsorbed at the maximum (Fig. 3.2). The basal spacing of Ca^{2+}-montmorillonite containing ~0.30 g g^{-1}, dried at room temperature or at 70°C, was 1.8—2.0 nm which on thorough drying (over P_2O_5 at 70°C under vacuum for 72 h) fell to 1.47 nm. This would indicate a collapse from a 1 nm deep layer to a fully extended monolayer of PVA molecules, sandwiched between two opposing silicate layers. If 0.13 g g^{-1} were apportioned to external crystal surfaces as was done for the cesium clay, the difference of (0.30—0.13) = 0.17 g g^{-1} would represent the monolayer capacity. The basal spacing of the thoroughly dried Na^+-montmorillonite complex did not fall below 3 nm, giving an interlayer separation (Δ-value) of ~2 nm. By analogy with the calcium system, it may be inferred that this value represents four layers of fully extended PVA molecules, two on each interlayer surface. This view is further supported by the fact that the calculated interlayer amount of (0.80—0.13) = 0.67 g g^{-1} is very close to four times the monolayer capacity of 0.17 g g^{-1}.

The reduction in interlayer separation of the Ca^{2+}-montmorillonite complex from ~1 nm for the air-dry material to ~0.5 nm for the well-dried specimen, indicated that the displacement of interlayer water by the intercalating polymer was far from being complete. This residual water was presumably associated with the divalent cation, being involved in water bridge formation. Since no such reduction was observed with the Na^+-montmorillonite complex, we may infer that the hydroxyl groups of the polymer were capable of displacing the water molecules around the exchangeable interlayer Na^+ ions. Direct cation—dipole interactions might likewise be expected to occur with Cs^+-montmorillonite because the Cs^+ ion is essentially nonhydrated (Nightingale, 1959). Similar bonding mechanisms have been postulated by Theng and Scharpenseel (1975) for the adsorption of humic acid by montmorillonite saturated with monovalent and divalent cations (cf. Chapter 12).

It is also interesting to note that for dilute (<1% w/v) aqueous suspensions of Na^+-montmorillonite, the amount adsorbed increases with molecular weight (Greenland, 1963; Schamp et al., 1975). A similar dependence of adsorption on molecular weight has been reported for kaolinite— (Kashmoula and Schurz, 1975), illite— (Schamp et al., 1975), and other colloidal solid—PVA systems, such as silver iodide (Fleer et al., 1972), antimony sulphide (Baran et al., 1976), and hydrous oxides of aluminium and iron (Kavanagh et al., 1976). On the other hand, for the comparable calcium and cesium suspensions as well as for more concentrated sodium clay suspensions, Greenland (1963) noted a reduction in uptake as polymer size increased. These observations must again be related to surface accessibility; as access becomes restricted due to a closer packing of the silicate layers or clay crystals, the low molecular weight is adsorbed in preference to the larger, bulkier species. As remarked on in section 2.2, this behaviour is characteristic of the interactions of polymeric solutes with porous solids.

As compared with the "G" series of polyvinyl alcohols described above, the "N" series (Table 3.1) generally showed higher uptakes under comparable experimental conditions. This may be ascribed to the tendency of the "N" polymers to form micelles in solution, leading to the creation of a more extensive network of clay—polymer bonds. This suggestion accords with the data of Toyoshima (1973) who found that adsorption by a kaolinite of polyvinyl alcohols with varying extents of hydrolysis reached a minimum at ~90% hydrolysis ($x \sim 10$), a point which coincided approximately with a maximum in solution viscosity (Fig. 3.5).

The reactions of polyvinyl alcohol with other clay minerals, such as halloysite, illite, and kaolinite, are essentially similar to those obtained with montmorillonite. Thus, adsorption is reduced with an increase in suspension concentration although the extent of reduction is not as great as for montmorillonite. In other words, much stronger suspensions are required before uptake is substantially decreased.

The isotherms are again of the H-type and once adsorbed, the polymer is very resistant to being removed from the complex by washing with water (Kuznetsova and Serb-Serbina, 1968; Greenland, 1972b; Kashmoula and Schurz, 1975).

The plateau values, obtained by Greenland (1972b) for a number of clay and soil minerals using GM-14 as the adsorbate, are summarised in Table 3.2. As a comparison, Kashmoula and Schurz reported values of 0.014 and 0.013 g/g for a kaolinite using NH-17 and NH-05, respectively (cf. Fig. 3.5). Apart from Na^+-montmorillonite, none of the clay and mineral species listed in Table 3.2, are capable of intercalating polyvinyl alcohol. As compared with Na^+-montmorillonite, the accessible surface of these minerals is smaller and the amount adsorbed per unit weight of adsorbent is accordingly lower.

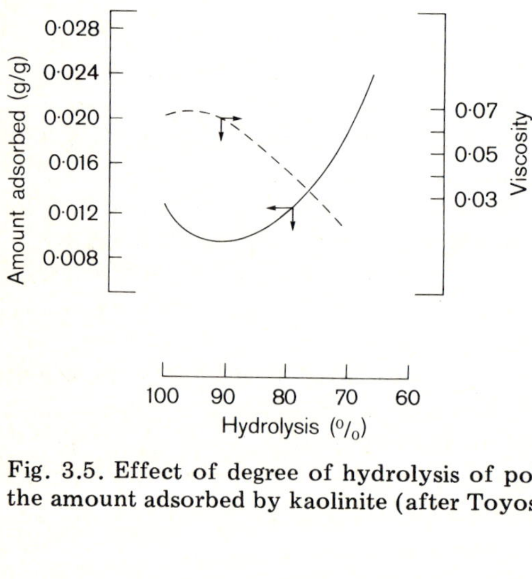

Fig. 3.5. Effect of degree of hydrolysis of polyvinyl alcohol on solution viscosity and on the amount adsorbed by kaolinite (after Toyoshima, 1973).

TABLE 3.2

Surface area of some clays, oxides, and hydroxides together with their respective plateau value for the adsorption of polyvinyl alcohol (GM-14; molecular weight 70 000), expressed on the basis of either per unit weight or per unit area of adsorbent (from Greenland, 1972b)

Clay/mineral	Specific surface area ($m^2\ g^{-1}$)	Maximum adsorption	
		($g\ g^{-1}$)	($mg\ m^{-2}$)
Na^+-montmorillonite	725	0.80	1.11 (~1)
Cs^+-montmorillonite	115	0.14	1.21
Fithian illite	93	0.12	1.29
Grundite illite	105	0.13	1.25
Willalooka illite	132	0.14	0.7
Malone kaolinite	17	0.015	1.06
Rocky Gully kaolinite	36	0.034	0.98
Djebel Debar halloysite	60	0.02	0.33
Gibbsite	48	0	0 (2.1—2.7)
Boehmite	108	0	0
Alumina			
Unignited	256	0	0
Ignited at 800° C	220	0	0
Silica			
Unignited	155	0	0
Ignited at 800° C	127	0.094	0.74

Values in brackets are from Kavanagh et al. (1975, 1976).

However, if this amount is expressed on a per unit area basis, that is, in terms of adsorption density (Γ), the plateau value for the 2 : 1 type layer silicates is seen to be remarkably similar, ranging from 1.1 to 1.3 mg m^{-2}. On the other hand, the kaolinite group of minerals give smaller values which, furthermore, show marked variations among individual species. The explanation for this behaviour is not obvious but differences in surface properties and accessibility are probably involved. Thus, Panasevich et al. (1976) have shown that the adsorption of PVA onto kaolinite depended on the size and porosity, that is, the degree of dispersion of the clay aggregates.

Of particular interest is the absence of any measurable adsorption of PVA by the various oxides and hydroxides of aluminium. Similarly, silica failed to adsorb PVA although the corresponding freshly ignited sample did take up an appreciable quantity of the polymer. In sharp contrast, Kavanagh et al. (1975, 1976) were able to measure substantial adsorption of PVA onto their gibbsite specimens even at low solute concentrations, an adsorption density as high as 7 mg m^{-2} being found at a PVA concentration of ~1.5 g l^{-1} (cf. Fig. 3.1).

Greenland (1972a, b) has tentatively suggested that the unreactivity of aluminol and silanol surfaces towards PVA is due to the inability of the polymer to displace the strongly hydrogen-bonded water molecules which are associated with these surfaces. This suggestion is in line with the observation by Griot and Kitchener (1965b) who also failed to find adsorption of polyacrylamide onto silica after it has been aged, a treatment which led to hydration of the surface. On the other hand, the siloxane surface (produced by igniting silica at 800°C), like the basal surfaces of 2 : 1 type layer silicates, does not form strong hydrogen bonds with water and polar organic species. Any water present at the siloxane surface can presumably be readily displaced by the polymer which then becomes adsorbed.

The data of Kavanagh et al. (1975, 1976) indicated, however, that the presence of water at the gibbsite surface did not preclude PVA from being adsorbed. Indeed, prior air-drying or heating to 100°C of the mineral was found to decrease, rather than increase, adsorption. This they ascribed to a reduction in surface accessibility resulting from a more extensive face-to-face aggregation of the mineral particles.

These workers also noted that adsorption of PVA by gibbsite, goethite, illite, and montmorillonite (in the sodium form) led to a decrease in the electrophoretic mobility of the particles, in agreement with the findings of Maslenkova et al. (1965) for polyacrylamide, and of Birkner and Edzwald (1969) in the case of polyethylene oxide. This behaviour was ascribed to the physical displacement of the plane of shear by the adsorbed polymer rather than to changes in the ionic distribution within the electrical double layer (cf. Fig. 2.6). From the measurement of zeta potentials in the presence and absence of adsorbed PVA, they were able to deduce an adsorbed layer thickness in excess of 10 nm, which was an order of magnitude greater than the value derived by Greenland (1963) for montmorillonite. Accordingly, unlike Greenland's system, only a small fraction (~5%) of the polymer segments could be in contact with the surface, the rest being present in the form of loops and tails. Similarly, Panasevich et al. (1976) have calculated a minimum thickness of 1.55 nm on the assumption that all of the kaolinite surface was accessible to PVA, with adsorbed layer thicknesses greater than 3.4 nm being possible.

The question naturally arises as to why there is such a disparity between the two sets of results. In searching for some plausible explanation, it would seem to us that subtle differences in the way these workers brought the reactants together might be significant. Thus, in Greenland's (1972b) case, the PVA solutions were added to air-dry samples of the adsorbent whereas Kavanagh et al. (1975) made their additions to stirred suspensions of the mineral.

Relevant in this connection is the study of Jankovics (1965) on the interactions of polyacrylamide with calcium phosphate. He made a distinction between the "dry" and the "wet" method of complex formation, which corresponded to that used by Greenland and by Kavanagh et al., respectively, and also considered the effect of agitation on flocculation. The wet method gave a consistently higher rate of adsorption than the dry technique. By promoting floc disruption, agitation increased the extent of surface accessible to the polymer. In addition, stirring tended to distribute the solute evenly over the accessible surface.

On the basis of Jankovics' observations, the method used by Kavanagh et al. will conceivably yield higher adsorption densities as well as thicker adsorption layers, as compared with Greenland's approach. Other factors may, of course, be involved and further work into the behaviour of PVA at clay and metal oxide surfaces is clearly warranted.

Reference has already been made to the reactions of polyacrylamide with some minerals. Indeed, uncharged polyacrylamide (PAM) has been used extensively both as a flocculant of mineral suspensions and a soil aggregate stabiliser. It has the advantage over PVA in that negatively as well as positively charged functional groups can be introduced into the polymer backbone. In this section we are only concerned with the interactions of the nonionic species with clays.

The factors affecting the clay—PAM interaction have been studied in some detail by Schamp and Huylebroeck (1972, 1973). For this purpose they synthesised PAM with a range of molecular weight ($140—1150 \times 10^3$) by radical polymerisation of the monomer in water at different temperatures. In preparing the complexes, aqueous solutions of PAM were added to clay suspensions (in water) of varying strength.

As would perhaps be expected, the behaviour of PAM at clay surfaces has many common features with that of PVA. Thus, the polymer was strongly and virtually irreversibly adsorbed. Like PVA, adsorption was influenced by the final suspension concentration, polymer molecular weight, and to a small extent, by temperature. The nature of the exchangeable cation also affected adsorption more by altering the particle arrangement and hence, the porosity and surface accessibility, than by modifying the surface reactivity of the clay. Both the suspension and cation effects were more markedly displayed with montmorillonite than with either illite or kaolinite. These points are summarised in Table 3.3.

The effect of polymer molecular weight on adsorption (column 5) is more clearly illustrated in Fig. 3.6 which shows that whereas the maximum uptake for Na^+-montmorillonite increases that for the hydrogen clay decreases with rising molecular weight. The curves for H^+-illite and H^+-kaolinite also show a sharp decline initially but then level off when the molecular weight approaches 1000×10^3.

These observations may again be explained in terms of surface accessibility. Like polyvinyl alcohol, PAM can apparently gain unrestricted access to

TABLE 3.3

Data for the adsorption of uncharged polyacrylamide from aqueous solutions at 298 K by different clay minerals (from Schamp and Huylebroeck, 1973)

Clay mineral species	Exchangeable cation	Polymer molecular weight ($\times 10^3$)	Clay suspension concentration (% w/v)	Maximum adsorption [a] (mg g^{-1})
Kaolinite	Na^+	140	5	34
	Ca^{2+}	140	5	30
	H^+/Al^{3+}	140	5	26 (33)
	H^+/Al^{3+}	140	2.5	27
	H^+/Al^{3+}	140	1.7	27
	H^+/Al^{3+}	250	5	17
	H^+/Al^{3+}	250	2.5	16
	H^+/Al^{3+}	250	1.7	16
	H^+/Al^{3+}	1150	5	7
	H^+/Al^{3+}	1150	2.5	9
	H^+/Al^{3+}	1150	1.7	12
Illite	H^+/Al^{3+}	128	2.5	41 (46)
	H^+/Al^{3+}	250	2.5	34
	H^+/Al^{3+}	530	2.5	14 (28)
Montmorillonite	Na^+	250	0.05	757
	Na^+	250	0.025	787
	H^+/Al^{3+}	140	0.25	328
	H^+/Al^{3+}	250	0.17	330 (270)

[a] Estimated from the respective Langmuir plots of the isotherm data; values in brackets refer to adsorption at 348 K.

all the surfaces of Na^+-montmorillonite. This is so because in the range of suspension concentrations used, the clay crystal would show extensive interlayer swelling with large distances separating individual silicate layers. On the other hand, the interlayer expansion of H^+-montmorillonite would be relatively limited due to the partial occupancy of exchange sites by Al^{3+} ions released from the structural framework by proton attack. The association of several unit layers into a crystal and the turbostratic arrangement of the crystals give rise to a porous clay matrix. On this basis, Schamp and Huylebroeck (1973) have made a distinction between "surface" and "pore" adsorption.

This concept of an aggregated clay system is in fact very much like that proposed earlier by Aylmore and Quirk (1960) who envisaged a clay mass as being made up of regions of oriented clay crystals, termed "domains", which are arranged in turbostratic array. For this reason, we shall use the terms inter- and intra-domain adsorption rather than surface and pore

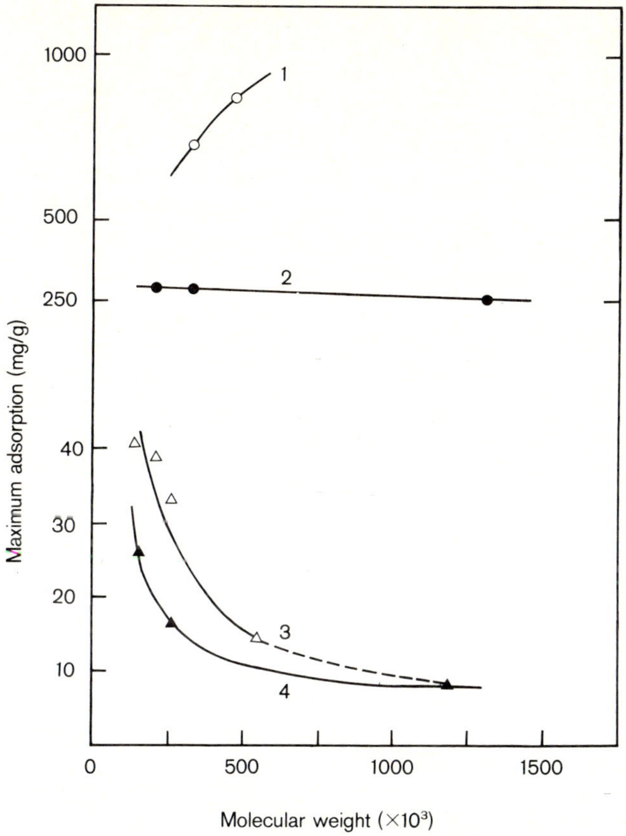

Fig. 3.6. Influence of molecular weight on the adsorption of polyacrylamide by different clay mineral species. Curve 1, Na^+-montmorillonite; curve 2, H^+-montmorillonite; curve 3, H^+-illite; curve 4, H^+-kaolinite (after Schamp and Huylebroeck, 1973).

adsorption. Access to intra-domain surfaces will clearly be more difficult to achieve as the polymer molecular weight or coil size increases. Indeed, access will become impossible when a certain molecular size is exceeded. For H^+-illite and H^+-kaolinite, this limiting size appears to correspond to a molecular weight of $\sim 1000 \times 10^3$.

The results of kinetic measurements, shown in Fig. 3.7, tend to accord with the view that there are two types of adsorbing surfaces in non-expanding clay systems. Thus, adsorption of the high molecular weight PAM (curve 2) by H^+-kaolinite is complete in less than 2 h whereas that of the low molecular weight material (curve 1), which presumably can penetrate the intra-domain surfaces, keeps on rising gently with time. The difference in adsorption between curves 1 and 2 may therefore be assumed to represent intra-

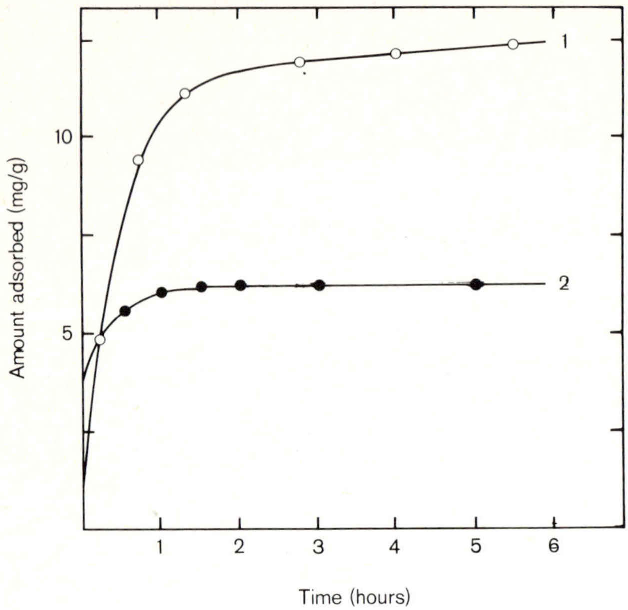

Fig. 3.7. Kinetics of polyacrylamide adsorption by H^+-kaolinite as a function of molecular weight (M). Curve, 1, $M = 140 \times 10^3$; curve 2, $M = 1150 \times 10^3$. In both instances, 10 mg of polymer in 10 ml water was mixed with 500 mg clay (after Schamp and Huylebroeck, 1973).

domain adsorption, a process which is shown to obey second-order kinetics. This would indicate that the velocity of the reaction at any given instant is proportional to the concentration of the polymer and the accessible sites on the clay surface.

The above analysis does not apply to Na^+-montmorillonite where adsorption is relatively rapid and decreases with rising temperature, reminiscent of the behaviour of simple (non-polymeric) solutes. As already mentioned, adsorption here increases with molecular weight. This again is similar to what is normally observed with small organic species. It is also in line with theoretical predictions (cf. section 2.3) implying that the proportion of polymer segments in trains does not vary greatly with molecular weight.

Interestingly, the temperature coefficient for H^+-kaolinite is positive. This may mean that as loop size increases with temperature, there is no corresponding increase in the area per attached segment. Why this is so is not clear but an increase in the rate of pore diffusion with temperature may offset the contrary effects of exothermicity and polymer solubility.

An interesting feature of the polyacrylamide—clay interaction is the occurrence of a minimum in the adsorption isotherms which for a given polymer molecular weight is independent of clay mineral species and of final clay suspension strength (Fig. 3.8). Furthermore, the onset of this minimum occurs at the same initial polymer concentration (the "critical" concentration). Since this concentration is proportional to the inverse square root of molecular weight, Schamp and Huylebroeck (1972, 1973) have postulated that the polyacrylamide solution undergoes a phase transition at this point. Below and above the critical concentration, the polymer coils are thought to be separate from each other, whereas at the transition point the coils intertouch, filling out the whole volume. Consequently, the surface requirement of the coils is larger at the critical concentration than would otherwise have been and adsorption decreases accordingly. The occurrence of a minimum is not confined to polyacrylamide—clay systems, being also observed during the adsorption of polyacrylamide onto charcoal and of polyethyleneimine onto clay.

Like polyvinyl alcohol, polyacrylamide can form interlayer complexes with montmorillonite. Using a 1% suspension of the clay and a 0.2% aqueous solution of the polymer, Tanihara and Nakagawa (1975) were able to inter-

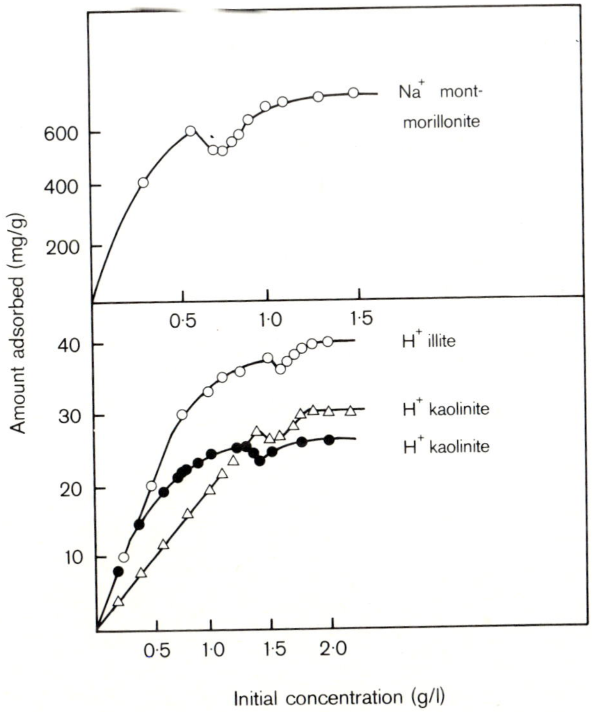

Fig. 3.8. Isotherms for the adsorption of polyacrylamide by different clay minerals showing the occurrence of a minimum at a given ("critical") concentration. Data for illite and kaolinite refer to PAM with a molecular weight of 140×10^3; for montmorillonite of 250×10^3 (after Schamp and Huylebroeck, 1973).

calate a monolayer of PAM into montmorillonite saturated with Na^+, Ca^{2+}, and Mg^{2+} ions. The strong binding and thermal stability of adsorbed polyacrylamide in the divalent cation clay systems were explained in terms of the formation of coordination complexes between the exchangeable cations and the amide groups of the polymer.

The importance of entropy effects in the interaction of uncharged polymers with clays has been demonstrated by Parfitt and Greenland (1970a) using a thermodynamic approach. These workers determined the isotherms at 275 and 298 K for the adsorption of polyethylene glycols (PEGs), which are low molecular weight polyethylene oxides, by montmorillonite saturated with Cs^+, Na^+, Ca^{2+}, and Al^{3+} ions.

Fig. 3.9 shows the data for Ca^{2+}-montmorillonite at 298 K. For polyethylene glycol with an average molecular weight of 200, 300, and 400 (denoted by PEG 200, PEG 300, and PEG 400, respectively), the isotherms were linear over the concentration range of 0—1 g/100 ml. The same was true for the curves at 275 K but the slopes here were steeper, indicating that adsorption was enhanced by a decrease in temperature. On the other hand, PEG 600 and PEG 1500 gave rise to L-type isotherms and for PEG 20 000 the curve was of the H-type. The isotherms for PEG 300 with Cs^+-, Na^+-, and Al^{3+}-montmorillonite at 275 and 298 K were either linear or slightly curvilinear; those with Na^+- and Ca^{2+}-illite were straight lines. The change from C- to L-type isotherms probably coincides with the transition from a linear to a random coil conformation of PEG in solution.

Since the initial slope of the isotherms is a measure of solute—surface interaction, the data of Fig. 3.9 indicate that the affinity of PEG for the clay increases with molecular weight. This view is further supported by the observation that more than 80% of the adsorbed PEG 300 could be removed

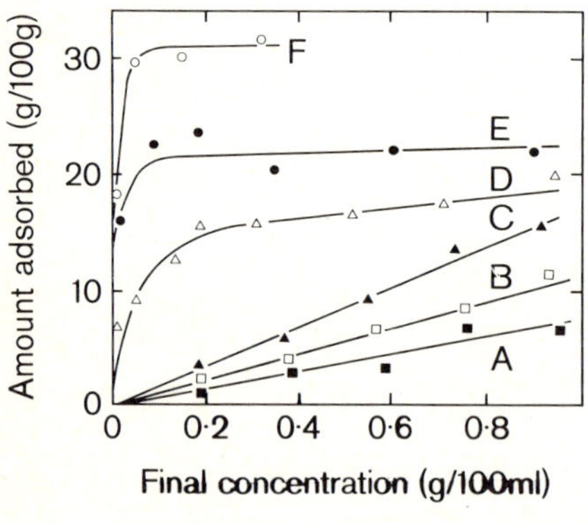

Fig. 3.9. Isotherms for the adsorption of polyethylene glycol (PEG) at 298 K by Ca^{2+}-montmorillonite as a function of molecular weight. Curve A, PEG 200; curve B, PEG 300; curve C, PEG 400; curve D, PEG 600; curve E, PEG 1500; curve F, PEG 20 000 (after Parfitt and Greenland, 1970a).

by washing the complex with water whereas this treatment failed to desorb any PEG 20 000. The effect of molecular weight on adsorption affinity is also borne out by thermodynamic measurements (Table 3.5) and is in keeping with theoretical predictions (cf. section 2.3).

Where C-type or linear isotherms obtain, the adsorption process may be analysed in terms of a constant partition of solute between the surface and the bulk phase (Greenland et al., 1962; 1962; Theng, 1972; Theng and Scharpenseel, 1975), that is

$$K_m = (\mathrm{PEG})_s/(\mathrm{PEG})_b$$

K_m is the partition coefficient; $(\mathrm{PEG})_s$ and $(\mathrm{PEG})_b$ refer to the concentration of the surface (adsorbed) and bulk (solution) species, respectively. In order to evaluate $(\mathrm{PEG})_s$ the volume in which the adsorbed polymer is contained must be known.

The basal spacing of the parent calcium clay in water is 1.9 nm. Taking a value of 0.95 nm for the thickness of an individual montmorillonite layer (cf. Fig. 1.5) gives an interlayer separation (or Δ-value) of 0.95 nm which is equivalent to twice the Stern layer thickness. On intercalation of PEGs the basal spacing of Ca^{2+}-montmorillonite tends to expand slightly and in the process water is desorbed. The volume occupied by adsorbed PEG is then estimated from the total surface area of the crystals (~800 $m^2\ g^{-1}$) and the extent of interlayer swelling. These calculations are set out in Table 3.4.

TABLE 3.4

Basal spacings and properties of moist complexes between Ca^{2+}-montmorillonite and a series of polyethylene glycols (PEG) together with the volume of water desorbed from the interlayer space (from Parfitt and Greenland, 1970a)

Molecular weight of PEG	Basal or $d(001)$ spacing (nm)	Weight of PEG adsorbed (g/100 g)	Volume of PEG adsorbed [a] ($cm^3\ g^{-1}$)	Increase in interlayer volume [b] ($cm^3\ g^{-1}$)	Volume of water desorbed ($cm^3\ g^{-1}$)
200	1.80	7.0 [c]	0.070	−0.04	0.110
300	1.90	13.5 [c]	0.114	0	0.114
400	1.90	14.6 [c]	0.124	0	0.124
600	1.97	19.0 [d]	0.160	0.028	0.132
1500	2.01	22.2 [d]	0.190	0.044	0.146
20 000	2.23	31.0 [d]	0.290	0.132	0.158

[a] Taking 61×10^{-24} cm^3 as the volume per segment or monomer.

[b] Calculated from $\frac{(d(001) - 1.9)}{2} \times 10^{-7} \times 800 \times 10^4\ cm^3\ g^{-1}$

[c] At the upper limit of isotherm linearity.

[d] At plateau adsorption.

Once K_m is known, the free energy change (ΔG^m) of partition can be estimated from $-\Delta G^m = RT \ln K_m$, where R is the gas constant and T the absolute temperature. The corresponding enthalpy change (ΔH^m) can be derived from the adsorption data of two temperatures using the Van 't Hoff equation, allowing the entropy change (ΔS^m) to be calculated by means of the relationship $\Delta G^m = \Delta H^m - T\Delta S^m$. The thermodynamic data for Ca^{2+}-montmorillonite adsorbing PEGs of molecular weights 200, 300, 400, and 600 are summarised in Table 3.5.

Although these data refer to non-standard conditions, they provide a useful clue to the mechanism underlying the adsorption process. It is noted, for example, that appreciable uptake can occur despite the generally small enthalpy change. Indeed, ΔH^m tends to diminish with increasing molecular weight, becoming vanishingly small for PEG 600. It seems clear that, with the exception of PEG 200, the driving force of the reaction as indicated by the ΔG^m values, is largely entropic. The entropy term increases with molecular weight and this is associated with the displacement of an increasing amount of water from the clay surface on adsorption since the water so desorbed gains one translational degree of freedom (cf. Fig. 2.1).

The total entropy change may be regarded as the sum of two terms, a conformational term (ΔS^m_c) and a "hydration" term (ΔS^m_H). The former is lost on adsorption and this tends to inhibit uptake. On the other hand, ΔS^m_H arising from (i) loss of water from the polymer, (ii) release of water from the clay, and (iii) entropy of dilution of the bulk phase, promotes adsorption. The data of Table 3.4 indicate that (ii) probably makes the most important contribution to ΔS^m_H and thus to ΔS^m. Glass (1968a, b) came to entirely analogous conclusions for the adsorption of polyethylene oxide and polyvinyl alcohol at the aqueous—air interface.

The effect of the exchangeable cations on adsorption is illustrated by the data for PEG 300, summarised in Table 3.6. The cesium-saturated clay

TABLE 3.5

Changes in non-standard free energy (ΔG^m), enthalpy (ΔH^m), and entropy (ΔS^m) for the adsorption of polyethylene glycols (PEG) by Ca^{2+}-montmorillonite (from Parfitt and Greenland 1970a)

Molecular weight of PEG	$-\Delta G^m$ (at 298 K) (cal mol^{-1})	$-\Delta G^m$ (at 275 K) (cal mol^{-1})	$-\Delta H^m$ (cal mol^{-1})	ΔS^m (entropy units)
200	1890	1870	1960	−0.2
300	2070	2000	1470	2.0
400	2340	2240	1500	2.8
600	3340	3040	0	11.1

TABLE 3.6

Some isotherm and thermodynamic data for the adsorption of polyethylene glycol (molecular weight 300) by montmorillonite and illite saturated with different cations (from Parfitt and Greenland, 1970a)

Clay	Cation	Amount adsorbed [a] (g/100 g)	$-\Delta G^{m}_{298}$ (cal mol^{-1})	$-\Delta G^{M}_{275}$ (cal mol^{-1})	$-\Delta H^{m}$ (cal mol^{-1})	ΔS^{m} (e.u.)
Montmorillonite	Cs^+	5.5	2920	2680	0	9.7
	Na^+	9.1	2280	2150	860	4.6
	Ca^{2+}	6.0	2070	2000	1470	2.0
	Al^{3+}	5.8	1920	1910	2100	—0.6
Illite	Na^+	0.90	2060	—	—	—
	Ca^{2+}	1.05	2180	—	—	—

[a] At a concentration of 0.6 g/100 ml in solution and 298 K.

stands out as having the largest $-\Delta G^{m}$ value which arises entirely from the large gain in ΔS^{m} by the system. This finding seems surprising since the Cs^+ ion is very weakly hydrated, if at all (Nightingale, 1959). However, the behaviour of Cs^+-montmorillonite is not strictly comparable to that of the other numbers of the series. Thus, although the cesium ion system adsorbed almost as much PEG 300 as did Ca^{2+}- and Al^{3+}-montmorillonite and about two-thirds as much as the sodium clay, no intercalation of the polymer by Cs^+-montmorillonite could be detected by X-ray diffractometry. Adsorption by the cesium clay appeared to be confined to external crystal surfaces with an area of some 150 $m^2\ g^{-1}$, instead of the 800 $m^2\ g^{-1}$ available to the polymer in the case of the other cationic forms of the clay. As a result, the adsorption density for the cesium sample would be 3—5 times greater than for the sodium, calcium, and aluminium ion systems, on the assumption that the adsorbed polymer adopts a similar conformation on all surfaces. Accordingly, a larger volume of water would have been desorbed per unit area of Cs^+-montmorillonite as compared with the other samples. In addition, this water is presumably readily displaceable since it is only weakly adsorbed onto the surface.

By the same token, the sequence of $-\Delta G^{m}_{Na} > -\Delta G^{m}_{Ca} > -\Delta G^{m}_{Al}$ is related to the ease with which water can be displaced and the series $\Delta S^{m}_{Na} > \Delta S^{m}_{Ca} > \Delta S^{m}_{Al}$ reflects the relative amount of water desorbed. It is interesting to note that the observed sequence for $-\Delta G^{m}$ is the reverse of that for the polarising power of the cations (cf. Fig. 12.4), indicating that cation—dipole interactions are not important. This may be due partly to steric factors and partly to the fact that the exchangeable polyvalent cations retain all of their hydration shells. Certainly, the presence of polyethylene glycol at the surface does not seem to affect the hydration and swelling properties of montmorillonite (Parfitt and Greenland, 1970c).

As is the case with Cs^+-montmorillonite, adsorption (of PEG 300) by Na^+- and Ca^{2+}-illite was confined to external crystal surfaces (~90 m^2 g^{-1}). However, the amount of polymer adsorbed by illite at a given solute concentration in solution was disproportionately low (Table 3.6). This is ascribable to the relatively high surface charge density of illite which may be expressed in terms of the cation concentration in the Stern layer. For illite this concentration amounts to ~5.7 N as compared with ~2.5 N for montmorillonite. This would mean that water is more strongly held onto the illite surface; adsorption is accordingly depressed since the polymer has to compete with water for surface sites. A similar "salting-out" effect has been invoked by Rubio and Kitchener (1976) in rationalising the effect of pH and neutral salts on the adsorption—flocculation behaviour of silica—polyethylene oxide systems.

By analogy with the montmorillonite—polyvinyl alcohol system, basal spacings of ~2 nm (Table 3.4) may be explained in terms of the intercalation of two layers of flatly adsorbed PEG molecules together with some residual water. The formation of such double-layer complexes has earlier been reported by Schott (1964) for polyethylene glycols with a molecular weight in excess of 1000. For lower molecular species (e.g. PEG 300), Parfitt and Greenland (1970c) obtained spacings of 1.3—1.4 nm in the absence of interlayer water. This would indicate the presence of a single layer of adsorbed molecules in the interlayer space, that is, between every two opposing silicate surfaces. The corresponding double-layer complexes gave basal spacings of 1.71—1.74 nm. Single-layer complexes of Na^+-, Ca^{2+}-, and Mg^{2+}-montmorillonite with polyethylene oxide were also obtained by Tanihara and Nakagawa (1975).

The respective Δ-values for the single- and double-layer complexes of 0.35—0.45 nm and 0.76—0.79 nm indicate a certain amount of "keying" or recession of the adsorbed molecules into the ditrigonal depressions of the silicate surface (Schott, 1964; Theng, 1974). In addition, the high molecular weight species must undergo extensive uncoiling and stretching on intercalation. Interestingly, such an effect also obtains with chromium and silica surfaces for which an adsorbed layer thickness of <3 nm has been derived (Killmann and Von Kuzenko, 1974; Killmann and Winter, 1975).

From diffusion, viscosity, and sedimentation measurements, Chew and Couper (1976) have deduced values of 3.0—7.7 for the root-mean-square end-to-end distance of PEG coils with molecular weights from 1000 to 6000 (in aqueous solutions at 298 K). Since the maximum extent of interlayer expansion for Ca^{2+}-montmorillonite is ~1 nm, it seems probable that a certain amount of uncoiling precedes intercalation. Conformational changes of this kind have been noted previously with polyvinyl alcohol. The data of

Tanihara and Nakagawa (1975) for polyacrylamide and of Francis (1973) for polyvinyl pyrrolidone (discussed below) suggest that these polymers behave similarly. Apparently, such modifications are of general occurrence when uncharged polymeric compounds adsorb onto clay mineral surfaces.

Surprisingly, the presence of polyethylene glycol in the interlayer space has little effect on the hydration and swelling properties of montmorillonite (Parfitt and Greenland, 1970c). Thus, the basal spacing of dehydrated Ca^{2+}-montmorillonite containing 10% PEG 300 increases in a step-wise manner from ~1.3 to ~1.9 nm as water vapour is adsorbed at increasing relative humidity. The corresponding complex with Na^{+}-montmorillonite shows extensive (region 2) swelling when exposed to water vapour at R.H. > 50%. These observations suggest that the water initially adsorbed (i.e. at low R.H.) tends to associate more with the exchangeable cation than with either the silicate surface or the interlayer polymer.

This view is further supported by infrared spectroscopic measurements on the hydrated complexes in that the surface oxygens as well as the ether oxygens of adsorbed polyethylene glycols, are only weakly hydrogen bonded to the interlayer water. Thus, with the possible exception of the Al^{3+}-montmorillonite complex, the principal hydroxyl stretching bands of adsorbed water occur in the region of 3400—3460 cm^{-1} indicating that this water exists, or is hydrogen bonded to the same extent, as liquid (bulk) water. In the aluminium clay system, the hydroxyl stretching band of interlayer water is displaced to lower frequencies by as much as 300 cm^{-1}. This is a reflection of the high polarising power of the Al^{3+} ion. In this instance, the ether oxygens of the polymer are possibly associated with the cation through a water bridge.

The use of polyvinyl pyrrolidone as an aid in separating soil clays by density gradient centrifugation (Francis et al., 1970) has led Francis (1973) to study the interaction of this polymer with some representative clay mineral species. Polyvinyl pyrrolidone (PVP) is soluble in both water and ethanol, but more so in the former solvent. The material used by Francis probably had a wide distribution of molecular weight (Killmann and Von Kuzenko, 1974), the average of which was 40 000. In addition, it was not entirely non-ionic but carried a certain amount of positively charged functional groups.

No isotherms were given except to say that the data conformed to the empirical Freundlich equation (cf. equation 7.12). However, the course of adsorption with time suggests that the isotherms are of the H-type as might be expected by analogy with the behaviour of polyvinyl alcohol and (high molecular weight) polyethylene glycol (cf. Figs. 3.1 and 3.9). Adsorption of PVP by illite and kaolinite was apparently very low and the polymer was weakly held since three washings of the complexes with water or ethanol could remove all or nearly all of the adsorbed material.

The amount adsorbed by montmorillonite (at a concentration of 1 g/100 ml in aqueous solutions) ranged from ~11 to ~80 g/100 g, depending on the nature of the exchangeable cation. As with PVA and PEG previously, the sodium saturated clay had the highest adsorptive capacity. Undoubtedly, this was again attributable to the ability of Na^+-montmorillonite to swell extensively in water and so expose a relatively large accessible surface to the polymer. This view is further supported by the observation that as much as four times more PVP may be taken up from an aqueous medium than from an ethanolic solution of comparable strength although the opposite tendency would be expected on the basis of relative solubility of PVP in these two solvents (cf. section 2.2). This behaviour is to be explained in terms of the effect of the dielectric constant (ϵ) on the diffuse double layer thickness ($1/\kappa$), developed on interlayer surfaces (cf. section 1.2). Since $1/\kappa \propto \sqrt{\epsilon}$ (equation 1.5) and $\epsilon_{ethanol} < \epsilon_{water}$, the diffuse double layers in ethanol would be relatively thinner, making these surfaces less accessible to the polymer. Indeed, it was possible to vary the adsorptive capacity of Na^+-montmorillonite by varying the extent of interlayer expansion. This could be achieved by treating the parent clay with ethanol before exposing it to the PVP solution (Levy and Francis, 1975a).

The presence in PVP of positively charged groups was inferred from the appearance in the bulk solution of the exchangeable cations, initially present at the clay surface, as PVP was being adsorbed. From the proportion of cations so released in function of the amount of polymer adsorbed, the excess positive charge at pH ~4 was calculated as ~11 meq per 100 g polymer. Clearly, cation exchange or charge neutralisation plays an important part in the adsorption process and this may account for the positive enthalpy change of ~6 kcal mol^{-1} (cf. Tables 3.5 and 3.6).

Another interesting feature, related to the partial polycationic character of the PVP used was that adsorption led to an apparent reduction in the exchange capacity of montmorillonite. That is to say, some of the cations remaining after PVP adsorption were not exchangeable. This behaviour should be contrasted with that of polyvinyl alcohol whose adsorption by montmorillonite does not affect cation exchangeability (Emerson and Raupach, 1964). The observed effect in the case of PVP is reminiscent of that shown by large organic cations capable of "covering up" some of the exchange sites at the clay surface (Hendricks, 1941).

Since the cover-up effect was greater for montmorillonite saturated with small monovalent cations (e.g. Na^+) than for divalent ion systems, both steric and kinetic factors seemed to be involved. On losing its hydration shell, the small Na^+ ion may possibly be recessed into the ditrigonal depression of the silicate surface whereas the divalent ions which tend to remain hydrated are likely to be located near the centre of the interlayer space. This suggestion is in line with the desorption behaviour of ^{85}Sr and ^{109}Cd introduced into the exchange complex of Na^+- and Ca^{2+}-montmorillonite, in the absence and presence of PVP. Thus, for the calcium clay the amount of ^{85}Sr and ^{109}Cd retained against washing with water, was the same whether or not PVP was present in the interlayer space. On the other hand, much more of these radionuclides was desorbed from the PVP-free sodium clay as compared with the sample containing interlayer polymer (Levy and Francis, 1975a).

In this connection, the basal spacing of the PVP complex with Na^+- and Ca^{2+}-montmorillonite (after heating at 393 K) is ~2.7 and ~1.6 nm, corresponding to an interlayer separation or Δ-value of ~1.75 and ~0.65 nm, respectively. If the thickness of the vinyl-pyrrolidone monomer is taken as 0.56 nm, the observed Δ-values seem to indicate the presence of three layers and one layer of PVP molecules in the interlayers of the sodium- and calcium-saturated samples, respectively.

However, on the basis of the adsorption data and by analogy with the polyvinyl alcohol systems described earlier, the basal spacing data are more consistent with the formation of double- and single-layer complexes. If so, the average depth of the adsorbed layer on each interlayer surface of Na^+-montmorillonite is ~0.88 nm, giving a value of ~0.7 for the fraction (p) of segments in direct contact with the surface. This is appreciably higher than for polyvinyl alcohol (p ~ 0.5) and probably reflects the slight positively charged character of the PVP sample used. The PVP chain in the interlayer space of divalent ion montmorillonite is even more extended, due presumably to the fact that the swelling in water of the parent clay is limited to the fact that the swelling in water of the parent clay is limited to ~1.9 nm. By the same token the lack of interlayer swelling in the vermiculite sample used (irrespective of the nature of the saturating cation) prevents PVP from being intercalated.

Attempts by Francis (1973) to characterise the probable bonding mechanisms using infrared spectroscopy have met with only limited success. This might be due to the generally weak segment surface interaction, the presence of cationic groups in the molecule, and the polydispersity of the polymer sample. The spectra gave no indication of significant hydrogen bonding interactions involving the carbonyl groups of the adsorbed polymer. The suggestion was made, however, that the polymer might interact with the clay surface by means of C—H···O—Si bonding but the evidence for this claim was not unequivocal. Furthermore, this type of bonding has been shown on many previous occasions (see Theng, 1974) to be of little, if any, importance in clay—organic reactions.

As already alluded to, Na^+- and Ca^{2+}-montmorillonite complexes with polyvinyl pyrrolidone gave rise to well-defined $d(001)$ peaks at widely separated positions and showing an integral series of higher order reflections. This fact has been used by Levy and Francis (1975b) to demonstrate the occurrence of cation demixing in montmorillonite crystals. To this end, they replaced the Na^+ ions in the fully (sodium-) saturated clay with increasing proportions of Ca^{2+} ions and exposed the sodium/calcium samples so obtained to a solution of PVP. The X-ray data indicated the formation of two distinct zones within a single crystal whose interlayers dominantly contained either Na^+ ions ($d(001) = 2.68$ nm) or Ca^{2+} ions ($d(001) = 1.6$ nm). More common is the observation that the intercalation of organic cations into homoionic forms of montmorillonite leads to the creation of organic cation-rich and inorganic-cation rich interlayers which are randomly interstratified within a single crystal (Barrer and Brummer, 1963; Theng et al., 1967). Mixed alkylammonium-exchanged montmorillonite likewise shows cation segregation into randomly ordered layers (McBride and Mortland, 1973). When this occurs there is a continuous increase in basal spacing with increasing amounts of the exchanging species introduced. This behaviour is not confined to the intercalation of small organic molecules. The X-ray data of Greenland (1963) and Toyoshima (1973) for polyvinyl alcohol (cf. Fig. 3.4) and of McLaren et al. (1958) for lysozyme (cf. Fig. 7.3) with montmorillonite indicate the occurrence of layer segregation in these systems.

On the other hand, with vermiculite McBride and Mortland (1973) observed segregation into crystallites, analogous to that described by Levy and Francis (1975b) for the Na^+/Ca^{2+}/PVP montmorillonite system. The behaviour shown by vermiculite was ascribed to clay inhomogeneity and this might equally well apply to the sample used by Levy and Francis.

In conclusion, we wish to comment on the effectiveness of uncharged polymers as flocculating agents of clay suspensions. Since non-ionic macromolecules exist as random coils in solution, their ability to form interparticle bridges is very limited (Kuzkin et al., 1964; Vincent, 1974). Only in situations where the clay particles are separated by distances approximating the mean equivalent spherical diameter of the polymer, can appreciable interparticle bonds be formed. This would account for the increase in flocculating effectiveness of uncharged polymers with increasing molecular weight (M) (Linke and Booth, 1960; Vincent, 1974) since, to a first approximation, the coil radius is proportional to $\sqrt{M}$ (Schamp and Huylebroeck, 1972). Flocculation through interparticle bridging will clearly be promoted by increasing the clay concentration or soild content per unit volume and/or polymer molecular weight.

The dependence of adsorption on the concentration of clay particles in the suspension, in the case of polyvinyl alcohol acting on montmorillonite, has already been described (cf. Fig. 3.1). This observation may partly be explained in terms of the relative extent of interparticle bridge formation. With increasing solid content, the mean distance of separation between particles will be reduced and consequently there is a greater probability for bridging to occur. Although several clay particles may initially be attached by only a few, well-separated polymer segments, the subsequent collapse and spreading of the molecule over the surface would draw these particles together, thereby reducing the surface area accessible to the following molecules. Similarly, the decrease in adsorption as the clay concentration is reduced follows the sequence GL-05 < GM-14 < GH-20 presumably because the lowest molecular weight species (cf. Table 3.1) would form the fewest interparticle bonds (Greenland, 1963). By the same token, the uptake by Na^+-montmorillonite of polyacrylamde increases with molecular weight (cf. Fig. 3.6).

We have mentioned earlier that although neutral salts generally promote clay flocculation by uncharged polymers, more extensive interparticle bridging seems to occur when the salt is added after than before the polymer. To account for this behaviour, Emerson and Raupach (1964) have suggested that in the absence of electrolytes a loose network of polymer—clay links is formed. When salt is now added to the system, the clay particles tend to align themselves, allowing the PVA chains to spread out over a greater expanse of surface. This gives rise to stronger flocs and a marked increase in viscosity. On the other hand, if salt were added before the polymer, the latter would encounter a system in which the particles are already mutually aligned and in relatively close proximity to each other. Adsorption is more restricted and hence, interparticle bridging is less extensive than in the previous instance. It is important to note, however, that the salt effect described above refers to concentrations which are much lower (0.003 M NaCl) than those required to induce particle coagulation in the absence of polymer (0.025 M).

REFERENCES

Aylmore, L.A.G. and Quirk, J.P., 1960. Domain or turbostratic structure of clays. Nature (London), 187: 1046—1048.

Baran, A.A., Kocherga, I.I., Solomentseva, I.M. and Kurilenko, O.D., 1976. The stabilization of hydrophobic sols by water-soluble polymers. 1. Poly (vinyl alcohol) adsorption and adsorbed layer structure on colloidal antimony sulfide particles. Kolloidnyi Zhurnal, 38: 16—24.

Barrer, R.M. and Brummer, K., 1963. Relations between partial ion exchange and interlamellar sorption in alkylammonium montmorillonites. Transactions of the Faraday Society, 59: 959—968.

Birkner, F.B. and Edzwald, J.K., 1969. Nonionic polymer flocculation of dilute clay suspensions. Journal of the American Water Works Association, 61: 645—651.

Chew, B. and Couper, A., 1976. Diffusion, viscosity, and sedimentation of poly (ethylene oxide) in water. Journal of the Chemical Society, Faraday Transactions 1, 72: 382—388.

Dowdy, R.H., 1972. Effects of hydroxyl-containing organics on the strength—energy characteristics of montmorillonite. Soil Science Society of America Proceedings, 36: 162—166.

Emerson, W.W., 1956. Synthetic soil conditioners. Journal of Agricultural Science (Cambridge), 47: 117—121.

Emerson, W.W., 1960. Complexes of calcium montmorillonite with polymers. Nature (London), 186: 573—574.

Emerson, W.W. and Raupach, M., 1964. The reaction of polyvinyl alcohol with montmorillonite. Australian Journal of Soil Research, 2: 46—55.

Fleer, G.J., Koopal, L.K. and Lyklema, J., 1972. Polymer adsorption and its effect on the stability of hydrophobic colloids. Kolloid Zeitschrift und Zeitschrift für Polymere, 250: 689—702.

Fontana, B.J., 1963. The configuration of an adsorbed polymeric dispersant by infrared studies. The Journal of Physical Chemistry, 67: 2360—2362.

Francis, C.W., 1973. Adsorption of polyvinyl pyrrolidone on reference clay minerals. Soil Science, 115: 40—54.

Francis, C.W., Tamura, T., Bonner, W.P. and Amburgey, W., Jr, 1970. Separation of clay minerals and soil clays using isopycnic zonal centrifugation. Soil Science Society of America Proceedings, 34: 351—353.

Garvey, M.J., Tadros, Th.F. and Vincent, B., 1976. A comparison of the adsorbed layer thickness obtained by several techniques of various molecular weight fractions of poly (vinyl alcohol) on aqueous polystyrene latex particles. Journal of Colloid and Interface Science, 55: 440—453.

Glass, J.E., 1968a. Adsorption characteristics of water-soluble polymers. I. Poly (vinyl alcohol) and poly (vinyl pyrrolidone) at the aqueous—air interface. The Journal of Physical Chemistry, 72: 4450—4458.

Glass, J.E., 1968b. Adsorption characteristics of water-soluble polymers. II. Poly (ethylene oxide) at the aqueous—air interface. The Journal of Physical Chemistry, 72: 4459—4467.

Greenland, D.J., 1963. Adsorption of polyvinyl alcohols by montmorillonite. Journal of Colloid Science, 18: 647—664.

Greenland, D.J., 1972a. Interactions between organic polymers and inorganic soil particles. Mededelingen Fakulteit Landbouwwetenschappen Rijksuniversiteit Gent, 37: 897—914.

Greenland, D.J., 1972b. Adsorption of polyvinyl alcohols by oxides and clays with non-expanding lattices. Mededelingen Fakulteit Landbouwwetenschappen Rijksuniversiteit Gent, 37: 915—922.

Greenland, D.J., Laby, P.H. and Quirk, J.P., 1962. Adsorption of glycine and its di-, tri-, and tetra-peptides by montmorillonite. Transactions of the Faraday Society, 58: 829—841.

Griot, O. and Kitchener, J.A., 1965a. Role of surface silanol groups in the flocculation of silica suspensions by polyacrylamide. I. Chemistry of the adsorption process. Transactions of the Faraday Society, 61: 1026—1031.

Griot, O. and Kitchener, J.A., 1965b. Surface silanol groups in the flocculation of silica suspensions by polyacrylamide. II. Surface changes of silica suspensions on ageing. Transactions of the Faraday Society, 61: 1032—1038.

Hendricks, S.B., 1941. Base exchange of the clay mineral montmorillonite for organic cations and its dependence upon adsorption due to van der Waals forces. The Journal of Physical Chemistry, 45: 65—81.

Howard, G.J. and McConnell, P., 1967. Adsorption of polymers at the solution—solid interface. I. Polyethers on silica. The Journal of Physical Chemistry, 71: 2974—2981.

Iler, R.K., 1975. Coacervates of polyvinyl alcohol and colloidal silica. Journal of Colloid and Interface Science, 51: 388—393.

Jankovics, L., 1965. Effect of agitation and molecular weight on polymer adsorption and deflocculation. Journal of Applied Polymer Science, 9: 545—552.

Kashmoula, T. and Schurz, J., 1975. Die Eigenschaften von Polyvinylalkohol in verschiedenen Lösungsmitteln und seine Wechselwirkung mit Kaolin. Das Papier, 29: 45—48.

Kavanagh, B.V., Posner, A.M. and Quirk, J.P., 1975. Effect of polymer adsorption on the properties of the electrical double layer. Faraday Discussions of the Chemical Society, No. 59: 242—249.

Kavanagh, B.V., Posner, A.M. and Quirk, J.P., 1976. The adsorption of polyvinyl alcohol on gibbsite and goethite. The Journal of Soil Science, 27: 467—477.

Killmann, E. and Von Kuzenko, M., 1974. Ellipsometrische Charakterisierung der Adsorptionsschichten von Polyäthylenglycol, Polyvinylpyrrolidon und Polystyrol an Chromoberflächen. Die Angewandte Makromolekulare Chemie, 35: 39—55.

Killmann, E. and Winter, K., 1975. Kalorimetrische Untersuchungen zur Adsorption von Polyäthylenglykolen und niedermolekularen Substanzen an Siliziumdioxid. Die Angewandte Makromolekulare Chemie, 43: 53—73.

Kuzkin, S.F., Nebera, V.P. and Zolin, S.N., 1964. Aspects of the theory of suspension flocculation by polyacrylamides. Proceedings of the 7th International Mineral Processing Congress, New York, N.Y., Part 7: 347—357.

Kuznetsova, L.E. and Serb-Serbina, N.N., 1968. Influence of addition of polyvinyl alcohol on structure formation in aqueous dispersions of clay minerals. Kolloidnyi Zhurnal, 30: 853—859.

Levy, R. and Francis, C.W., 1975a. Interlayer adsorption of polyvinyl pyrrolidone on montmorillonite. Journal of Colloid and Interface Science, 50: 442—450.

Levy, R. and Francis, C.W., 1975b. Demixing of sodium and calcium ions in montmorillonite crystallites. Clays and Clay Minerals, 23: 475—476.

Linke, W.F. and Booth, R.B., 1960. Physical and chemical aspects of flocculation by polymers. Transactions of the American Institute of Mining, Metallurgical, and Petroleum Engineers, 217: 364—371.

Maslenkova, G.L., Revut, N.B. and Rusakova, G.N., 1965. Effect of exchangeable cations on the processes that take place in soil minerals during formation of synthetic soil structure. Pochvovedenie, No. 6: 73—79.

McBride, M.B. and Mortland, M.M., 1973. Segregation and exchange properties of alkylammonium ions in a smectite and vermiculite. Clays and Clay Minerals, 21: 323—329.

McLaren, A.D., Peterson, G.H. and Barshad, I., 1958. The adsorption and reactions of enzymes and proteins on clay minerals. IV. Kaolinite and montmorillonite. Soil Science Society of America Proceedings, 22: 239—244.

Mortland, M.M., 1970. Clay organic complexes and interactions. Advances in Agronomy, 22: 75—117.

Nightingale, E.R., Jr, 1959. Phenomenological theory of ion solvation. Effective radii of hydrated ions. The Journal of Physical Chemistry, 63: 1381—1387.

Norrish, K., 1954. The swelling of montmorillonite. Discussions of the Faraday Society, No. 18: 120—134.

Panasevich, A.A., Klimova, G.M. and Tarasevich, Yu.I., 1976. Study of poly (vinyl alcohol) adsorption on kaolinite. Kolloidnyi Zhurnal, 38: 1188—1191.

Parfitt, R.L. and Greenland, D.J., 1970a. The adsorption of poly (ethylene glycols) on clay minerals. Clay Minerals, 8: 305—315.

Parfitt, R.L. and Greenland, D.J., 1970b. Adsorption of polysaccharides by montmorillonite. Soil Science Society of America Proceedings, 34: 862—866.

Parfitt, R.L. and Greenland, D.J., 1970c. Adsorption of water by montmorillonite—poly (ethylene glycol) adsorption products. Clay Minerals, 8: 317—324.

Rubio, J. and Kitchener, J.A., 1976. The mechanism of adsorption of poly (ethylene oxide) flocculant on silica. Journal of Colloid and Interface Science, 57: 132—142.

Schamp, N. and Huylebroeck, J., 1972. Physico-chemical interaction of polyacrylamide on clay particles. Mededelingen Fakulteit Landbouwwetenschappen Rijksuniversiteit Gent, 37: 923—937.

Schamp, N. and Huylebroeck, J., 1973. Adsorption of polymers on clays. Journal of Polymer Science, Symposium No. 42: 553—562.

Schamp, N., Huylebroeck, J. and Sadones, M., 1975. Adhesion and adsorption phenomena in soil conditioning. In: Soil Conditioners. Soil Science Society of America Special Publication No. 7, pp. 13—23.

Schott, H., 1964. Interaction of nonionic detergents and swelling clays. Kolloid Zeitschrift und Zeitschrift für Polymere, 199: 158—169.

Tanihara, K. and Nakagawa, M., 1975. Flocculation treatment of waste water containing montmorillonite. IV. Interlamellar complex formation between various ion forms of montmorillonite and poly (ethylene oxide) or polyacrylamide. Nippon Kagaku Kaishi, No. 5: 782—789.

Theng, B.K.G., 1972. Adsorption of ammonium and some primary n-alkylammonium cations by soil allophane. Nature (London), 238: 150—151.

Theng, B.K.G., 1974. The Chemistry of Clay—Organic Reactions. Adam Hilger Ltd., 343 pp.

Theng, B.K.G. and Scharpenseel, H.W., 1975. The adsorption of ^{14}C-labelled humic acid by montmorillonite. Proceedings of the International Clay Conference, Mexico City, pp. 643—653.

Theng, B.K.G., Greenland, D.J. and Quirk, 1967. Adsorption of alkylammonium cations by montmorillonite. Clay Minerals, 7: 1—17.

Toyoshima, K., 1973. Reactions of polyvinyl alcohol with clay. In: C.A. Finch (Editor), Polyvinyl Alcohol. Properties and Applications. John Wiley and Sons, London, pp. 331—338.

Varfolomeeva, E.K., Zgadzai, L.K., Vlasov, V.V. and Sunchaleeva, S.S., 1977. Complexes of activated montmorillonites with poly (vinyl alcohol). Kolloidnyi Zhurnal, 39: 252—257.

Vincent, B., 1974. The effect of adsorbed polymers on dispersion stability. Advances in Colloid and Interface Science, 4: 193—277.

Chapter 4

NEGATIVELY CHARGED POLYMERS (POLYANIONS)

4.1. INTRODUCTION

The reactions of anionic polymers with clays began to be systematically studied in the early nineteen-fifties, research being stimulated by the idea that synthetic polyanions might simulate the action of negatively charged soil polymers, such as humic substances and polyuronides, in stabilising natural soil aggregates. It was no coincidence that two types of synthetic anionic polymers, known under the trade name of "Krilium", were marketed by the Monsanto Chemical Company at about this time. The polymers in question were the sodium salt of hydrolysed polyacrylonitrile (HPAN) and the predominantly calcium-saturated copolymer of vinyl acetate and maleic acid (VAMA), the structures of which are shown in the Appendix.

That "Krilium" did have a beneficial effect on soil physical conditions was soon demonstrated by a number of workers using a variety of methods. These early results were published in a series of papers presented at a symposium on "Improvement of Soil Structure" (Bear, 1952). The efficacy of Krilium, in this sense, is well illustrated by the data of Rennie et al. (1954) which are reproduced in Fig. 4.1. It is apparent that Krilium is at least as effective as polysaccharides which are considered to be the most important organic aggregate stabilising agents in soil (cf. section 10.4).

The concept that polyanions can flocculate clay suspensions by forming interparticle bonds goes back to 1952 when Ruehrwein and Ward (1952) proposed that a single polymeric chain was capable of attaching itself to more than one particle surface. Since then a large volume of literature has accumulated in support of the view that interparticle bridging is the mechanism by which polymers, in general, flocculate particulate colloidal dispersions (cf. section 2.4). Perhaps the most direct and compelling evidence for the bridging concept is provided by examining polyanion-flocculated clay systems under the electron microscope. Thus, using Krilium-type polymers, Beutelspracher (1955), Fischer and Rentschler (1957) and Jordine (1963) have been able to demonstrate visibly that clay particles so flocculated are linked together by the polymer which appears as slender, stretched-out threads. It ought to be interpolated here that the stabilisation of soil and clay aggregates in the field, be it by synthetic or naturally occurring polymers, is not due so much to the flocculation of soil/clay particles which may be suspended in the soil solution, as to the strengthening of an already existing structure by the adsorbed polymer (cf. Chapter 6).

Because much of the information on the clay—polyanion interaction relates to the effect of adsorbed polymers on clay colloid stability, it is often

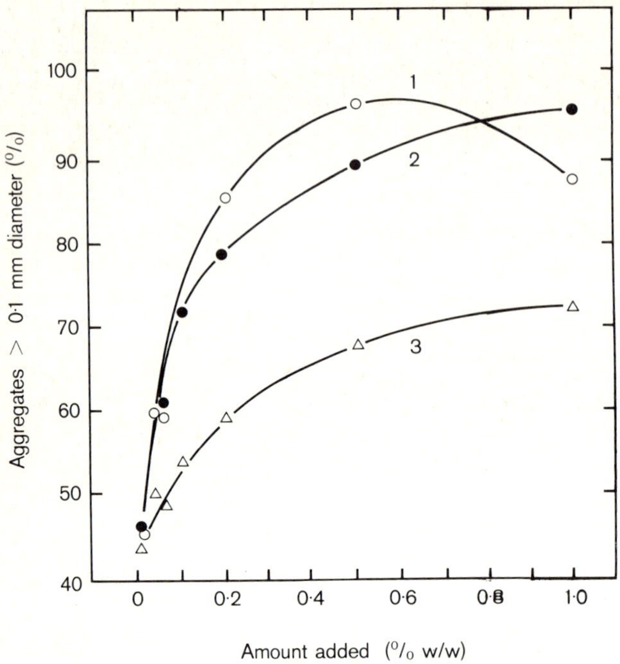

Fig. 4.1. The effectiveness of "Krilium" as compared with polysaccharides in stabilising soil aggregates. Curve 1, microbially synthesised polysaccharides; curve 2, "Krilium"; curve 3, soil polysaccharides. The relatively low efficacy of polysaccharides extracted from soil probably reflects the fact that these materials are composed of relatively short chain molecules (after Rennie et al., 1954).

difficult to treat the adsorption process separately from that of flocculation. It should be pointed out, however, that although polyanion adsorption must precede interparticle bridging by the polymer, adsorption does not always result in flocculation. Indeed, conditions which favour adsorption may work against flocculation. This is so because the former process is largely determined by the chemical constitution of the polymer and the net segment—surface interaction energy (cf. section 2.3) whereas flocculation is controlled more by the effective length of the polyanion chain than by the amount adsorbed (Michaels and Morelos, 1955; Flaig and Söchtig, 1959; Caskey et al., 1974).

The influence of chain length (i.e., molecular weight) on flocculation has been examined in some detail by Bergmann and Fiedler (1956) using polyacrylic acid and suspensions of kaolinite and montmorillonite. As would be expected, a good correlation was observed between these two parameters. However, the importance of chain length diminished as suspension strength increased, that is, as the mean distance between suspended particles was reduced. Similarly, in comparing the action of polyacrylamide on kaolinite and calcium carbonate suspensions with that of polyacrylic acid, Sakaguchi and Nagase (1966) noted that the former polymer attained its optimum effectiveness when the root-mean-square end-to-end distance approached the mean particle separation which, under the conditions used, was of the order of 100 nm. On the other hand, polyacrylic acid was effective at lower molecular weights (viscosities) because its chains were relatively extended due to the mutual repulsion of carboxylate groups.

Thus, factors which tend to promote adsorption such as low solution pH and high ionic strength, would also lead to chain coiling and hence, reduced flocculation. Conversely, under conditions which favour maximum chain extension and interparticle bridging, the polymer may be so highly ionised that it is largely repelled by the negatively charged particle surface. The relationship between adsorption and flocculation may be illustrated by the data of Slater et al. (1969) for the fluorite—hydrolysed polyacrylamide system (cf. Fig. 2.5). As the pH was decreased from 6 to 1, for example, the amount adsorbed increased three-fold but flocculation was only slightly reduced. On the other hand, on raising the pH from 9 to 12, there was little change in adsorption although flocculation was markedly reduced as was the apparent viscosity. Relevant in this context is the work of Michaels (1954), who found that there was an optimum degree of hydrolysis above and below which flocculation was reduced (cf. Fig. 4.3).

4.2. FORMATION AND PROPERTIES OF COMPLEXES

Because of their negative surface charge, clay minerals would be expected to repel rather than attract polyanions when most of their anionic groups are dissociated or fully ionised. That little, if any, uptake occurs under such conditions is generally borne out by experiment. Thus, in the absence of neutral salts, Ruehrwein and Ward (1952) found very little adsorption of sodium polymethacrylate by kaolinite. Similarly, Michaels and Morelos (1955) failed to detect any uptake of hydrolysed polyacrylamide and sodium polyacrylate by kaolinite above pH 7—8 since both polymers were then nearly completely ionised. As the solution pH was progressively lowered, ionisation became increasingly suppressed. This reduced the electrostatic repulsion between polymer and clay surface and increased chain coiling, giving rise to enhanced adsorption. By the same token, Parfitt and Greenland (1970) noted that poly(galacturonic acid) was not adsorbed by Al^{3+}-montmorillonite at pH 6 whilst the sodium- and calcium-saturated clays actually showed negative adsorption. Here again, appreciable (positive) uptake occurred with all three samples on lowering the solution pH (cf. section 10.3.2).

On the other hand, Bidwell et al. (1970) were able to measure some adsorption of a low molecular weight ($M \sim 1500$) polyacrylate by kaolinite at a pH as high as 7.3. However, the polyanion sample used apparently contained an appreciable proportion (~35%) of unionised carboxyl groups at this pH. This and its relatively small size possibly allowed the molecule to approach the clay surface sufficiently closely for bonding to take place. Joyce and Worrall (1970), also using polyacrylate and kaolinite, have reported similarly.

Besides being influenced by solution pH, the adsorption of polyanions by clays is enhanced by the presence of neutral or indifferent electrolytes, added in concentrations below what is required for particle coagulation (Ruehrwein and Ward, 1952; Ahlrichs, 1962; Mortensen, 1962; Zavorskhina and Kaganskaya, 1967). This observation may be explained in terms of charge screening effected either through cation binding or, in the case of polyvalent and transition cation salts, through coordination to and complexation with, the anionic groups of the polymer. These interactions, together with the induced tendency for the chain to coil, favour the adsorption process. In addition, there will be a slight compression of the diffuse double layers around the clay particles by which the polymer—clay repulsion is further reduced.

The question as to where and how negatively charged polymers are attached to clay mineral surfaces, is still not fully resolved. X-ray diffraction analysis of polyanion—clay complexes has consistently and repeatedly indicated that anionic polymers do not penetrate the interlayer space of montmorillonite-type crystals (Ruehrwein and Ward, 1952; Hagin and Bodman, 1954; Emerson, 1955; Holmes and Toth, 1957; Ahlrichs, 1962). To our knowledge, the only notable exception to this "rule" is shown by fulvic and humic acids in that these substances can be intercalated into montmorillonite at pH < 4 when they exist and behave as virtually uncharged species (Schnitzer and Kodama, 1966; Martin Martinez and Perez Rodriguez, 1969; cf. section 12.2). The failure of synthetic polyanions to form interlayer complexes with expanding 2 : 1 type phyllosilicates has led to the widely accepted view that the active sites on clay surfaces, in general, are mainly located at the crystal edges and identifiable with $Al \cdot OH$ or its protonated form, $Al \cdot OH_2^+$. Although such sites clearly play an important part in the adsorption by kaolinite-type minerals (cf. Fig. 1.3) they may not be of great significance to polyanion uptake by 2 : 1 type layer silicates.

Electrostatic attraction or "anion exchange" (Ruehrwein and Ward, 1952; Packter, 1957; Warkentin and Miller, 1958; Mortensen, 1962; Bidwell et al., 1970), hydrogen bonding (French et al., 1954; Michaels and Morelos, 1955; Emerson, 1956; Kohl and Taylor, 1961), or a combination of both (Ahlrichs, 1962; Van der Watt and Bodman, 1962) together with physical (Van der Waals) interactions (Joyce and Worrall, 1970) have all been postulated as bonding mechanisms. By analogy with the behaviour of simple anions, ligand exchange may be added to the list (Parfitt, 1978).

Although a great many experimental data are in line with the edge-site hypothesis, the external basal (planar) surfaces of clays are by no means inactive. Indeed, as we shall show below, there is adequate evidence to suggest that the anionic and polar groups of the polymer can interact with the exchangeable cations located at these surfaces (Mortensen, 1962; Van der Watt and Bodman, 1962; Roberts et al., 1974; Theng and Scharpenseel, 1975; Theng, 1976). It can therefore be said that the types of bonding involved in the adsorption of synthetic anionic polymers are, in all probability, similar to those postulated for humic substances (cf. section 12.2), summarised in Table 12.3.

The participation of crystal edges in polyanion uptake is supported by the finding that the amount adsorbed correlates with the measured positive edge charge ("anion exchange capacity") of the clay (Ruehrwein and Ward, 1952) and with the extent of the estimated edge area (Warkentin and Miller, 1958). Also in keeping with the edge-site hypothesis is Mortensen's (1962) observation that kaolinite loses its ability to adsorb hydrolysed polyacrylonitrile (HPAN) after pretreatment with an aluminium complexing agent. Similarly, the rheological measurements of Packter (1957) and Van der Watt and Bodman (1962) indicate that, under conditions conducive to interparticle bridging, polyanion adsorption (by montmorillonite) leads to a marked increase in the viscosity and yield stress of the system. This is ascribed to formation of edge-to-edge bonds giving rise to a ribbon-like structure with a high degree of dissymmetry, that is, a large surface diameter to particle thickness ratio. On the other hand, in situations where the polyanion merely neutralises the positive edge charge of the clay (without forming interparticle bonds), the viscosity and yield stress are reduced presumably because of the disruption of edge-to-face particle association which exists in the original (untreated) suspension (Street, 1957; Zavorskhina and Kaganskaya, 1967; Joyce and Worrall, 1970).

In this context, it is relevant to mention the measurements of Rausell-Colom and MacEwan (1964) on the swelling of Na^+-montmorillonite—polyanion complexes in water and electrolyte solutions. As indicated in section 1.2 (see also Table 1.4), the forces opposing the osmotic (region 2) swelling are largely in the nature of a frictional resistance arising from edge-to-face particle association. Although this type of bonding would be disrupted when polyanions, such as Krilium, adsorb onto the edge surfaces, this effect is more than offset by the attendant interparticle bonding by the polymer. Consequently, the presence of Krilium was found to increase the forces resisting interlayer expansion and so reduce the extent of swelling.

If polyanion adsorption were largely restricted to the crystal edges of clays, we might also expect that those minerals which show interlayer swelling in water would take up more of the polymer as compared with their non-swelling counterparts. This is so because, unlike the adsorption of small, simple anionic species, steric and accessibility factors come into play during the interaction of polyanions and clay minerals. In accord with this expectation, Ahlrichs (1962) and De and Jain (1964) have observed that the adsorption of Krilium-type polymers by 2 : 1 type layer silicates decreased in the order of montmorillonite > vermiculite > biotite > pyrophyllite. This observation further argues for the reactivity of the edge surface towards polyanions since such minerals as biotite and pyrophyllite have few, if any, exchangeable cations with which the functional groups of the polymer may interact.

Similarly, Kohl and Taylor (1961) have interpreted their infrared data for the adsorption of HPAN and VAMA by Ca^{2+}- and H^+/Al^{3+}-montmorillonite in terms of hydrogen bonding between the carbonyl groups of the polyanion and edge aluminols. This suggestion was based on the downward shift of the carbonyl group frequency near 1720 cm^{-1} by some 15—20 wavenumbers when complex formation occurred. In the absence of independent measurements, however, the adsorption sites cannot be unequivocally and exclusively assigned to the edge surface. Thus, hydrogen bonding of the carbonyl groups to water molecules in the primary hydration shell of the exchangeable cations is likely to contribute significantly to the observed shift in carbonyl absorption band. The difficulty in distinguishing between crystal edge aluminium and exchangeable aluminium, using infrared techniques, applies particularly to acid washed montmorillonite (Ahlrichs, 1962).

The active participation of the exchangeable cations in polyanion adsorption by clays is well illustrated by the work of Mortensen (1957, 1962) using HPAN and kaolinite saturated with different cations. As indicated in Fig. 4.2, uptake increased in the order of $Na^+ < K^+ < NH_4^+ < H^+ < Ba^{2+} < Ca^{2+} < Th^{4+}$. With the exception of H^+, this was also the sequence in which these

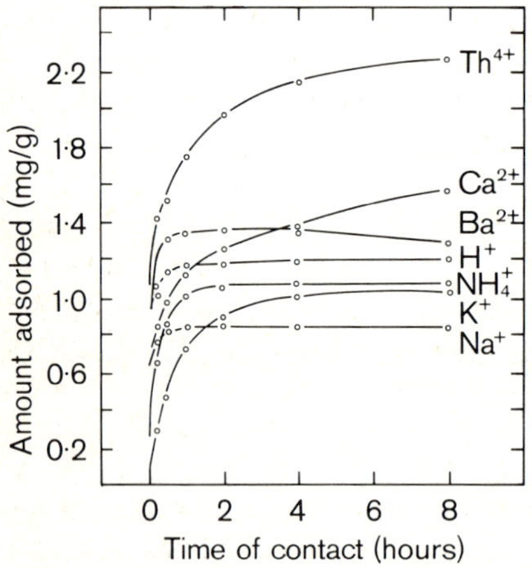

Fig. 4.2. The adsorption of hydrolysed polyacrylonitrile (HPAN) by kaolinite as influenced by the nature of the exchangeable cation at the clay surface (after Mortensen, 1962).

cations reduced the zeta potential of the clay. On this basis, Mortensen suggested that as ζ became increasingly less negative, the polyanion was able to approach the clay surface more closely and this enhanced the probability for surface—polymer bonding to be established. This attachment might be effected by means of anion- and ligand-exchange as well as through hydrogen bonding interactions, all of which primarily involved edge aluminol groups. Polyvalent exchangeable cations, present at the basal surfaces of the clay, might also act as bridges between the polyanion and the mineral surface.

That polyvalent cations, such as Ca^{2+}, promote the adsorption of polyanions by clays and, indeed, are necessary for effective flocculation to occur, has subsequently been demonstrated by Black et al. (1965) for the interaction of hydrolysed polyacrylamide with dilute kaolinite suspensions. The favourable effect of Ca^{2+} ions was ascribed to the reduction of surface—polymer and particle—particle repulsion resulting from the compression of diffuse double layers. A similar mechanism has been proposed by Sommerauer et al. (1968) for the interaction of some anionic polymers (polyacrylic acid, hydrolysed polyacrylamide, and polystyrene sulphonate) with AgBr sols in the presence of small amounts ($\leqslant 5 \times 10^{-4}$ M) of Ca^{2+} ions. Unlike the situation with clay minerals, however, the question of where the adsorption sites may be located does not arise with the spherical AgBr particles and formation of $RCOO^{-} \cdots Ca^{2+} \cdots$ surface or $RSO_3^{-} \cdots Ca^{2+} \cdots$ surface type of complexes proposed by Sommerauer et al. is likely to be the principal mode of attachment betweeen polymer and the negatively charged sol particles. Likewise, Slater et al. (1969) have found that the adsorption of an anionic polyacrylamide by silica particles was markedly improved by the addition of copper sulphate to the medium (over a narrow range of pH and concentration). These workers also inferred that the Cu^{2+} ions acted as bridges joining the carboxylate groups of the polymer to the negatively charged silica surface.

It is interesting to compare Mortensen's (1957, 1962) data with those obtained by Theng and Scharpenseel (1975) and Theng (1976) for the adsorption of humic and fulvic acids by montmorillonite. In fact, if the plateau values for the HPAN—kaolinite system (Fig. 4.2) were to be plotted against the ionic potential of the respective cations, the points for Ba^{2+}, Ca^{2+}, and Th^{4+} fall on a line with a positive slope whilst those for Na^{+}, K^{+}, and NH_4^{+} lie on another line of which the slope is negative. The close resemblance of these plots to those shown by the humic acid—montmorillonite system (cf. Fig. 12.4) would indicate that a similar mechanism is operative in both systems. That is to say that, like humic and fulvic acids, the functional ($C=O$, $C\equiv N$) groups of HPAN are probably associated with the exchangeable cations through a water bridge and through direct ion—dipole interactions in the case of polyvalent and monovalent cation clay systems, respectively (cf. Table 12.3). Whatever the underlying mechanism may be, however, it seems clear that the basal surfaces of clays play an important part in polyanion adsorption since it is here that the cation exchange sites are located.

Rather puzzling was the fact that prior treatment of the parent clay with strong (0.5 M) solutions of the sodium salt of different anions also led to enhanced adsorption of HPAN. The extent to which adsorption was enhanced decreased in the order $F^- > OH^- > H_2PO_4^- > Cl^- > CH_3COO^- > NO_3^-$ (Mortensen, 1962). By analogy with the promoting effect of cations on HPAN uptake, we might expect that anions would produce the opposite effect. Slater et al. (1969), for example, noted that anionic polyacrylamide at pH 6.4 was capable of flocculating galena but failed to do so in the presence of 1 mM sodium sulphide which presumably induced a large negative potential on the particle surface. It is known, however, that under the drastic conditions used by Mortensen, extensive structural decomposition would occur through rupture of Si—O—Si and Al—O—Si bonds (e.g. Fijal et al., 1976). It seems reasonable to suppose that the extent of this degradation follows the same sequence so that the observed rise in adsorption is largely due to the resulting increase in particle surface area exposed to the polyanion. On the other hand, the increase in HPAN adsorption on addition of dilute, indifferent electrolytes to the medium, may be ascribed to the attendant compression of diffuse double layers (cf. Fig. 1.13 C). In addition, the polyanion chain will be more coiled at the high ionic strength and this will decrease the surface requirement of the polymer.

Mortensen (1962) also reported that adsorption of HPAN onto kaolinite could be markedly reduced by raising the pH of the medium from 4 to 9. This observation may be explained in terms of the increased extension of the polyanion chain induced by intramolecular charge repulsion since a rise in pH would increase the degree of ionisation of the molecule (cf. Fig. 2.3). Although an increase in chain extension at high pH would favour interparticle bond formation, this tendency would be somewhat offset by the attendant enhancement of particle dispersion due to the progressive disruption of edge-to-face particle association and the increase in electrostatic repulsion between the polymer and the particle.

Michaels (1954) was able to get around this difficulty by synthesising a series of polyacrylamides with different degrees of hydrolysis (β), testing their flocculating properties on aqueous clay/soil suspensions. The flocculation results together with the probable conformations of the polymer in solution as a function of β, are shown in Fig. 4.3. Under the conditions used, optimum flocculation (as measured by the reduction in sediment density) occurs when $\beta \sim 30\%$. At this point, there is apparently a good compromise between chain extension, on the one hand, and polymer—particle repulsion, on the other. At very low or very high values of β, polyacrylamide is a poor flocculant of dilute clay/soil suspensions because its chain is highly coiled or it is effectively repelled by the negatively charged particle. Similarly, Williams and Ottewill (1971) have shown that fully hydrolysed polyacrylamide, which is the same as polyacrylic acid, failed to flocculate negatively charged AgI sols.

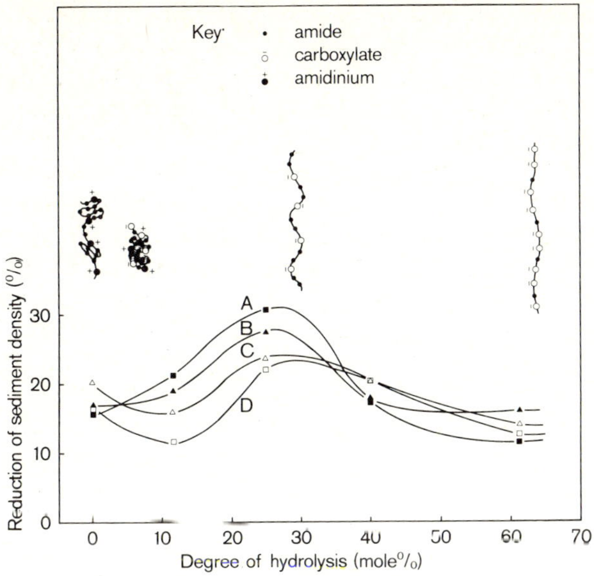

Fig. 4.3. Relationship between the flocculating efficiency (measured in terms of the percentage reduction in sediment density) and the degree of hydrolysis for polyacrylamide acting on aqueous suspensions of a sandy soil. The probable chain conformations in solution of the polymer at different degrees of hydrolysis are also indicated (after Michaels, 1954).

Another approach which has been shown to work well with saponite suspensions at high pH is to add a coagulant (e.g. $MgSO_4$) to the stable sol, followed by an anionic polyacrylamide (O'Gorman and Kitchener, 1973). In this way, the repulsion between particles and between particle and polymer is reduced whilst the benefit of high chain extension is being retained. A variant of this is to adsorb hydrolysed aluminium ions onto the clay surface prior to adding the polyanion, as Roberts et al. (1974) have done in flocculating kaolinite suspensions with polyacrylamide. The beneficial effect of adsorbed polyhydroxy aluminium compounds on flocculation is presumably similar to that exerted by polyvalent exchangeable cations.

Because of coulombic repulsion, polyanions are relatively weakly adsorbed by clays although this situation, as we have seen, can be largely reversed by the presence of positively charged species at the particle surface. By the same token, however, anionic polymers are effective flocculants of dilute clay suspensions since their extended chain conformation in solution is more or less retained (in the form of long tails) on adsorption. On this premise we might expect the flocculating effectiveness of polyanions to increase with chain length, that is, with molecular weight. That this is so has been demonstrated by Michaels (1954) for the flocculation of kaolinite suspensions by hydrolysed polyacrylamide. Likewise, Smith and Volk, as cited by Walles (1968), have found that the time required to flocculate montmorillonite suspensions by sodium polystyrene sulphonate is sharply decreased with an increase in polymer molecular weight up to a limiting value (Fig. 4.4). Interestingly, Montgomery and Hibbard (1955) came to similar conclusions in the case of the soil aggregating activity of polystyrene sulphonates.

We may perhaps ask at this juncture what influence, if any, chain branching would have on the flocculating effectiveness of anionic polymers in general. In attempting to answer this question, Caskey et al. (1974) have prepared linear and branched polyacrylamides of varying molecular weight and degree of hydrolysis (β), assessing their respective effectiveness with aqueous (3% w/v) suspensions of kaolinite.

In accord with previous reports, flocculation by the uncharged or non-hydrolysed, linear forms increased with molecular weight and, for a given molecular weight, the anionic polymers were superior to their non-ionic counterparts in flocculating the clay suspension. Again, as Michaels (1954) has found, the linear polymers were most effective for $\beta \sim 25\%$, that is, when the chains attained their optimum extension. Interestingly, the branched polymers were less effective than the linear types at all levels of molecular weight and degree of hydrolysis. Entirely analogous results were obtained by Anthony (1973) for the flocculation of domestic sewage by linear and branched anionic polyacrylamides. Somewhat related to this was the

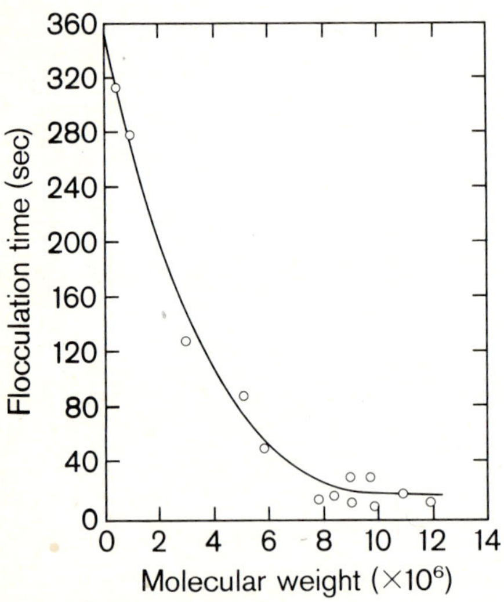

Fig. 4.4. Relationship between polymer molecular weight and flocculation time for the flocculation of aqueous montmorillonite suspensions by sodium polystyrene sulphonate (after Walles, 1968).

observation of Montgomery and Hibbard (1955) that the (soil) aggregating activity of polystyrene sulphonates decreased as the number of (sulphone) cross-links increased. More surprisingly, however, was the finding that less clay was flocculated under conditions when the individual side chains of the branched polymers were optimally extended than when they are fully coiled.

In order to rationalise these observations, Caskey et al. (1974) proposed that chain branching created a physical or steric barrier to interparticle bridging, an effect which became more pronounced as the branches became more extended. This would imply that in the case of branched polymers steric stability could be achieved at relatively low amounts adsorbed. By contrast, linear anionic polymers do not exert their protective action or give rise to steric stabilisation until much higher levels of adsorption have been reached. Indeed, at low concentrations linear polyanions may render the clay suspension more sensitive to coagulation by simple, neutral electrolytes (cf. Fig. 2.4 C).

The ability of anionic polymers to sensitise and protect clay suspensions at low and high amounts adsorbed, respectively, has been demonstrated by Van Olphen (1963) using sodium carboxymethylcellulose (CMC) and Na^+-montmorillonite. He noted that there was a narrow range of polyanion concentration (1×10^{-4} to 10×10^{-4} g/100 ml) over which the flocculation value (i.e. the amount of NaCl required to cause flocculation) was less than that shown in the absence of CMC. Thus, in a plot of CMC concentration (or [CMC]) against flocculation value, a well-defined minimum occurred at [CMC] $\sim 3 \times 10^{-4}$ g/100 ml, corresponding to the point at which the montmorillonite suspension was most sensitive to being coagulated by salt. For [CMC] $> 10 \times 10^{-4}$ g/100 ml, on the other hand, the flocculation value increased more or less linearly with polyanion concentration. In this range of concentration, the suspension was apparently protected or sterically stabilised by the presence around the particle surfaces of a rather dense polymer layer consisting of extended loops and tails (cf. Fig. 2.4 F).

We ought to interpolate here that no interparticle bridging is indicated in the idealised diagram for the sensitisation process (Fig. 2.4 D). However, as Van Olphen has suggested, some bridging probably occurs even at low amounts of polyanion adsorbed although no visible flocculation is to be detected in the absence of electrolytes. Another point of note is that, for a given polyanion concentration, the flocculation value is smaller when the electrolyte is added before the polyanion than when the order of addition is reversed. Furthermore, the flocs produced in the former instance are more stable to redispersion (Ruehrwein and Ward, 1952; Warkentin and Miller, 1958). This observation may be compared with the behaviour of uncharged polymers towards clay suspensions. Here the extent of interparticle bridging is apparently increased when the electrolyte is added after the polymer, as Emerson and Raupach (1964) have found for the montmorillonite—polyvinyl alcohol system (cf. section 3.2).

It seems clear that, under certain conditions, polyanions can act as effective flocculating agents of dilute clay suspensions, despite the fact that both the polymer and the suspended particle carry a net negative surface charge. Indeed, because of their extended chain conformation in solution, anionic polymers are, in this respect, generally more effective than their non-ionic counterparts of comparable molecular weight (Michaels, 1954; Sakaguchi and Nagase, 1966). However, since many variables affect the adsorption—flocculation process, these conditions should be assessed separately for each system.

REFERENCES

Ahlrichs, J.L., 1962. Bonding of organic polyanions to clay minerals. Dissertation Abstracts, 22: 2121.

Anthony, A.J., 1973. The effects of branching and other physical properties of anionic polyacrylamides on the flocculation of domestic sewage. Ph.D. Thesis, Virginia Polytechnic Institute and State University, Blacksburg, Virginia, 111 pp.

Bear, F.E. (Editor), 1952. Synthetic Soil Conditioners. Soil Science, 73: 419—492.

Bergmann, W. and Fiedler, H.J., 1956. Der Einfluss synthetischer heteropolarer Linearkolloide unterschiedlicher Kettenlänge auf Kaolin- und Bentonitaufschlämmungen. Zeitschrift für Pflanzenernährung, Düngung, Bodenkunde, 72: 114—136.

Beutelspracher, H., 1955. Wechselwirkung zwischen anorganischen und organischen Kolloiden des Bodens. Zeitschrift für Pflanzenernährung, Düngung, Bodenkunde, 69: 108—115.

Bidwell, J.I., Jepson, W.B. and Toms, G.L., 1970. The interaction of kaolinite with polyphosphate and polyacrylate in aqueous solutions — some preliminary results. Clay Minerals, 8: 445—459.

Black, A.P., Birkner, F.B. and Morgan, J.J., 1965. Destabilization of dilute clay suspensions with labelled polymers. Journal of the American Water Works Association, 57: 1547—1560.

Caskey, J.A., King, P.H. and Martin, J.T., 1974. Effect of polyacrylamide molecular structure on flocculation efficiency. American Institute of Chemical Engineers Symposium Series, 70: 120—128.

De, S.K. and Jain, R.K., 1964. Some aspects of krilium adsorption by clay minerals and their derivatives. Journal of the Indian Chemical Society, 41: 379—383.

Emerson, W.W., 1955. Complex formation between montmorillonite and high polymers. Nature (London), 176: 461.

Emerson, W.W., 1956. Synthetic soil conditioners. Journal of Agricultural Science (Cambridge), 47: 117—121.

Emerson, W.W. and Raupach, M., 1964. The reaction of polyvinyl alcohol with montmorillonite. Australian Journal of Soil Research, 2: 46—55.

Fijal, J., Klapyta, Z. and Zabinski, W., 1976. Study of the modification of surface properties of clay mineral. Przeglad Geologiczny, 24: 257—259.

Fischer, E.W. and Rentschler, W., 1957. Zur Verbesserung der Bodenstruktur durch synthetische Hochpolymere. Zeitschrift für Pflanzenernährung, Düngung, Bodenkunde, 76: 232—244.

Flaig, W. and Söchtig, H., 1959. Über die Wechselwirkung zwischen Kaolinit and Polyanionen. Zeitschrift für Pflanzenernährung, Düngung, Bodenkunde, 87: 44—57.

French, R.O., Wadsworth, M.E., Cook, M.A. and Cutler, I.B., 1954. The quantitative application of infrared spectroscopy to studies in surface chemistry. The Journal of Physical Chemistry, 58: 805—811.

Hagin, J. and Bodman, G.B., 1954. Influence of the polyelectrolyte CRD-186 on aggregation and other physical properties of some California and Israeli soils and some clay minerals. Soil Science, 78: 367—378.

Holmes, R.M. and Toth, S.J., 1957. Physico-chemical behaviour of clay—conditioner complexes. Soil Science, 84: 479—487.

Jordine, E.St. A., 1963. Fine structure of polyelectrolytes and clay—organic complexes. Journal of Electron Microscopy, 12: 236—239.

Joyce, I.H. and Worrall, W.E., 1970. The adsorption of polyanions by clays and its effect on their physical properties. Transactions of the British Ceramic Society, 69: 211—216.

Kohl, R.A. and Taylor, S.A., 1961. Hydrogen bonding between the carbonyl group and Wyoming bentonite. Soil Science, 91: 223—227.

Martin Martinez, F. and Perez Rodriguez, J.L., 1969. Interlamellar adsorption of a black-earth humic acid on Na^{+}-montmorillonite. Zeitschrift für Pflanzenernährung und Bodenkunde, 124: 52—57.

Michaels, A.S., 1954. Aggregation of suspensions by polyelectrolytes. Industrial and Engineering Chemistry, 46: 1485—1490.

Michaels, A.S. and Morelos, O., 1955. Polyelectrolyte adsorption by kaolinite. Industrial and Engineering Chemistry, 47: 1801—1809.

Montgomery, R.S. and Hibbard, B.B., 1955. Theoretical aspects of the soil-conditioning activity of polymers. Soil Science, 79: 283—292.

Mortensen, J.L., 1957. Adsorption of hydrolyzed polyacrylonitrile on kaolinite: I. Effect of exchange cation and anion. Soil Science Society of America Proceedings, 21: 385—388.

Mortensen, J.L., 1962. Adsorption of hydrolyzed polyacrylonitrile on kaolinite. Clays and Clay Minerals, 9: 530—545.

O'Gorman, J.V. and Kitchener, J.A., 1973. The flocculation and de-watering of Kimberlite clay slimes. International Journal of Mineral Processing, 1: 33—49.

Packter, A., 1957. Interaction of montmorillonite clays with polyelectrolyte. Soil Science, 83: 335—343.

Parfitt, R.L., 1978. Anion adsorption by soils and soil materials. Advances in Agronomy, 30: 1—50.

Parfitt, R.L. and Greenland, D.J., 1970. Adsorption of polysaccharides by montmorillonite. Soil Science Society of America Proceedings, 34: 862—866.

Rausell-Colom, J.A. and MacEwan, D.M.C., 1964. El hinchamiento de la montmorillonita sodica y del complejo montmorillonita—krilium en electrolitos. Anales de Edafologia y Agrobiologia, 23: 499—520.

Rennie, D.A., Truog, E. and Allen, O.N., 1954. Soil aggregation as influenced by microbial gums, level of fertility and kind of crop. Soil Science Society of America Proceedings, 18: 399—403.

Roberts, K., Kowalewska, J. and Friberg, S., 1974. The influence of interactions between hydrolyzed aluminium ions and polyacrylamides on the sedimentation of kaolin suspensions. Journal of Colloid and Interface Science, 48: 361—367.

Ruehrwein, R.A. and Ward, D.W., 1952. Mechanism of clay aggregation by polyelectrolytes. Soil Science, 73: 485—492.

Sakaguchi, K. and Nagase, K., 1966. The flocculation of kaolin and precipitated calcium carbonate by polymers. Bulletin of the Chemical Society of Japan, 39: 88—91.

Schnitzer, M. and Kodama, H., 1966. Montmorillonite: Effect of pH on its adsorption of a soil humic compound. Science, 153: 70—71.

Slater, R.W., Clark, J.P. and Kitchener, J.A., 1969. Chemical factors in the flocculation of mineral slurries with polymeric flocculants. In: Clays and Other Colloidal Systems. Proceedings of the British Ceramic Society, No. 13: 1—12.

Sommerauer, A., Sussman, D.L. and Stumm, W., 1968. The role of complex formation in the flocculation of negatively charged sols. Kolloid Zeitschrift und Zeitschrift für Polymere, 225: 147—154.

Street, N., 1957. Effect of a polyanion on the rheology of a kaolinite suspension. Journal of Colloid Science, 12: 1—9.

Theng, B.K.G., 1976. Interactions between montmorillonite and fulvic acid. Geoderma, 15: 243—251.

Theng, B.K.G. and Scharpenseel, H.W., 1975. The adsorption of ^{14}C-labelled humic acid by montmorillonite. Proceedings of the International Clay Conference, Mexico City, pp. 643—653.

Van der Watt, H.v.H. and Bodman, G.B., 1962. Viscosimetric constants of suspensions of clay—polymer complexes. Clays and Clay Minerals, 9: 568—584.

Van Olphen, H., 1963. An Introduction to Clay Colloid Chemistry. Interscience, New York, N.Y., 301 pp.

Walles, W.E., 1968. Role of flocculant molecular weight in the coagulation of suspensions. Journal of Colloid and Interface Science, 27: 797—803.

Warkentin, B.P. and Miller, R.D., 1958. Conditions affecting formation of the montmorillonite—polyacrylic acid bond. Soil Science, 85: 14—18.

Williams, D.J.A. and Ottewill, R.H., 1971. Stability of silver iodide sols in the presence of poly(acrylic acids) of various molecular weights. Kolloid Zeitschrift und Zeitschrift für Polymere, 243: 141—147.

Zavorskhina, N.A. and Kaganskaya, K.A., 1967. Action of hydrolyzed polyacrylamide on clay suspensions. Kolloidnyi Zhurnal, 29: 797—802.

Chapter 5

POSITIVELY CHARGED POLYMERS (POLYCATIONS)

5.1. INTRODUCTION

As compared with the negatively charged and uncharged polymers, described in the preceding chapters, the interactions of polycations with clay minerals have not received a great deal of attention. This seeming lack of interest may be ascribed to the fact that soil organic matter or humus which is intimately associated with the clay fraction, largely consists of negatively charged polymers together with minor amounts of uncharged and amphoteric substances. By and large, this is a reflection of the overlying vegetation from which humus is ultimately derived and of the pathways through which plant residues are decomposed and transformed into humic substances (cf. Chapter 12).

In recent years, however, there has been an upsurge of interest in the behaviour of positively charged synthetic polymers at the aqueous/solid interface, research being stimulated by the potential of polycations as aggregating agents of negatively charged colloids. Most of the information available on the polycation—solid interaction is therefore concerned more with the effect of these polymers on dispersion stability than with their adsorption behaviour.

The use of positively charged macromolecules as aggregants has been studied in such systems as aqueous dispersions of metal oxides and hydroxides (Linke and Booth, 1960; Kane et al., 1964a, b, c; Shyluk, 1964; Dixon et al., 1967; Posselt et al., 1968; Iler, 1971), of clay minerals (Ruehrwein and Ward, 1952; Shyluk, 1964; Black et al., 1965, 1966; Pressman, 1967; Narkis et al., 1968; Ueda and Harada, 1968a; Black and Vilaret, 1969; Yorke, 1973; Kimura et al., 1976), of polymer fibres and latices (Gregory, 1969, 1973; Goossens and Luner, 1976; Lindström and Söremark, 1976), and of humic substances (Narkis et al., 1968; Rebhun et al., 1969). It is in this type of application that polycations really come into their own in that they are extremely effective, even at low concentrations (<5 p.p.m.) and in the absence of neutral electrolytes. Their high effectiveness in this sense is related to the ability of polycations to act as both a coagulant and flocculant (cf. section 2.4). The former effect is exerted through charge neutralisation with the attendant compression of the diffuse double layers on the particle surface through which the height of the energy barrier to coagulation (V_m) is reduced (cf. Fig. 1.15). Indeed, for low to medium molecular weight ($M < 4.5 \times 10^4$) species, the flocculating action effected through interparticle bridge formation may only be of secondary importance. Thus Goossens and Luner (1976) found little evidence for interparticle bridging on adding an ionene polymer ($M \sim 6 \times 10^4$) to a suspension of microcrystalline cellulose. In many instances, however, charge neutralisation alone cannot adequately account for the interaction process. Certainly, at high polycation concentrations and molecular weights, a certain amount of flocculation or interparticle bridging must be invoked in order to rationalise the experimental data (e.g. Shyluk, 1964; Black et al., 1965, 1966; Narkis et al., 1968; Iler, 1971).

5.2. FORMATION AND PROPERTIES OF COMPLEXES

The ability of different polyelectrolytes to aggregate suspended particles has been investigated by Kane et al. (1964a, b, c) and Dixon et al. (1967) using silica suspensions. They were able to demonstrate that aggregation and good filtration by polycations occurred for molecular weights and polymer dosages which were at least one order of magnitude smaller than those shown by anionic polymers. Earlier, Flaig et al. (1962) have reported similarly for the behaviour of aqueous kaolinite suspensions towards different polymers of comparable molecular weight in that the flocculating efficiency increased in the order polyanion < non-ionic polymer < polycation (Fig. 5.1).

As early as 1952 Ruehrwein and Ward put forward the suggestion that the aggregation of suspended clay particles by polyelectrolyte addition was brought about by adsorption of, and interparticle bridging by, the polymer. Such a mechanism would account for the observation that the aggregation of kaolinite suspensions by sodium polymethacrylate was promoted by the presence of NaCl at low concentrations. Although neutral electrolytes would reduce intramolecular charge repulsion, giving rise to a less stretched-out conformation, they would also bring the clay particles closer together. Adsorption as well as interparticle bridging would thus be enhanced. Unlike

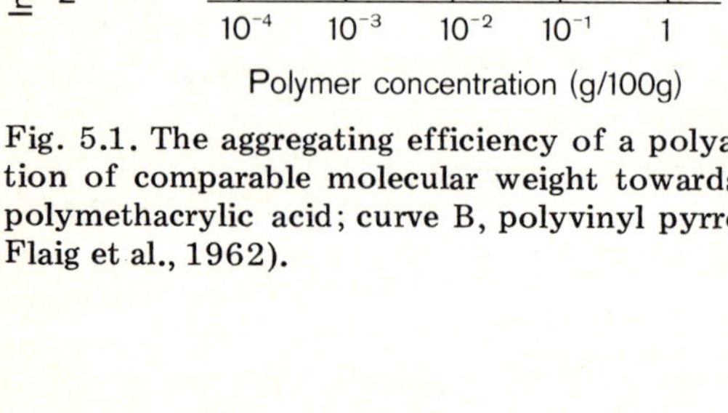

Fig. 5.1. The aggregating efficiency of a polyanion, an uncharged polymer, and a polycation of comparable molecular weight towards aqueous kaolinite suspensions. Curve A, polymethacrylic acid; curve B, polyvinyl pyrrolidone; curve C, polyethyleneimine (after Flaig et al., 1962).

polymethacrylate, the cationic polymer β-dimethyl-aminoethylmethacrylate hydroacetate (DMAEM) was capable of penetrating the interlayer space of montmorillonite. In doing so, the basal spacing of the resultant complexes increased with the amount adsorbed. This would indicate that DMAEM was intercalated by exchanging for the Na^+ ions and that such an interchange gave rise to Na^+-rich and polycation-rich layers which were randomly interstratified within a single crystal (cf. Figs. 3.4 and 7.3). The extent of interlayer expansion (~0.5 nm) at the maximum corresponds to the presence of a single layer of molecules, adsorbed in a flat, extended conformation between every two opposing silicate layers. This is in keeping with theoretical predictions since, for an ion exchange process, the net segment—surface interaction energy is greatly in excess of 1 kT (cf. section 2.3).

A more compelling evidence for the charge neutralisation concept and the attendant chain collapse is provided by the studies of Ueda and Harada (1968a). Using Na^+-montmorillonite with a cation exchange capacity (CEC) of 62.4 meq/100 g and a diallyldimethylammonium chloride—SO_2 copolymer of molecular weights 16.7×10^4 (copolymer I) and 1.4×10^4 (copolymer II), these workers followed the changes in CEC and anion exchange capacity (AEC) of the complex as adsorption progressed. The results, summarised in Table 5.1, show that with rising amounts (of copolymer I) adsorbed, the AEC (column C) increased whilst the CEC (column B) declined. Since the parent clay sample did not possess a measurable AEC, the development of an AEC by the complex was ascribed to the adsorption of polymer segments in the form of loops and tails whilst the decrease in CEC was due to neutralisation of the negative surface charge by train segments. On this premise, the difference between the amount of polymer adsorbed (column A) and the AEC represents the quantity which is actually in contact with the surface (column D). The ratio of D/A would then give the proportion of segments adsorbed or bound in trains (p). The virtual constancy of the sum (B + D) which approaches the total cation exchange capacity of the clay, lends further support to the proposed mechanism.

Of particular interest is that up to a surface coverage of ~0.5, all or nearly all of the polymer segments are attached to the surface ($p \sim 1$); even at full coverage only some 25% of the segments is contained in loops and tails. A consequence of chain flattening (high p values) is that the adsorption process is "irreversible" with respect to water and simple electrolytes, no detectable desorption being observed on washing the complexes with up to 1.5 M $CaCl_2$ solutions. Although interlayer adsorption was likely to occur, basal spacing measurements were not carried out to verify this point which would have served to clinch the argument for the extended surface conformation inferred from chemical analysis.

As compared with copolymer I, more of the lower molecular weight copolymer II was taken up at comparable concentrations in solution. Further, very little looping occurred with copolymer II ($p > 0.9$) even at high surface

TABLE 5.1

Data for the adsorption of diallyldimethylammonium chloride—SO_2 copolymer I by montmorillonite at a suspension concentration of 1.5 g/100 ml and 303 K (from Ueda and Harada, 1968a)

Amount added (g/100 g)	Amount adsorbed		CEC (meq/100 g) B	AEC (meq/100 g) C	Amount in trains (meq/100 g) D = A − C	Sum of B + D (meq/100 g)	Proportion of segments in trains (p) D/A
	(g/100 g)	(meq/100 g) A					
0	0	0	62.4	0	0	62.4	1
2.67	1.93	8.56	49.5	0	8.56	58.1	1
5.33	4.18	18.5	44.6	0	18.5	63.1	1
10.0	7.57	33.6	30.7	0.97	32.6	63.3	0.97
13.3	9.18	40.7	24.5	3.85	36.8	61.3	0.90
20.0	9.82	43.5	23.9	5.34	38.2	62.1	0.88
26.7	10.8	47.9	21.4	6.68	41.2	62.6	0.86
33.3	10.5	46.6	23.7	6.37	40.2	63.9	0.86
43.3	13.0	57.7	15.8	13.2	44.5	60.3	0.77
56.7	13.2	58.5	14.4	15.1	43.4	57.8	0.74
73.7	14.2	63.0	12.8	15.5	47.5	60.3	0.75
93.3	14.5	64.3	11.1	18.2	46.1	57.2	0.71

coverage. In being adsorbed or intercalated, the smaller size polymer presumably encountered less steric hindrance, enabling a more intimate polymer—surface association to be established. An enhanced adsorption with a decrease in polymer molecular weight and charge, has also been reported by Lindström and Söremark (1976) for the interactions of cationic polyacrylamides with cellulose.

The effect of molecular shape (rather than size) on uptake is reflected in the observation that the adsorption of copolymer I by montmorillonite increases in the presence of neutral salts (Table 5.2). This may be ascribed to the greater tendency of the polycation to coil as the ionic strength of the medium is increased. As a result, the surface requirement of the polymer is lessened and more can be accommodated at the clay surface (column A). Apparently, the coiled conformation is largely retained on adsorption since the amount of bound segments (column D) remains more or less invariant with surface coverage. Peyser and Ullman (1965) have observed similarly for the adsorption of poly-4-vinylpyridinium chloride onto negatively charged porous glass surfaces.

As mentioned at the beginning of this chapter, research into the polycation—solid interaction, in general, has emphasised the aggregating effect of polycations on aqueous particulate dispersions. One of the more detailed and systematic attempts in this direction is that of Black et al. (1965, 1966) who studied the destabilisation of dilute (<150 mg l^{-1}) suspensions of kaolinite and montmorillonite by ^{14}C-labelled N-substituted piperidinium

TABLE 5.2

Adsorption of diallyldimethylammonium chloride—SO_2 copolymer I by montmorillonite in the presence of sodium chloride solutions of varying strengths (from Ueda and Harada, 1968a)

NaCl conc. (M)	Amount adsorbed [a] (g/100 g) A	Amount adsorbed [a] (meq/100 g)	AEC (meq/100 g) B	Amount in trains (meq/100 g) C	CEC [b] (meq/100 g) D
0	9.66	42.8	5.03	37.8	24.6
0.02	11.4	50.4	13.3	37.1	25.3
0.05	12.2	54.1	17.7	36.4	26.0
0.15	13.5	59.8	21.8	38.0	24.4
0.50	13.5	59.8	23.0	36.8	25.6
1.25	13.4	59.5	21.4	34.4	28.6

[a] Based on a suspension concentration of 1.5 g clay per 100 ml solution and an amount of polymer added of 20.8 g per 100 g clay.

[b] Calculated from 62.4—C where 62.4 is the cation exchange capacity of the parent montmorillonite sample.

chloride polymer of molecular weight $>5 \times 10^4$. The approach used was to mix the polycation solution with the clay suspension under controlled conditions of pH, ionic strength and stirring. The extent of aggregation ("destabilisation") and adsorption was assessed by measuring the residual turbidity and polymer concentration of the supernatant liquid, respectively, after a given time of settling. Concurrently, the electrophoretic mobility of the clay—polymer system was determined.

The results for kaolinite are shown in Fig. 5.2 where the electrophoretic mobility of the particles is given as a function of polymer dosage (concentration) for three levels of (initial) clay suspension strength. For mobility we may substitute zeta potential (cf. equation 7.3) which, in turn, is a measure of the potential at the particle surface. The corresponding adsorption isotherms (not shown) have the same appearance as the mobility curves, indicating that adsorption increases with polymer concentration until a point is reached at which the residual turbidity of the supernatant solution is minimal. This point is assumed to coincide with that of maximal aggregation, the concentration at which is referred to as the optimum polymer dosage (cf. section 2.4). The approximate position of this optimum (P_m) is indicated by arrows.

It might be added that polycations are generally very rapidly adsorbed by clays in accordance with a cation exchange mechanism. Further, it is usually observed that P_m becomes broader and less well defined as the clay content is increased. That is to say, there is a range of polymer concentra-

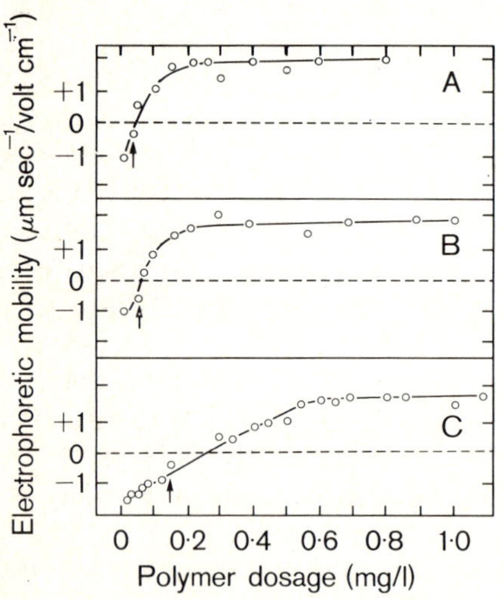

Fig. 5.2. The electrophoretic mobility of kaolinite suspensions as a function of polycation (N-substituted piperidinium chloride polymer) concentration in the system. The arrows indicate the optimum polycation dosage, that is, the concentration at which maximum aggregation occurs (after Black et al., 1966).

tions — rather than a single value — over which turbidity removal is most effective. Besides being dependent on the particle surface area, the width of this aggregation "zone" may increase with polymer molecular weight (Gregory, 1969).

With reference to Fig. 5.2, it seems clear that charge reduction through polycation adsorption plays an important part in the aggregation process. As adsorption progresses, the mobility and hence the zeta potential of the particles becomes less negative, whilst above a certain dosage there is actually a reversal of surface charge. With low molecular weight polycations and cationic surfactants, maximal aggregation (minimal residual turbidity) commonly occurs at $\zeta \sim 0$ (Ottewill and Rastogi, 1960; Kane et al., 1964c; Narkis et al., 1968). On this basis, Dixon et al. (1967) have suggested that aggregation of negatively charged particles by polycations is the result of charge reversal, followed by interparticle bridging. With high molecular weight polycations, however, P_m is frequently less than that required to bring the zeta potential to zero (Vincent, 1974). In this situation, interparticle bridging probably occurs concurrently with charge reduction; in any case, before all of the negative surface charge has been neutralised (Heller, 1966; Gregory, 1969). However, since the effect of surface area and of polymer molecular weight may not operate in the same direction, a balance must be struck between these two parameters if polycations are to be applied to full advantage. Using kaolinite samples of different particle size (surface area) and polydiallyldimethylammonium chloride (polyDADMAM) fractions Yorke (1973), for example, has found that the aggregating efficiency of this polycation decreases below and above a certain molecular weight.

The importance of surface area is also illustrated by the data of Black et al. (1965, 1966). Thus, for kaolinite the electrophoretic mobility corresponding to P_m ranges from 0 to −0.7 μm s^{-1}/Volt cm^{-1} (Fig. 5.2) whereas for montmorillonite with its larger surface area and higher adsorptive capacity and in which intercalation may occur, this range is between −0.3 and +0.7 μm s^{-1}/Volt cm^{-1} (Table 5.3). As would be expected P_m increases with rising initial clay concentrations. Of greater interest is the observation that a strict proportionality exists between solid content in the suspension and P_m in the case of kaolinite. The correlation is again not so well displayed by montmorillonite even in the absence of any intercalation. A linear relationship between solid content (or particle surface area) and P_m has also been reported by Linke and Booth (1960) for polyacrylamide and by Kane et al. (1964a) for polyethyleneimine with silica suspensions. Similarly, Iler (1971) has found that the amount of a polycation required to aggregate a unit weight of silica varies inversely with particle diameter when this is smaller than 40 nm. Another noteworthy feature is that the actual amount of polymer adsorbed tends to decrease as the suspension strength increases (Table 5.3). This accords with what we have observed earlier for the adsorption of

TABLE 5.3

Some data on the interactions of a polycation (^{14}C-labelled N-substituted piperidinium chloride) with dilute suspensions of kaolinite and montmorillonite (from Black et al., 1966)

Clay species	Initial suspension strength or solid content (mg l^{-1})	Extent of polycation adsorption [a] (mg g^{-1})	Optimum polymer dosage, P_m (mg l^{-1}) × 10^{-3}	Electrophoretic mobility at P_m (μm s^{-1}/Volt cm^{-1})	P_m/solid content (×10^{-3})
Kaolinite	73.2	8.1	143	−0.7	1.95
	29.8	12.0	62	0.0	2.08
	14.9	18.4	30	−0.4	2.01
Montmorillonite	144.0	18.5	890	+0.5	6.18
	66.8	29.2	670	+0.7	10.03
	34.4	30.4	147	−0.3	4.27

[a] Estimated from applying the linear transform of the Langmuir equation to the adsorption data.

polyvinyl alcohol by kaolinite and montmorillonite (cf. Fig. 3.1) and may be explained in similar terms.

The relationship between optimum polymer dosage and particle surface area has been further investigated by Black and Vilaret (1969) using kaolinite, latex, and silica of different particle sizes and poly-DADMAM as the aggregant. The results for kaolinite, shown in Fig. 5.3, clearly indicate that the dosage required for maximum aggregation to occur is closely correlated with the surface area of the particles which, in turn, is a function of particle size. They also noted that the rate of aggregation and the size of the flocs produced increased with a decrease in mean particle size.

In this connection, Shyluk (1964) has pointed out that at low shear rates and when the molecular weight of the polymer (poly(1,2-dimethyl-5-vinylpyridinium methyl sulphate)) is sufficiently high, the rate of floc formation is initially greater than that of floc disruption. This gives rise to the creation of large flocs which enclose an appreciable amount of active (adsorbing) sites. As agitation continues the rate of floc formation declines because the polycation supply in the medium becomes depleted and/or more of the "hidden" surface is exposed, resulting in the formation of smaller and denser flocs. Similarly, Goossens and Luner (1976) have found that the polycation concentration required to aggregate cellulose suspensions increases with the duration and intensity of agitation.

Most studies on the clay—polycation interaction have been carried out at more or less constant, near neutral pH conditions, using polymers with a somewhat wide molecular weight and charge density distribution. The degree of polymerisation (n) and of charge substitution are, of course, intrinsic

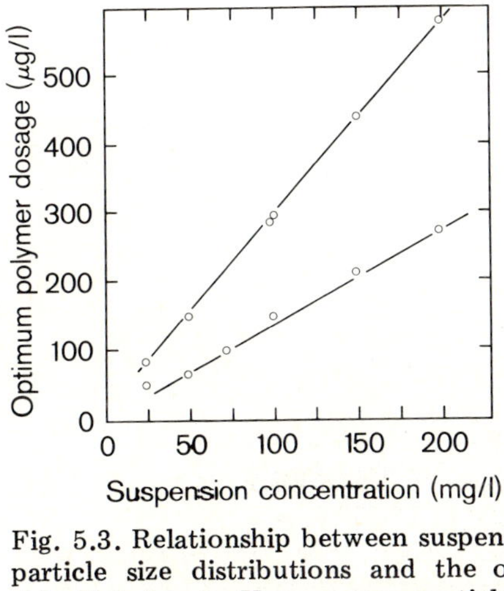

Fig. 5.3. Relationship between suspension concentration of kaolinite with two different particle size distributions and the optimum polymer (polydiallyldimethylammonium chloride) dosage. Upper curve, particles with an average equivalent spherical diameter of 0.26 mμ; lower curve, 0.69 mμ (after Black and Vilaret, 1969).

properties of the polymer concerned and some of their effects on dispersion stability have already been described. On the other hand, the pH of the medium (as well as its ionic strength) influences clay colloid stability by modifying the charge characteristics of both the polycation (through its effect on the degree of dissociation) and the clay.

In order to separate these effects, Ueda and Harada (1968b) have examined the behaviour of different cationic polymers towards kaolinite suspensions in function of *n* and solution pH. To this end, they prepared what are essentially polymers or copolymers of a secondary amine hydrochloride (diallylamine HCl—SO_2 or DAA—HCl—SO_2 copolymer), a tertiary amine hydrochloride (diallylmethylamine HCl—SO_2 or DAMA—HCl—SO_2 copolymer), and a quaternary ammonium chloride (diallyldimethyl—ammonium—SO_2 or DADMAM—SO_2 and diallyldiethyl—ammonium—SO_2 or DADEAM—SO_2 copolymers), ranging in molecular weight from 1.4×10^4 to 33×10^4. The extent of aggregation was assessed by measuring the change in transmittance of the supernatant liquid. Aggregating efficiency was expressed in terms of an "aggregation value" which was defined as the amount of polymer required to increase the transmittance (at 660 mμ) of a 2% (w/v) kaolinite suspension by 50%.

Ueda and Harada (1968b) first demonstrated that the amount of polymer (DADMAM—SO_2) required to produce maximum transmittance (minimum turbidity) for a given initial clay concentration (2% w/v suspension) increased from ~4 to ~22 p.p.m. as *n* decreased from 1430 to 64. Thus, the higher the molecular weight, or the longer the chain, the less material is needed to achieve a comparable extent of aggregation.

This behaviour is perhaps better illustrated by plotting the variation in aggregation value (*c*) as a function of *n*, as shown in Fig. 5.4 from which the relationship

$$c = k/n^{0.734}$$

may be derived, where *k* is a constant. This observation was explained in terms of an extension of the Schulze—Hardy rule (cf. section 1.2), that is, charge neutralisation was regarded as being the prime mechanism underlying aggregation. It seems likely, however, that interparticle bridging also plays a significant role in the process, particularly for the high molecular weight species ($n > 200$). Further evidence in support of this view is given below.

The effect of pH on aggregation is shown in Fig. 5.5 for four different polymers having a similar degree of polymerisation ($n \sim 200$). Below pH 7 all four polycations were equally effective in aggregating kaolinite suspensions but under alkaline conditions differences were noted, becoming more pronounced with increasing pH. As indicated above, this observation may be ascribed to the influence of pH on the degree of dissociation (α) of the polycation, on the one hand, and on the state of dispersion, on the other.

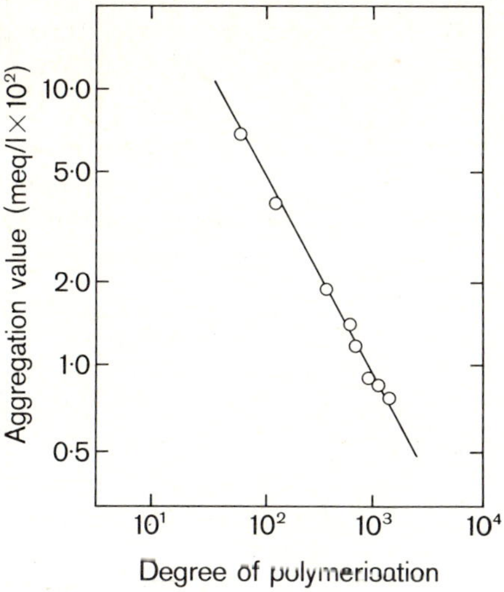

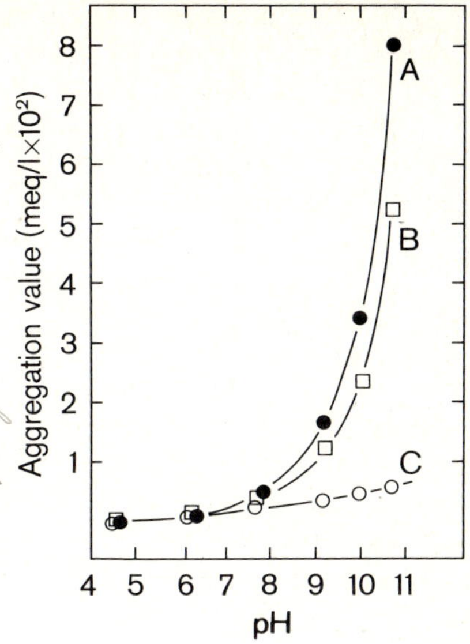

Fig. 5.4. The aggregating ability of polydiallyldimethylammonium chloride—SO_2 copolymer towards kaolinite suspensions as a function of degree of polymerisation. The data refer to a suspension concentration of 2% w/v and at pH 7.1 (after Ueda and Harada, 1968b).

Fig. 5.5. The effect of suspension pH on the aggregating ability of some cationic polymers with a mean degree of polymerisation of 224. Curve A, polydiallylmethylamine HCl—SO_2 copolymer; curve B, polydiallylamine HCl—SO_2 copolymer; curve C, polydiallyldimethylammonium—SO_2 and polydiallyldiethylammonium—SO_2 copolymers (after Ueda and Harada, 1968b).

Being quaternary ammonium derivatives, the degree of dissociation of poly-DADMAM—SO_2 and poly-DADEAM—SO_2 is more or less independent of solution pH over the range of pH 4.5—10.5 and hence, their effectiveness as aggregants remains high. Linke and Booth (1960) have reported similarly for the aggregation of silica by polyacrylamide containing 11% quaternary ammonium groups. By contrast, the values of α for poly-DAA—HCl—SO_2 and poly-DAMA—HCl—SO_2 fall off sharply on increasing the pH from 6 to 8, as indicated by titration studies. Indeed, for pH >9, these polymers exist as uncharged, randomly coiled molecules in solution, incapable of acting as a coagulant and losing much of their flocculating (bridging) properties. It is interesting to note that although their effective charge at pH 7 is about half that at pH 5, their aggregation value is thereby only slightly decreased (Curves A and B). This observation argues in favour of interparticle bridging as being a significant mechanism.

As indicated in Fig. 1.3, the edge surface of kaolinite crystals tends to become increasingly negative as the solution pH is raised above 7. Besides disrupting any edge-to-face association that may occur, the resulting mutual repulsion would increase the mean particle separation, reducing the rate of particle collision and aggregation. In this instance, the effect of pH on particle dispersion is apparently not as important as compared with that on the charge characteristics of the polymer (Curve C).

Positively charged polymers have been used to good effect in removing turbidity and "colour" from surface waters (Cohen et al., 1958; Dixon, 1967; Pressman, 1967). These characteristics arise from the presence in the water of suspended clay and mineral particles, together with humic substances whether they be in solution or associated with particle surfaces (cf. Chapter 12). Such organic materials and their mineral complexes tend to "interfere" with the aggregating action of polycations in that the clay and the organic acids compete with each other for the polymer (Narkis et al., 1968). Thus, when a polyethyleneimine solution is added to a clay or organo-clay suspension containing (free) humic acid in solution, the polycation apparently reacts with the highly mobile, negatively charged humic acid before acting on the clay. Since a portion of the added polymer is used to neutralise the charge on the free or adsorbed humic acid, larger concentrations or dosages of the polycation are required for aggregation to occur.

By the same token, the high adsorptive capacity of montmorillonite for a wide variety of organic compounds (Theng, 1974), including the reaction products of polycations with dissolved organics, makes montmorillonite well suited for removing contaminants from municipal waste water. The application of montmorillonite in combination with cationic polymers to secondary effluents therefore lifts the efficiency of turbidity removal and decreases the optimum polycation dosage (Rebhun et al., 1969).

"Competition" of a different kind has been described by Roberts et al. (1974) in the aggregation of kaolinite suspensions by a cationic polyacrylamide. In this instance, the competing or interfering substance is the positively charged hydroxy-aluminium ion. When this polymeric inorganic ion is added to the system in sufficiently large amounts (~30 p.p.m.), its adsorption by kaolinite reverses the negative surface charge of the clay particles. The polycation is now effectively repelled from the particle surface and its flocculating efficiency is thereby markedly reduced.

REFERENCES

Black, A.P. and Vilaret, M.R., 1969. Effect of particle size on turbidity removal. Journal of the American Water Works Association, 61: 209—214.

Black, A.P., Birkner, F.B. and Morgan, J.J., 1965. Destabilization of dilute clay suspensions with labelled polymers. Journal of the American Water Works Association, 57: 1547—1560.

Black, A.P., Birkner, F.B. and Morgan, J.J., 1966. The effect of polymer adsorption on the electrokinetic stability of dilute clay suspensions. Journal of Colloid and Interface Science, 21: 626—648.

Cohen, J.M., Rourke, G.A. and Woodward, R.L., 1958. Natural and synthetic polyelectrolytes as coagulant aids. Journal of the American Water Works Association, 50: 463—478.

Dixon, J.K., 1967. Flocculation. Encyclopedia for Polymer Science and Technology, 7: 64—78.

Dixon, J.K., La Mer, V.K., Li, C., Messinger, S. and Linford, H.B., 1967. Effect of the structure of cationic polymers on the flocculation and the electrophoretic mobility of crystalline silica. Journal of Colloid and Interface Science, 23: 465—473.

Flaig, W., Beutelspracher, H. and Söchtig, H., 1962. Ein Beitrag aus dem Bereich der molekularen Dimension zur Morphologie des Bodens. Zeitschrift für Pflanzenernährung, Düngung, Bodenkunde, 98: 225—231.

Goossens, J.W.S. and Luner, P., 1976. Flocculation of microcrystalline cellulose suspensions with cationic polymers: effect of agitation. The Journal of the Technical Association of the Pulp and Paper Industry (Tappi), 59: 89—94.

Gregory, J., 1969. Flocculation of polystyrene particles with cationic polyelectrolytes. Transactions of the Faraday Society, 65: 2260—2268.

Gregory, J., 1973. Rates of flocculation of latex particles by cationic polyelectrolytes. Journal of Colloid and Interface Science, 42: 448—456.

Heller, W., 1966. Effects of macromolecular compounds in disperse systems. Pure and Applied Chemistry, 12: 249—274.

Iler, R.K., 1971. Relation of particle size of colloidal silica to the amount of a cationic polymer required for flocculation and surface coverage. Journal of Colloid and Interface Science, 37: 364—373.

Kane, J.C., La Mer, V.K. and Lindford, H.B., 1964a. The filtration of amorphous and sized crystalline silica suspensions flocculated by high polymers. The Journal of Physical Chemistry, 68: 2273—2277.

Kane, J.C., La Mer, V.K. and Lindford, H.B., 1964b. The effect of solid content on the adsorption and flocculation behaviour of silica suspensions. The Journal of Physical Chemistry, 68: 3539—3544.

Kane, J.C., La Mer, V.K. and Lindford, H.B., 1964c. Filtration and electrophoretic mobility studies of flocculated silica suspensions. Journal of the American Chemical Society, 86: 3450—3453.

Kimura, C., Serita, H. Murai, K. and Takahashi, Y., 1976. Flocculation effect on kaolin suspension of quaternary poly(2- or 4-vinylpyridine). Yukagaka, 25: 424—428.

Lindström, T. and Söremark, C., 1976. Adsorption of cationic polyacrylamide on cellulose. Journal of Colloid and Interface Science, 55: 305—312.

Linke, W.F. and Booth, R.B., 1960. Physical chemical aspects of flocculation by polymers. Transactions of the American Institute of Mining, Metallurgical and Petroleum Engineers, 217: 364—371.

Narkis, N., Rebhun, M and Sperber, H., 1968. Flocculation of clay suspensions in the presence of humic and fulvic acids. Israel Journal of Chemistry, 6: 295—305.

Ottewill, R.H. and Rastogi, M.C., 1960. Stability of hydrophobic sols in the presence of surface-active agents. II. Stability of silver iodide sols in the presence of cationic surface-active agents. Transactions of the Faraday Society, 56: 866—879.

Peyser, P. and Ullman, R., 1965. Adsorption of poly-4-vinylpyridine onto glass surfaces. Journal of Polymer Science, Part A, 3: 3165—3173.

Posselt, H.S., Reidies, A.H. and Weber, W.J., Jr, 1968. Coagulation of colloidal hydrous manganese dioxide. Journal of the American Water Works Association, 60: 48—68.

Pressman, M., 1967. Cationic polyelectrolytes as prime coagulants in natural water treatment. Journal of the American Water Works Association, 59: 169—182.

Rebhun, M., Narkis, N. and Wachs, A.M., 1969. Effect of polyelectrolytes in conjunction with bentonitic clay on contaminants removal from secondary effluents. Water Research, 3: 345—355.

Roberts, K., Kowalewska, J. and Friberg, S., 1974. The influence of interactions between hydrolyzed aluminum ions and polyacrylamides on the sedimentation of kaolin suspensions. Journal of Colloid and Interface Science, 48: 361—367.

Ruehrwein, R.A. and Ward, D.W., 1952. Mechanism of clay aggregation by polyelectrolytes. Soil Science, 73: 485—492.

Shyluk, W.P., 1964. Poly(1,2-dimethyl-5-vinylpyridinium methyl sulfate). Part II. Polymer properties. Journal of Applied Polymer Science, 8: 1063—1076.

Theng, B.K.G., 1974. The Chemistry of Clay—Organic Reactions. Adam Hilger Ltd., London, 343 pp.

Ueda, T. and Harada, S., 1968a. Adsorption of cationic polysulfone on bentonite. Journal of Applied Polymer Science, 12: 2395—2401.

Ueda, T. and Harada, S., 1968b. Effect of the structure of cationic polysulfone on the flocculation of kaolinite. Journal of Applied Polymer Science, 12: 2383—2393.

Vincent, B., 1974. The effect of adsorbed polymers on dispersion stability. Advances in Colloid and Interface Science, 4: 193—277.

Yorke, M.A., 1973. Particle size vs. molecular weight relationships in cationic flocculation. In: N.M. Bikales (Editor), Water-Soluble Polymers. Polymer Science and Technology Series, Vol. II. Plenum Press, New York, N.Y., pp. 93—104.

Chapter 6

SOME PRACTICAL APPLICATIONS OF THE CLAY—POLYMER INTERACTION

6.1. INTRODUCTION

Although this monograph is primarily concerned with the chemistry of clay—polymer reactions, a description of the more important applications of the clay—polymer interaction, however brief, should be included in a summary of this type. After all, the impetus behind our enquiry into the mechanisms underlying the interaction process was the importance and technological potential of clay—polymer systems in agriculture and industry. Whether polymers are added to clay (soil) in order to improve its physical condition, or whether clays are incorporated into a polymer matrix so as to modify its properties in a given direction, the mineral and organic components must form some kind of bond or complex if their union is to have more than a transitory existence.

In the majority of cases the polymer, dissolved in a suitable solvent, is added to dry clay or to a clay suspension and complex formation then essentially involves the adsorption of the organic phase onto the inorganic particle surface with or without the participation of interparticle bonding. Such systems have been described in some detail in the preceding chapters. In some instances the polymer is introduced in the form of an emulsion, a solid, or a melt which is insoluble in the continuous phase. Here the principal mode of attachment is one of adhesion although this process may be regarded as an extension or amplification of adsorption. An outstanding example of adhesive interactions is the reinforcement of elastomers and plastics by finely divided clays and inorganic powders ("fillers") which are dispersed in the organic matrix. Physical (Van der Waals) attraction plays a prominent part in adhesion although chemical interactions and bridging of the filler surface to the polymer network are also involved (Zisman, 1965; Dannenberg, 1975). Thus, the compatibility of the mineral with the organic phase, in terms of wetting and dispersion, may be substantially increased by the prior adsorption of a monolayer of a suitable organic substance over the particle surfaces or by the covalent addition of some simple organics (alcohols, organosilanes) to the reactive surface groups of the mineral.

The practice of adding organic materials to soil in order to improve its structure and so maintain a workable level of fertility is, of course, as old as agriculture itself. The improvement in "tilth" caused by organic amendments is basically due to an increase in the extent of soil aggregation and in the water-stability of the aggregates formed. This is brought about by the adsorption onto and the linking together of clay particles or domains (making up the pore walls within an aggregate) by soil organic matter. In view of their abundance and resistance to microbial decomposition, humic substances undoubtedly play an important part in aggregate stabilisation (cf. Chapter 12). However, evidence has accumulated to indicate that, among the many and varied organic constituents of soil, the polysaccharides and notably the anionic polyuronides, are the single most important class of compounds which are involved in this process (cf. Chapter 10).

These observations, together with the advances in polymer technology made during the immediate post-war years, provided the initial stimulus for using synthetic polyanions as "soil conditioners". By analogy with their behaviour towards dilute clay suspensions, described in Chapter 4, the aggregate stabilising action of such synthetic anionic polymers as "Krilium" was ascribed to and identified with their ability to flocculate clay particles. However, as already remarked on, aggregate stabilisation should not be confused with flocculation although both processes involve the adsorption of and interparticle bridging by, the polymer molecule. In the former situation the polymer is added to and stabilises an existing structure in which the particles are already in close proximity, however weak and unstable this arrangement may be. On the other hand, flocculation is a process by which particles, initially separated by relatively large distances and without having any structural order, are linked together by the polymer.

This distinction is also at the basis of the difference in the aggregating effectiveness between polyelectrolytes and uncharged polymers. As we have seen, polyanions are effective flocculants of dilute particulate dispersions because their chains are extended in solution. Since relatively few segment—surface contacts are established, their flocculating action has been likened to the formation of a "string of beads" (Greenland, 1963). By the same token, the polyanions do not confer great strength or stability to clay aggregates unless the system contains polyvalent cations or other positively charged species which can act as "anchor points" to the surface. In soils, the hydrated oxides of iron and aluminium, particularly when they exist as an amorphous film, coating the clay/soil particles, can perform this function. At first sight it might appear that polycations would be effective aggregate stabilisers but this does not accord with practice. Apparently, the rate of collapse of the polymer onto the negatively charged clay/soil surface is so great that only a very limited amount of interparticle bonds can be achieved. Because uncharged polymers exist as randomly coiled molecules in solution, their effectiveness as flocculants of dilute clay suspensions is not as great as polyanions of comparable molecular weight (cf. Chapter 4). However, when uncharged polymers, such as polyvinyl alcohol, are introduced into dry clay or clay/soil aggregates, they can impart great strength and stability to the system (Emerson, 1956; Williams et al., 1967a,b, 1968; Dowdy, 1972, 1975). This is so because the molecules tend to uncoil on contact with the solid and spread out like a "coat of paint" over adjacent clay/soil particle surfaces (Greenland, 1963).

There is good evidence to suggest that besides forming interparticle bridges, the polymer molecule must also be capable of penetrating the pore space of the aggregate if effective and long-lasting stability is to be achieved. Thus, in addition to those factors which affect the clay—polymer interaction, in general (cf. Chapter 3), polymer diffusion into and disposition within the aggregate, as well as pore size distribution and physical or steric effects, are important considerations (Quirk and Panabokke, 1962; Williams et al., 1967a, 1968; Quirk and Williams, 1974). Indeed, as Gardner (1972) has pointed out, the beneficial effects of soil conditioners on such processes as plant growth, soil erodibility, and water movement are only indirect, being the result of their ability to alter and stabilise particular porosity configurations within the aggregate.

Although the results of early experiments with synthetic polyelectrolytes were encouraging (e.g. Bear, 1952; Hedrick and Mowry, 1952; cf. Fig. 4.1), the popularity of krilium-type polymers as soil conditioners steadily declined from 1954 to the early nineteen-sixties. This was chiefly due to the high cost/benefit ratio in many large-scale agricultural operations (Gardner, 1972; Carr and Greenland, 1975). Over the past decade, however, there has been a resurgence of interest in the soil conditioning and stabilising potential of synthetic polymers as attested by the convening of a series of symposia on this topic (De Boodt, 1972; Soil Science Society of America, 1975; De Boodt and Gabriels, 1976).

Several reasons may be advanced for this phenomenon. One is the pressure to bring into production so-called marginal and problem soils as a result of an ever-increasing human population, particularly in the developing countries. Another is the need to prevent the structural deterioration of intensively farmed soils so as to maintain adequate structure and crop production. Thirdly, the application of soil conditioners to arable land will reduce the pollution of water resources resulting from surface erosion and run-off into lakes, rivers, and reservoirs (Carr and Greenland, 1975; De Boodt, 1975, 1976).

These developments, however, would not have been possible had it not been for the availability of a cheaper and wider range of synthetic polymers. Among these, the uncharged, linear, water-soluble polyvinyl alcohols and polyacrylamides have proved to be very effective as soil aggregate stabilisers. A number of water-insoluble polymeric and bituminous compounds, applied in the form of emulsions, have also gained wide-spread usage as aggregate stabilising agents, particularly for coarse-textured soils and sands (Gabriels, 1972; Muljadi et al., 1972; Vandevelde and De Boodt, 1972). In this instance, stabilisation is essentially effected by the formation of localised adhesive links between the solid particles as the emulsion droplets ("micelles") at the surface progressively retract to interparticle points of contact during evaporation (Rigole and De Bisschop, 1972). This mode of action is reflected in the observation that the effectiveness of emulsified polymers is dependent on their droplet size (Schamp et al., 1975; Goor et al., 1976) rather than on their molecular weight as does that of water-soluble polymers (Williams et al., 1968; Carr and Greenland, 1972). Emulsified polymers may also be less effective than their water-soluble counterparts because relatively few interparticle links are formed in the former case (Sadones et al., 1975). These points are summarised in Table 6.1.

In addition to the three symposia volumes already mentioned, the subject of soil aggregation and the part played by naturally occurring and syn-

TABLE 6.1

Effectiveness (measured in terms of increases in hydraulic conductivity) of some water-soluble and emulsified polymers in stabilising natural soil aggregates as a function of polymer molecular weight and rate of application (from Sadones et al., 1975)

Treatment	Molecular weight (×10³)	Concentration (‰)	Hydraulic conductivity [a] (cm min⁻¹)
Untreated soil	—	—	0.1
Glass beads	—	—	6.7
Water-soluble polymers			
Polyvinyl alcohol	40	1; 2	0.4; 1.1
	80	0.5; 1; 2	0.9; 1.6; 2.5
	100	0.5; 1; 2	1.1; 1.9; 2.9
Polyacrylamide	100	0.5; 1; 2	0.2; 0.5; 0.8
	300	0.5; 1; 2	0.4; 0.9; 1.0
	1000	0.5; 1; 2	0.6; 0.9; 1.5
Polyvinyl pyrrolidone	40	0.5; 1; 2	0.2; 0.3; 0.4
	360	0.5; 1; 2	0.9; 1.9; 2.5
Polyvinyl alcohol urethenised	50	0.5; 1; 2	0.5; 1.0; 1.7
	120	0.5; 1	0.8; 1.4
Carboxymethyl cellulose		0.5; 1	0.5; 0.7
Polyethylene glycol	20	2	0.1
Polystyrene sulphonate	—	2	0.1
Emulsified polymers			
Styrene—acrylic acid ester copolymer	—	2.5; 5	0.5; 0.7
Vinylacetate—acrylic acid ester copolymer	—	2.5; 5	0.3; 0.5
Vinylacetate—ethylene copolymer	—	2.5; 5	0.5; 0.8
Vinylacetate—dibutyl-maleate copolymer	—	2.5; 5	0.2; 0.5

[a] <0.75 cm min⁻¹ poor, 0.75—1.0 moderate, 1.0—2.0 good, >2.0 excellent. These values were measured with 0.85—1.0 mm aggregates and the respective ratings were given in relation to wet sieving and rainfall simulation results.

thetic polymers in this process, have been summarised by a number of workers (Martin et al., 1955; Harris et al., 1966; Stepanova and Daulbayev, 1970; Brandt, 1972; De Boodt, 1972; Gardner, 1972; Kullmann, 1974; Carr and Greenland, 1975). These reviews, containing most of the relevant references to the original literature, should be consulted for more details than could possibly be included here.

Industrially, raw clays have been extensively used as fillers of polymer systems, their main function here being frequently to lower the cost of the final product or composite. For this reason, clays and other minerals have commonly been considered as mere extenders. The studies of Uskov et al. (1966a) and Solomon (1968), however, have clearly shown that clay minerals are by no means chemically inert. Besides interacting with the polymer matrix, clays can control and direct or serve as a substrate-template for, the polymerisation of organic monomers (Theng and Walker, 1970; Blumstein et al., 1971, 1974; Theng, 1974; Solomon, 1977). We would therefore prefer to use the term "filler" in the broad sense of a solid which is dispersed in the organic phase so as to modify the properties of the polymer system to meet specific industrial or commercial requirements. Because of their abundance, availability, low cost, and surface properties, clay minerals are admirably suited for this purpose.

Brief summaries of the industrial applications of clay—polymer systems and the clay—organic interaction, in general, have been given elsewhere (Haden, 1963; Jordan, 1963; Murray, 1963; Nahin, 1963; Theng, 1970). As might be appreciated, much of the information in this field has been published as patents and apart from some representative examples, this type of literature is not included here.

6.2. SOIL CONDITIONING

The effectiveness of polymers in improving soil physical conditions may be assessed in a variety of ways, among which water-stable aggregation is perhaps the single, most widely used index. This is so because water stability reflects the arrangement of the clay/soil particles and the pore geometry within the aggregate which, in turn, influence such properties as mechanical strength, hydraulic conductivity (permeability) and erodibility.

Experimentally, the degree of water stability can be evaluated on the basis of the changes induced in a given parameter relating to either the aggregate (e.g. stability on wet-sieving) or the supernatant liquid after a shaking ("dispersion") treatment (e.g. turbidity). Very rarely, however, have the different tests for water-stable aggregation been compared for a given system as Slater and Kitchener (1966) have done for the flocculating action of polymers on particulate dispersions (cf. Fig. 2.5). It has been pointed out (e.g. Brockman and Allenby, 1955; Hofman, 1976) that laboratory data obtained using the wet-sieving method, for example, are very much influenced by the manner of conditioner introduction, sample pretreatment (drying, storage), and moisture content. Indeed, the initial water content of the aggregate often plays a crucial role in aggregate stabilisation, particularly when emulsions are used as the stabilising agents (Gabriels and De Boodt, 1972; Rigole and De Bisschop, 1972; De Boodt and Gabriels, 1973). Wet-sieving may also yield misleading information when different soils are compared (Low, 1954) and its applicability to field situations has been questioned by Allison (1968). For that matter, however, great care should always be exercised in extrapolating data obtained in the laboratory to field conditions, if only because this type of information relates to short-term experiments which are carried out under more or less controlled conditions. A similar comment could be made as regards the extension of small-scale experiments to industrial operations.

There is little doubt, however, that the application to soils of various polymers, in the form of either solutions or emulsions, leads to marked increases in water-stable aggregation and other related soil properties such as hydraulic conductivity, mechanical strength, and resistance to water and wind erosion. In most cases the incorporation of polymers into soils also has a favourable effect on seed germination, seedling emergence, and crop yield. As already remarked on, these effects stem more from the improvement in the physical conditions of the treated soils than from the presence of the polymer as such (see Carr and Greenland, 1975). Although the polymers applied may provide a source of carbon and nitrogen for soil microorganisms and the growing crop, such direct effects are not very significant because of the small quantities involved.

The effectiveness of polyanions as soil conditioners tends to decrease and increase in the presence of neutral electrolytes and bases, respectively (Jones and Martin, 1957). This observation is in line with the effect of increasing ionic strength and pH on polymer conformation and hence, on its ability to form interparticle bonds. A similar pH effect has also been reported by Richard and Chausson (1957) and Richard (1959). By the same token, aggregation is enhanced by an increase in the molecular weight or degree of polymerisation of the polyanion (Nicholls and Davidson, 1957; Smith et al., 1958).

Because of their peculiar mode of action the uncharged polymers, notably polyvinyl alcohol (PVA), are, as remarked on earlier, very effective soil aggregate stabilisers. The interactions of PVA with natural soil aggregates and the water-stability of the treated aggregates have been examined in some detail by Williams et al. (1966, 1968). They were able to show that the diffusion of the polymer into the aggregate pores and its subsequent collapse and spreading over the surface of the clay particles or domains, making up the pore walls, are the controlling factors. Thus, the adsorption of PVA by aggregates (which had previously been wet) increased with a decrease in polymer molecular weight (M), presumably because the smaller molecules met with less steric hindrance in penetrating the pore space as compared with their bulkier counterparts. However, water-stable aggregation (as measured by the amount of <2 μm e.s.d. particles in the supernatant after shaking in water) increased with polymer molecular weight since relatively less of the larger size molecules was apparently required in order to achieve a comparable level of stability (Fig. 6.1).

These conclusions have subsequently been verified by Carr and Greenland (1972) who determined the extent to which PVA of varying molecular weight (M = 15 000, 70 000, and 100 000) penetrated an aggregate by sectioning the treated aggregate and measuring the carbon content of each section with an induction furnace. As might be expected, the amount adsorbed decreased with distance from the aggregate—PVA solution interface, that is, the aggregate face through which the polymer was introduced. They also found that the extent of penetration (over a given period) was greater for the lower molecular weight species. Of particular note was the finding that despite its limited penetration, the GH-20 sample (M = 100 000) was able to confer a high degree of stability to those parts of the aggregate which contained so little polymer that its amount escaped detection by the method of analysis used.

This observation indicates that it may be feasible to achieve substantial stabilisation by placing the polymer at strategic positions within an aggregate. In an attempt to demonstrate this point, Quirk and Williams (1974)

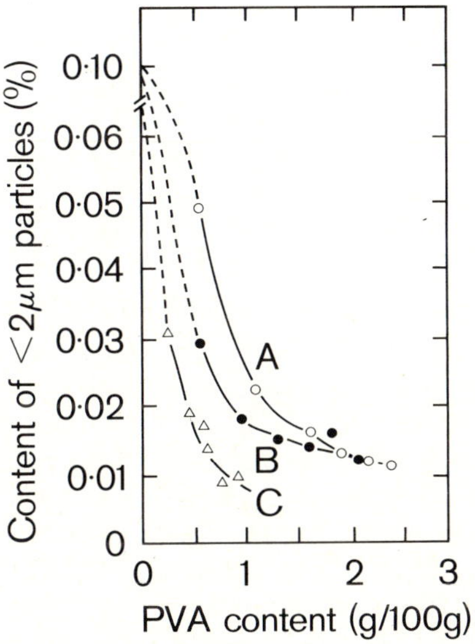

Fig. 6.1. The water stability of natural soil aggregates as a function of their content in polyvinyl alcohol of varying degrees of polymerisation (n). Curve A, GL-05 (n = 500); curve B, GM-14 (n = 1400); curve C, GH-20 (n = 2000) (after Williams et al., 1968).

have selectively introduced PVA into the different sized pores of natural, unstable soil aggregates, covering the pore-radius classes of 0—7.5, 7.5—15, 15—20 and 50—150 μm using the wetting-under-suction method and measuring the rate of water uptake by, and the mechanical strength of, the treated aggregates. The results showed that adsorption within the 15—50 μm radius pores produced the optimum effect in terms of the parameters used.

The effectiveness of synthetic polymers in increasing water-stable aggregation is clearly related to the strength with which these compounds are attached to the clay (soil) particle surface. In this context mention should be made of the suggestion by Koch et al. (1974) that coupling agents may be used to advantage in producing water-stable aggregates because of their ability to form chemical bonds with both the polymer and the surface. The agents they used were a variety of organosilanes with the general formula of $R(CH_2)_3Si(OC_nH_{2n+1})_3$ where R = NH_2, SH, Cl, $NH_2(CH_2)_2NH$. They were able to show that the mere addition of such compounds to soil (i.e. in the absence of extraneous polymers) led to significant increases in water-stable aggregation. From this they inferred that a bond was established between R and the functional groups of soil organic matter at one end, and between OH (formed by hydrolysis of (OC_nH_{2n+1})) and surface silanols at the other (cf. Fig. 6.4). However, as Koch et al. themselves have indicated, many of these silanes are capable of autopolymerisation and, when adsorbed onto a clay (soil) particle, may polymerise laterally along the surface. In this way, they can form strong interparticle bonds so that their mode of action is akin to that of non-ionic polymers.

The stability of an aggregate may also be assessed in terms of its mechanical strength, that is, of its ability to withstand compression by an externally applied load. A distinction should be made here between inter- and intra-aggregate strength (Carr and Greenland, 1975). The application of a conditioner to soil or clay systems should be aimed at increasing the strength within, rather than that between, aggregates since an increase in inter-aggregate strength may lead to surface crusting.

That polymer incorporation into an aggregate usually leads to substantial increases in the mechanical strength of the system, has been demonstrated on several occasions (Lambe and Michaels, 1954; Williams et al., 1967a; Quirk and Williams, 1974; Schamp et al., 1975). The data of Williams et al. (1967a) indicate that the tensile strength of PVA-treated soil cores increases to a maximum with the amount of polymer adsorbed and then tends to decline. This behaviour may be ascribed in terms of surface coverage by the polymer in that the strengthening effect is optimal when about half of the adsorbing sites at the aggregate surface is occupied by the polymer. Adsorption beyond this point presumably leads to a decrease in interparticle bonding efficiency. The tensile strength of soil cores is strongly affected by the amount of water present, decreasing with an increase in moisture content.

Dowdy (1972, 1975) has reported similarly for montmorillonite saturated with Na^+, Ca^{2+}, and Al^{3+} ions on treatment with PVA of different molecular weight. He also observed that the tensile strength of the clay films increased in the order of $Na^+ > Ca^{2+} >> Al^{3+}$, indicating that the size of the clay domains, linked together by the polymer, was an important factor. For a given level of PVA addition, the tensile strength of the film increased with polymer molecular weight (Fig. 6.2). Although no maximum occurred in the curve, there was a change in slope at ~7 g PVA/100 g clay. At this point, a considerable proportion of the added PVA entered the interlayer space and thus became no longer available for bridging. By the same token, very little, if any, increase in strength was produced by adding glucose to the clay, the molecules of which were too short to effect interdomain bonding.

Dowdy's interpretation is consistent with the swelling behaviour of PVA-treated montmorillonite cores (Theng et al., 1967). As shown in Fig. 6.3, the greatest reduction in swelling (as measured by the amount of water taken up at 15 mbar) occurred for amounts of PVA adsorbed which were too little (<0.01 g/g) to influence the extensive interlayer (osmotic) swelling proper-

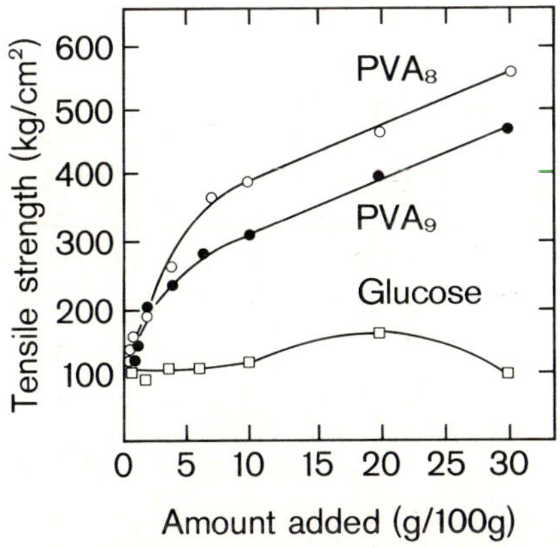

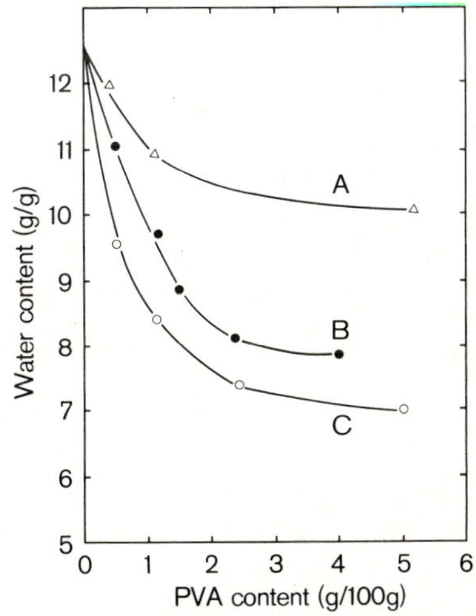

Fig. 6.2. The tensile strength of Na^+-montmorillonite as a function of the amount of polyvinyl alcohol (PVA) or glucose added. PVA_8 and PVA_9 refer to polymers with an average molecular weight of 75 000 and 14 000, respectively (after Dowdy, 1975).

Fig. 6.3. Relationship between the swelling of Na^+-montmorillonite as measured by its water uptake (at 15×10^{-3} bar suction) and the amount of polyvinyl alcohol (PVA) adsorbed. The polymer solutions were added to the clay to give a final suspension strength of 4% w/v. Curve A, GH-20 solution added to an aqueous clay dispersion; curve B, GH-20 solution added to air-dry clay; curve C, GL-05 solution added to air-dry clay (after Theng et al., 1967).

ties of Na^+-montmorillonite. This would suggest that the residual swelling occurred within domains which were rather loosely enclosed by the polymer chains. The adsorption of PVA in amounts greater than 0.01 g/g probably produced a larger number of bonds between domains which were already linked together and only a small increase in the total number of bonded particles (domains). The effectiveness of subsequent additions of polymer in reducing swelling was accordingly less. In keeping with this view, less swelling occurred with the cores prepared by adding the polymer to air-dry clay samples than when the polymer was added to aqueous dispersions of the clay. In the latter system, the added PVA was presumably expended in linking together a number of individual montmorillonite layers, giving rise to a more or less open structure.

Similarly, the reduction in surface area and the changes in pore size distribution (as measured by the nitrogen adsorption—desorption technique), which Williams et al. (1967b) have reported, may be explained in terms of the interlinking domain model. Interestingly, the volume of the <3 nm wide pores descreased by an amount which was greater than the volume of PVA adsorbed. From this they inferred that the polymer occupied sites at the pore peripheries, preventing the nitrogen molecules from gaining access into intradomain pores and surfaces. More recently, Epshtein (1976) reported that the adsorption of polyacrylamide by montmorillonite reduced the volume of 4—10 nm pores whilst pores of 11.5—17.5 nm diameter were "created", suggesting that polymer uptake induced a certain amount of particle (domain) rearrangment.

Water-soluble polymers and emulsions have also been used to good effect in stabilising soils for civil engineering and related non-agricultural applications, such as in the construction of airfields, canals, and highways and the preparation of building sites for housing and industrial purposes (Ingles, 1968; Moldenhauer and Gabriels, 1972; De Boodt, 1975). Here the cost factor is of relatively minor consideration since the expenditure involved is only a small proportion of the total capital investment and the area to be stabilised is generally rather restricted.

There is also considerable scope for using polymers and emulsions to control and reduce soil erosion by water and wind. Of the wide range of materials which have been tested for this purpose, using rainfall simulation techniques, PVA and polyacrylamide (PAM) are among the most effective in increasing soil stability and permeability (Blavia et al., 1971; Gabriels, 1972; Gabriels et al., 1973). Similarly, Stefanson (1973) has reported that the addition of only 0.005% w/w of PVA to the top 2 cm of various undisturbed soil cores nearly doubled the rainfall acceptance of the soil. By spraying aqueous PVA solutions at a rate of 8—10 g m^{-2}, Oades (1976) has been able to stabilise seedbeds (in the field) in that seedling emergence, rainfall infiltration, and water storage were significantly improved. These effects were ascribed to the prevention of crust formation by which soil loss by surface run-off was also substantially reduced. Besides PVA and PAM (Mausbach and Shrader, 1975), bitumen emulsions have been found to be effective in preventing crust formation for critical periods of time even at high intensities of simulated rainfall (Gabriels and De Boodt, 1975; Pla, 1975).

The potential application of different emulsions, polymers, and resins to control soil erosion by wind has been examined by Armbrust and co-workers (Armbrust and Dickerson, 1971; Armbrust and Lyles, 1975). They have been able to establish a number of criteria which have to be met before a given material is deemed suitable for this purpose. These are that the conditioner should completely cover the soil surface, have no deleterious effect on seedling emergence and plant growth, be able to prevent erosion initially and reduce it over the first two months, be cheap, and be simple to apply. Of the many compounds and formulations evaluated, only six were found to meet all of the above criteria. Polyvinyl alcohol was not among them because it failed to meet the last-mentioned criterion due to its slow rate of dissolution in water.

6.3. CLAYS AS FILLERS OF POLYMER SYSTEMS

As mentioned earlier, fillers may be regarded as finely divided solids which, on adding and bonding to the polymer matrix, modify the properties of the system in a certain (usually desirable) direction. The type and amount of a chosen filler to be incorporated into the system would clearly depend on the particular polymer and the purpose for which the filled product is designed. Fillers either as such or, frequently, after surface modification, should therefore be capable of fulfilling a variety of functions. Among the benefits which may be derived from using fillers are the improvements in strength, stability, rheology, insulating properties, economy, and cracking resistance of the system. Fuller details on the functions and uses of fillers, such as asbestos, calcium carbonate, carbon black, glass fibre, kaolinite, mica, and silica, in the polymer industry, have been given elsewhere (Zisman, 1965; Cook, 1966; Murphy, 1966; Smith, 1968; Dannenberg, 1975; Heitz, 1975).

All things being equal, the efficiency of a filler to improve the physico-mechanical properties of a given system is primarily determined by the degree of its dispersion in the polymer matrix. However, the surface of clays and inorganic mineral fillers, in general, is hydrophilic and this impedes their dispersion in, and rapid wetting by, the organic phase. To overcome this problem it is often necessary to make the surface hydrophobic and/or organophilic prior to blending or compounding the filler with the polymer. This may be achieved by one of several means.

By far the simplest way is to attach a suitable organic compound to the mineral surface (e.g. Murray, 1963). Such a treatment, however, may not be very effective for porous, high surface area solids. However, the adsorbed organic layer ("coat") may be displaced by solvents or during compounding (Dannenberg, 1975; Helmer et al., 1976). A more effective method of rendering clay surfaces compatible with the organic phase is to neutralise the acid-washed material with an organic base which is then held by the exchange complex of the mineral as the corresponding cation. Alternatively, stable hydrophobic clays can be prepared by replacing part or all of the exchangeable, inorganic cations with small organic cations or polycations (Carter et al., 1950; Gluesenkamp, 1957; Ruehrwein, 1957; Uskov, 1960a; Wilcox, 1961; Jordan, 1963; Nahin, 1963; Walker and Garrett, 1967).

A very successful method of modifying clay surfaces so as to make for strong polymer—filler interaction, is to react the minerals with a variety of organosilanes (Rehner et al., 1960; Libby et al., 1967; Ranney et al., 1973). Using an aminosilane, Libby et al. (1967) were able to attach the hydrolysed silanol group of the reagent to the kaolinite surface, leaving the pendent amino group (at the other end of the molecule) free to react with the rubber during vulcanisation. The resulting cross-linked rubber—aminosilane—clay network showed significant improvement in modules, abrasion resistance, and compression-set properties as compared with the unfilled system.

Following Dannenberg's (1975) scheme for silica the probable reaction mechanism, involving kaolinite, may be represented by the sequence depicted in Fig. 6.4.

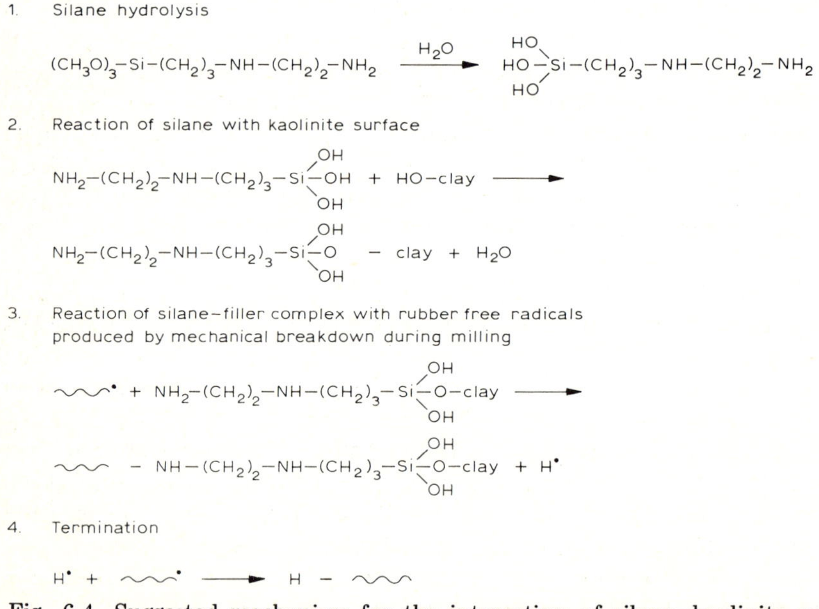

Fig. 6.4. Suggested mechanism for the interaction of silane—kaolinite complexes with rubber. ~~~ denotes a polymer chain and the dot a free radical species (after Dannenberg, 1975).

The process is therefore analogous to that proposed by Koch et al. (1974) for the stabilisation of soil aggregates (referred to in the preceding section) and by Plueddemann et al. (1962) for the coupling of glass fibres to polyesters by means of various reactive silanes.

Using a similar approach, Helmer et al. (1976) have modified kaolinites with a reaction product of tetraisopropyl titanate and oleic acid. The organo-titanium clays so obtained have proved to be effective in the filling of polyamides (nylon). They also show great promise as fillers of plastics, as the data of Table 6.2 indicate. It is interesting to note that elongation is generally reduced by the presence of the filler (cf. Table 6.5). Similarly, those polymer systems, which are designed to have a high impact strength, tend to be negatively affected by filling. On the other hand, the toughness and heat distortion properties of plastics are usually improved by incorporating the filler into the matrix.

A most effective method of achieving compatibility and hence, strong bonding between filler and matrix polymer, is to use minerals and clays onto whose surface a suitable polymeric species has previously been grafted by some means or other. Because of the potential industrial applications of such graft or rather, mixed graft polymers, the grafting of polymers onto inorganic substances in general, has received a great deal of attention. The extensive literature on this subject has been reviewed by Raff (1968). In this context, mention should also be made of the initiation of polymerisation by freshly formed surfaces of inorganic materials (Kargin and Platé, 1959, 1961) and clays (Uskov, 1960a,b) since this process may give rise to polymer grafting (see Jacovic, 1964). It is not intended here to go over these aspects in great detail, nor shall we discuss the ability of clays to influence and direct the polymerisation of adsorbed organic monomers, a topic which has been treated elsewhere (Solomon, 1968; Theng and Walker, 1970; Theng, 1974). Rather, we shall attempt to outline those features of the polymerisation process which lead to grafting, emphasising the applications of such mixed graft polymers as reinforcing fillers of plastics and rubber.

Grafting normally implies that the polymer is attached to the mineral by primary (covalent) bonds. However, as we have seen in Chapter 2, secondary interactions (e.g. Van der Waals, hydrogen bonding) may also lead to extremely strong bonding between polymer and solid surface. For this reason and because, in many instances, some homopolymer is formed concurrently, the occurrence of "true" grafting is often difficult to demonstrate directly. However, under certain conditions and using suitable criteria there seems little doubt that mixed graft polymers are formed when polymerisation is carried out in the presence of inorganic substances and clays.

The studies of Kargin and Platé (1959, 1961) have convincingly shown that fresh surfaces of metals, metal oxides, and inorganic salts, formed during their mechanical disintegration, can initiate the polymerisation of vinyl monomers. In the case of metals, the mechanism involves the transfer of an

TABLE 6.2

Modifications in some properties of polymers and plastics induced by the incorporation of an organo-titanium kaolinite filler, obtained by reacting the clay with tetraisopropyl titanate and oleic acid (from Helmer et al., 1976)

Property	Polymer or plastic										
	HDPE	Epoxy	Acetal	SAN	Nylon 6	Poly-propylene	Propylene copolymer	Rigid PVC	ABS	Impact PS	Phenolic
Tensile strength	+	+	0	—	0	+	+	0	+	+	0
Tensile yield	0	+	0	0	0	0	+	0	0	0	0
% Elongation	—	—	—	0	—	—	—	—	0	—	0
Tensile modulus	+	0	+	+	+	+	0	+	+	+	0
Hardness	0	+	0	0	0	0	0	0	+	0	0
Heat distortion	+	+	+	0	+	+	+	+	+	0	0
Melt flow	0	0	+	0	0	0	0	0	—	0	0
Izod impact	0	0	—	0	0	0	0	—	—	—	0
Optimum concentration	30	50	30	30	60	30	30	20	30	25	

Plus and minus signs denote a positive and negative change in property, that is, an advantage and disadvantage, respectively. Zero sign means that there is no significant alteration in property on incorporating the filler.
HDPE = high density polyethylene; SAN = styrene acrylonitrile polymer; PVC = polyvinyl chloride; ABS = acrylonitrile-butadiene-styrene resins; PS = polystyrene.

electron, excited from the surface layers by the kinetic energy of impact, to the adsorbed monomer by which it is transformed into a radical-anion (I). This species then grows by successive addition to other monomer units, either through an anionic (A) or a free radical (R) mechanism although the latter route is apparently preferred. For acetonitrile, for example, the process may be represented by the following diagram:

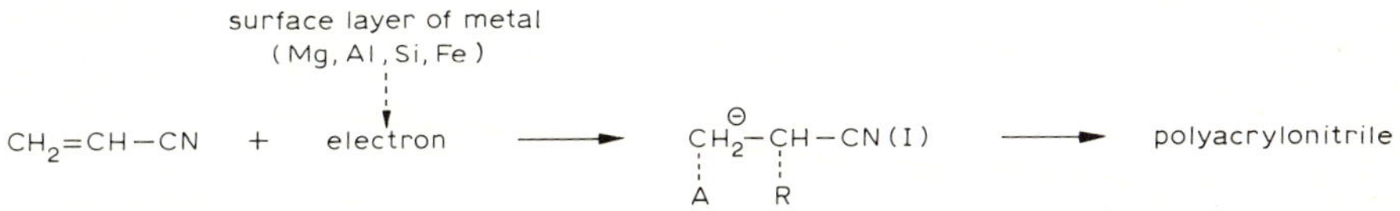

The proposed scheme is supported by the observation that the yield of polymer is related to the duration of grinding and the volume of steel balls in the mill whilst the addition of hydroquinone reduces, but does not completely inhibit, polymer formation.

In the case of polymerisation initiated by inorganic salts, grinding is thought to create ionic vacancies at the surface which serve as electron "traps". These electrons, subsequently emitted by ionisation, would then react with the monomers as indicated. The correlation between the intensity of electron emission and polymer yield, observed by Antonova et al. (1976), is consistent with the suggested mechanism.

What is important for the present discussion is that the polymer formed by this or similar means (Uskov, 1960b; Uskov et al., 1966b) may become grafted onto the mineral surface. Whether or not grafting occurs would ultimately depend on the strength of the mineral--polymer bond. As mentioned earlier, true grafting involves the formation of covalent bonds between the surface and the polymer. If, on the other hand, the surface—polymer bond is weak, the polymer could detach itself from the surface and little or no grafting would be observed. If the polymer can attach to the solid by a series of hydrogen bonds or other secondary interactions, both graft-like polymers and homopolymers would be yielded; the former being found in the vicinity of and the latter at some distance away from, the surface.

Kargin and Platé's (1959, 1961) work has been extended by Uskov and his co-workers to clay systems (Uskov, 1960a,b; Uskov and Kusnitsyna, 1960). Thus, when octadecylammonium (ODA)-montmorillonite was mechanically dispersed in the presence of methyl methacrylate by means of vibration milling, polymethyl methacrylate was formed. In this instance, little or no grafting occurred presumably because the clay surface had already been covered with a dense layer of long alkyl chains. However, the presence of dispersed clay particles in the matrix gave rise to a polymer which was harder and had a higher softening point than the corresponding unfilled material. Interestingly, when the same experiment was carried out using Na^+-

montmorillonite, considerable "grafting" (up to 22% by weight of clay) was observed as attested by the resistance of the polymer to prolonged extraction with boiling benzene. The rate of (radical) polymerisation and the yield of polymethyl methacrylate were further increased when benzoyl peroxide was added to the system and graft polymers as well as homopolymers were formed (Uskov et al., 1965). These observations were explained in terms of the adsorption onto the clay surface of the chemical initiator whose cleavage into radicals was promoted by the grinding process.

Uskov et al. (1961, 1963) then examined the effect of compounding Na^+- and ODA-montmorillonite with (preformed) polymethyl methacrylate and polystyrene on the properties of the filled systems. To this end, the clays were introduced into the polymer matrix by combined precipitation of the components from a general dispersing medium. Differential thermal analysis of the clay-filled products indicated that, in the case of ODA-montmorillonite, the mineral particles were densely screened by polymer molecules. Being mutually compatible, the clay surfaces were apparently "wet" by the organic phase. In the case of polypropylene containing dispersed organo-titanium kaolinite particles (cf. Table 6.2), the close adhesion between polymer and clay surface has been directly demonstrated by Helmer et al. (1976), using scanning electron microscopy. By contrast, the electron micrograph of the system filled with untreated kaolinite showed the clay particles to be surrounded by voids.

In common with the monomer-sodium clay system, the incorporation of Na^+-montmorillonite into polymethyl methacrylate resulted in "grafting" of the existing polymer onto the mineral surface in that ~20% of the material could not be extracted with benzene. On the other hand, polystyrene was completely removed by treating the polymer—clay system with benzene. This difference in behaviour between clay-filled polymethyl methacrylate and polystyrene was also reflected in their respective physico-mechanical properties. Thus, the hardness, glass transition temperature, yield point, and heat resistance of polymethyl methacrylate were markedly improved whereas those of polystyrene remained virtually unchanged as compared with the unfilled systems. Tarasenko et al. (1964) have observed similarly for the kaolinite-filled polymers.

To account for these observations, they suggested that hydrogen bonding occurred between the oxygens of polymethyl methacrylate and the hydroxyl groups at the clay surface whereas polystyrene could not be bonded in this manner. Although this suggestion seems feasible for kaolinite, their evidence for montmorillonite is not entirely convincing since the latter has very few exposed hydroxyl groups at its edge surface. It seems more probable that polymethyl methacrylate, like its monomer, attaches to the clay surface by means of ion—dipole interactions (Blumstein et al., 1971) whereas polystyrene, being essentially non-polar, is only weakly adsorbed. In addition, polymethyl methacrylate can conceivably penetrate the montmorillonite interlayers since its monomer does so (Blumstein, 1965) whereas styrene is not intercalated (Solomon and Rosser, 1965).

Rather than through mechanical disintegration of minerals in the presence of organic monomers, grafting of polymers onto clay surfaces is normally achieved by adding an organic initiating agent to or irradiating, the mineral—monomer system. Examples of each type, together with the potential uses of such mixed graft polymers as fillers of preformed polymer systems, will be given below. At this juncture, we wish only to mention the principles involved.

The method used by Poletukha et al. (1966) was to treat kaolinite with various organic initiating agents (e.g. tert-butylperacrylate, tert-butylperpropionate, β-hydroxyethyl-tert-butylperoxide). When such treated clays were then heated (above the decomposition temperatire of the respective initiator) in the presence of some organic monomers (e.g. methyl methacrylate, styrene), polymerisation occurred. An appreciable proportion of the polymer formed was found to be non-extractable and, on this basis, assumed to be grafted onto the clay. The amount of grafting was influenced by the nature of the initiator and the strength with which it is adsorbed onto the mineral surface.

A similar approach was earlier adopted by Dekking (1965) who introduced 2,2′-azobisisobutyramidine in the cationic form into the exchange sites of kaolinite and montmorillonite. As before, this complex decomposed to give free radicals (when gently heated), and was highly effective in initiating the polymerisation of such monomers as methyl methacrylate, acrylamide, vinyl acetate, and 4-vinyl pyridine. The observed increase in the rate of polymerisation, as compared with the clay-free system, was explained in terms of the higher rate of decomposition of the clay—initiator complex relative to that of the initiator by itself, and to the decrease in the rate of chain termination (Dekking, 1967). It seems probable, however, that the greater accessibility and availability to the monomers of the initiator, "extended" on the clay surface, is equally important since Ziegler—Natta catalysts, physically adsorbed on acidified montmorillonite (Friedlander, 1964) and on kaolinite surfaces, likewise initiate polymerisation (Theng and Walker, 1970).

Polyacrylic and polymethacrylic acids may also be grafted onto kaolinite surfaces by irradiating the mineral in the presence of a saturated vapour of the corresponding monomers, using γ-rays from a ^{60}Co source (Polushkin et al., 1976). Uskov et al. (1966b) have achieved similarly by irradiating different minerals (e.g. asbestos, glass fibre, kaolinite, silica gel) immersed in liquid methyl methacrylate and styrene. Like Dekking (1967), these workers observed an increase in polymerisation rate over the mineral-free system but ascribed this to the fact that monomer radiolysis was energetically more favoured in the adsorbed state than in bulk. In addition, the active radicals produced might be stabilised by adsorption onto the mineral surface.

Monomer adsorption, however, does not always lead to accelerated polymer formation. Indeed, Sugimoto et al. (1975) have noted the opposite during the radiation-induced polymerisation of 2-hydroxyethyl acrylate adsorbed onto Na^+-montmorillonite. In this instance, the reduction in polymerisation rate (as compared with the bulk) is probably due to the intercalation of a proportion of the monomer (Solomon and Loft, 1968; Theng and Walker, 1970) and this may conceivably impede the overall rate of polymer formation.

The paper industry is one of the largest users of clays, especially kaolinites, which serve as both filling and coating agents (Murray, 1963). The informative account of this topic by Morris and Brooks (1974) also contains a list of useful references to the literature.

Fillers are used to improve the brightness, colour, opacity, and printability of the sheet. In addition, fillers impart greater bulk and a better texture. Coating agents, on the other hand, are primarily applied to cover the depressions in the (uncoated) paper, making for a smooth surface which is receptive for printing. The coating process also gives a glossy and white finish to the product. Because of their controlled particle size distribution, low viscosity, non-abrasiveness, platy shape, and whiteness, kaolinites are eminently suitable as fillers and coating materials.

Various grades of kaolinites are available for either purpose, such qualities as brightness and particle size distribution being of particular importance. Prior to grading, the clays are fractionated by either flotation in air or sedimentation in water. In this way, impurities (e.g. sand and mica) are removed and the finer particles are reserved for the coating process. Table 6.3, taken from Morris and Brooks (1974), illustrates these points.

The filling process essentially consists of adding clays to a dilute suspension of the fibre in the beater, at a rate which depends on the ultimate use of the sheet (Table 6.4). In addition, a high refractive index filler, such as titanium dioxide, is usually included so as to give extra brightness and opacity.

In terms of quantity, some 70—90% of the dry weight of the coating material is made up of finely divided minerals (e.g. kaolinites), applied as an aqueous suspension, together with a suitable adhesive and an additive (dispersant). Starch and casein are the classic adhesives used (e.g. Kaliski, 1972) but, in recent years, various synthetic polymers, notably styrene—butadiene copolymers as also polyvinyl acetate, polyvinyl alcohol, and acrylics, have increasingly replaced the naturally occurring binding agents. As might be expected, a wide range of mixtures can be formulated depending on the proportions and types of clay, adhesive, and additive used together with the conditions under which the mixture is to be applied to the sheet. Not surprisingly, numerous patents for paper filling and coating compositions have been taken out, some of the more recent ones being those by Canard and Levy (1975), Davidson (1975), Davis et al. (1975), Matsumoto

TABLE 6.3

Filler and coating grades of clays (kaolinites) used in the paper industry (from Morris and Brooks, 1974

Grade	Brightness [a]	Particle size [b] (%)	High solids viscosity characteristics	Remarks
No. 1 coating clay	86—88	90	Excellent	
No. 2 coating clay	85—87	80	Excellent	
No. 3 coating clay	84—86	70	Excellent	
High glossing clay	86—90	98	Excellent	
High brightness	90+	—	Excellent	No. 1, No. 2, and high glossing grades available. Generally produced by special flotation procedures
Delaminated (large diameter platelets)	90	See remarks	Fair	Large diameter, thin (~0.25 μm) platelet. Higher opacity and sheet levelling properties permit lighter weight coatings
Delaminated (smaller diameter platelets)	88	100	Excellent	High glossing
High bulk clays	86—90	75	Poor	Entrapped air in "popcorn" shape yields opacity and fills sheet voids. Filler and/or coating use
English clays [c]	90	See remarks	Fair	Available in No. 2 and in high glossing grades
Wet washed filler	82—84	20—50	Fair	Various grades. Generally bleached for colour uniformity. Low grit content
Air floated filler	80—85	50—70	Variable	Not bleached. Grit content variable.

[a] Defined as the reflectance using ~457 nm light.
[b] Less than 2 μm equivalent spherical diameter.
[c] From Cornwall. The rest refers to the deposit near Macon, Georgia, U.S.A.

TABLE 6.4

Type and amount of filler used in different grades of paper (from Morris and Brooks, 1974)

Paper grade	Filler type	Amount added (% by weight)
Book and writing	High opacity pigment + clay	1—6
Book	High opacity pigment + clay	5—40
Offset paper	High opacity pigment + clay	0—10
Bible paper	Clay, calcium carbonate, high opacity pigment	20—30
Mimeograph paper	Clay, high opacity pigment, diatomaceous earth	20—25
Cigarette paper	Fine calcium carbonate	35—40
Envelope	Talc or clay, high opacity pigment	10—20
Newsprint	Clay or calcium carbonate	2—6

Note: high opacity pigment normally refers to titanium dioxide (rutile and anatase). By clay here is meant kaolinite.

and Ishida (1975) and Schroeder and Brehmer (1976).

Besides such compounds as calcium carbonate, carbon black and silica, clays enjoy a substantial application for the filling and reinforcement of natural and synthetic rubber (Cook, 1966; Cook and Pickering, 1968; Smith, 1968; Dannenberg, 1975). Indeed, next to the elastomer itself and its curing system, fillers are the most important ingredient of rubber and their use in rubber compounding is almost as old as the industry itself (Barrot, 1968; Smith, 1968).

The infrared study of Suito and Arakawa (1964) indicated that the interaction of kaolinite with rubber involved hydrogen bonding between the polymer and the hydroxyl groups on the mineral surface, particularly those located at crystal edges. This was inferred from the weakening of the OH absorption band at $\sim$3700 cm^{-1}, in the spectrum of kaolinite, on mixing the clay with the elastomer. In addition, prior heating of the mineral to temperatures above 673 K, which destroyed the crystal structure, led to the loss of its reinforcing ability.

As a rough guideline, the efficiency of a given filler in upgrading the mechanical properties of elastomers is inversely proportional to its ultimate particle size. This rule, however, is difficult to verify experimentally because of the lack of agreement as to the criteria used in measuring the degree of reinforcement (Smith, 1968). However, a reduction in particle size invariably leads to greater reinforcement as Cook and Pickering (1968) have found for rubber (Table 6.5) and Uskov et al. (1964) for amorphous polymers, using kaolinite and/or montmorillonite as fillers. This observation may be ascribed to the attendant increase in filler surface area (per unit weight) and hence, in

TABLE 6.5

The effect of filler particle size on the mechanical properties of a kaolinite-filled rubber compound [a] (from Cook and Pickering, 1968)

Property index [b]	Particle size fraction (μm)					
	<0.5	0.5—1.0	1.0—2.5	2.5—5.0	5.0—10.0	10.0—20.0
Modulus at 400%	100	96	86	62	37	37
Tensile strength	100	102	94	83	72	72
Elongation at break	100	106	104	112	119	119
Tear strength	100	79	79	54	52	50

[a] Made up of natural rubber (100), kaolinite (78), zinc oxide (5), stearic acid (4), sulphur (3), and mercaptobenzothiazole (1) where the numbers in parentheses denote parts by weight.

[b] Taking 100 for the results obtained using the finest particle size fraction.

the extent of polymer—filler interaction. In support of this hypothesis, De Nunzio and Held (1966) have found a good correlation between the surface area of a number of clays and the degree of reinforcement, as determined by tensile strength and 300% modulus measurements.

The shape of the filler particle also has a marked influence on its reinforcing ability, lamellar minerals being generally superior to their non-platy counterparts (Barrot, 1968). Cook and Pickering (1968) have noted similarly in that the spherical calcium carbonate (whiting) particles have a smaller stiffening effect on rubber than the platy kaolinite crystals of equivalent spherical diameter. If, in addition, some grafting occurs, the strength of the clay—polymer system is further increased, as Uskov et al. (1961) have shown for polymethyl methacrylate filled with Na^+-montmorillonite. Thus, the polymer of which about 10% is apparently grafted to the clay, has a strength of 28.5 kg mm^{-2} as compared with 24.3 kg mm^{-2} for the unfilled material. Calcium carbonate, used in the same ratio, actually reduces the strength to 23.5 kg mm^{-2}. By the same token, the yield point of the clay-filled polymer at 543 K is substantially higher than that of the polymer alone (503 K) whilst the system containing calcium carbonate has a yield point at 478 K.

The use of calcined kaolinite is preferred to that of unheated clay in the filling of elastomers which are to serve as cable compounds (Cook, 1966). Thus, the incorporation of calcined kaolinite into such materials as SB (styrene—butadiene) and butyl rubber (as well as polyvinyl chloride) yields products with markedly improved electrical properties. Calcined and acidic clays may also be used as fillers of EP (ethylene—propylene) rubber if a vulcanising aid (e.g. triallyl cyanurate, vinyl silane) is included in the formulation (Cook, 1966). In this regard, further improvements are obtained by using surface-modified clays, notably kaolinite, which have been reacted with organo-silicon compounds containing a vinyl functionality. Helmer et al. (1976) reported that EP rubber, filled with this type of clay, had a tensile strength which was 55% higher than the material containing untreated kaolinite.

An interesting approach to preparing organo-silicon modified kaolinites as fillers for rubber has been described by Kukharskaya and Skorik (1966). They added diethylpolysiloxane (liquid) to kaolinite, exposed the mixture to ultrasonic vibration, and removed the unbonded polymer by washing the system with toluene or tetrahydrofuran. When the mixed graft polymer was added to rubber and the blend vulcanised, a product was obtained which showed a substantial improvement in tensile strength over the compound filled with unmodified clay.

Untreated and surface-modified clays have also proved useful as fillers of polyethylene, polyvinyl chloride and other thermoplastics (Bostwick and Carey, 1950; Murray, 1963; Furter, 1964; Cook, 1966; Helmer et al., 1976; Schwarz, 1977). Kaolinites of fine particle size and high brightness (corresponding approximately to No. 2 coating clay of Table 6.3) are preferred. Coarser grades may be used in polyvinyl chloride formulations where hardness and toughness are required as, for example, in floor tiling. In addition to conferring hardness and rigidity to the product, clays give a smoother surface, good dimensional stability, and a high resistance to chemical attack. Untreated clays, however, confer little or no reinforcement to the system; they may, in fact, reduce the tensile strength and tear resistance of thermoplastics. Substantial reinforcement, however, can be achieved by using clays which have been treated by special means. The same is true for glass fibres which are more widely used than clays in filling thermoplastic polymers (Murphy, 1966).

A major advance in treating minerals, including clays, which results in filled thermoplastics of high strength, is the method described by Fallick et al. (1968). It involves dispersing the filler particles (e.g. kaolinite and asbestos) in the polymer melt, encapsulating them with a thin sheet of resin and firmly binding this encapsulant to both the mineral and the polymer matrix. By this means they were able to develop a family of mineral-reinforced linear polyethylene, termed "Ceraplasts", which showed considerable increases in tensile and impact strengths over conventionally filled polyethylene. This was ascribed to the presence of a modulus gradient in the encapsulant between the mineral and the polymer matrix.

That the mechanical strength and stability of such polymers as polycaproamide and polyethylene can be increased by adding untreated or surface-modified kaolinite to the polymer melt, has earlier been reported by Solomko et al. (1966, 1967). This finding was attributed to the fact that on crystallisation of the polymer, the clay particles tend to concentrate in the interspherulitic regions. This reduced the mobility of the polymer in these areas and also prevented the particles from regrouping in the crystalline phase. As a consequence, the average size of the spherulites was decreased, the reduction being greater as the size of the filler particles becomes smaller.

Reinforced polyethylene may be obtained by polymerising the monomer in the presence of a clay containing adsorbed, catalytically active transition metals (Howard, 1975). Alternatively, polyethylene can be reinforced by compounding the polymer with clay particles which have been coated with an olefinic substance and a free radical generator (Bixler and Fallick, 1969). A similar method was used by Baum (1969) who coated his clay samples with high boiling point ($>$413 K), oxygen-containing organic compounds and an organic peroxide which could act as a cross-linking agent between the modified mineral surface and the polyethylene chain.

Cross-linking or grafting of polyethylene and other thermoplastics (e.g. polypropylene, polyvinyl chloride) to clay surfaces has also been achieved by irradiation techniques (Nahin and Backlund, 1963; Nakada and Tagami, 1973). The extent of this cross-linking is greatly improved by coating the clay surface (kaolinite and montmorillonite) with organics having "polyethylene-like" chains (e.g. polyvinyl alcohol, polyethylene oxide, hexamethylene diamine). When such modified clays are exposed to γ-rays or fast electrons in the presence of polyethylene, products of great strength, excellent thermal stability, and resistance to organic solvents, are obtained (Nahin, 1965, 1966). Using a similar technique, Uskov et al. (1966b) were able to cross-link polystyrene to a layer of polymethyl methacrylate which had previously been grafted onto a mineral surface. Effective reinforcement of polyvinyl chloride was also claimed by Nakada and Tagami (1973) when a clay—polymethyl methacrylate composite, obtained by irradiating methyl methacrylate in the presence of the mineral, was incorporated into the system.

Using the above principles, Hawthorne et al. (1974) have prepared a series of kaolinite fillers which can be used to reinforce polyethylene. Their method consists of making the clay surface initially "acidic" (by heating the mineral at 383—393 K in air) and so capable of initiating the cationic polymerisation of various vinyl monomers (e.g. butadiene, isoprene, piperylene, propylene, styrene), and neutralising the remaining acidity with ammonia or an organic base. On compounding such treated clays with polyethylene, grafting takes place between the encapsulating polymer layer and the matrix polymer. This process is thought to be mediated by peroxidic groups in the encapsulant (generated by exposure of the coated clay to air) or by radical attachment at unsaturated centres. Polyethylene filled with surface-modified clays of this type show significant improvements in processability, tensile and impact strengths, as compared with the unfilled polymer or the material containing untreated fillers (Table 6.6).

Alternatively, kaolinite may be coated with diisopropoxyaluminium methacrylate polymer obtained by the in situ hydrolysis of adsorbed oxyaluminium derivatives in organic solvents (Hawthorne and Solomon, 1974). Such

TABLE 6.6

Some mechanical properties of high- and low-density polyethylene (HDPE and LDPE) containing 20% of different fillers (from Hawthorne et al., 1974)

Filler type	95% confidence limits for yield ($kg\ cm^{-2}$)		Elongation at break (%)		Impact strength (kg m/m)
	HDPE	LDPE	HDPE	LDPE	HDPE
None	289 ± 4	90 ± 3	1300	700	2.63
Untreated hydrite 10 (H 10) [a]	282 ± 5	96 ± 4	100	200	1.23
Freeport kaolin OX-3 [b]	285 ± 3	109 ± 1	40	140	1.30
Commercial reactive encapsulated kaolin	286 ± 3	96 ± 4	25	160	1.23
SRF Carbon black	318 ± 2	110 ± 1	50	84	1.00
H 10 + 4% isoprene [c]	300 ± 2	110 ± 3	90	210	—
H 10 + 4% piperylene [c]	300 ± 4	111 ± 2	140	140	1.32
H 10 + 4% styrene [c]	299 ± 4	111 ± 6	45	190	1.56

[a] From Georgia Kaolin Company.
[b] Coated with an organophilic compound, same average particle size as hydrite 10.
[c] Polymerisation terminated by ammonia; encapsulating polymer oxidised before use.

surface-modified clays, when blended with polyethylene, give rise to materials which display improved tensile and impact strengths whether or not peroxides have been added (during compounding) in order to promote grafting. Schwarz (1977) has claimed similarly for the properties of low density polyethylene filled with a block ethylene oxide—propylene oxide copolymer.

An added advantage of clay-filled over carbon black-filled thermoplastics is that the former materials, besides showing equal or superior mechanical properties to the latter, are white and may be given any desired colour by adding a suitable colouring agent.

We have already referred to the filling of polyamides with organo-titanium kaolinites (cf. Table 6.2), reported by Helmer et al. (1976). When a pendent amine group is introduced into the organic coating (cf. Fig. 6.4), the filler becomes even more compatible with the polymer matrix. Such mineral-filled nylons, which are commercially available, show significant reductions in water uptake. This would give them improved dimensional stability in situations where there are large fluctuations in relative humidity. Since these materials also have improved tensile and flexural properties, they have proved very useful in the manufacture of such objects as carburettor bodies, fan blades, and skate boards. As in the case of polyethylene (Howard, 1975), clay-filled polyamides may be obtained by introducing a catalytically active clay into the monomer. When montmorillonite treated with aminocaproic acid is used for this purpose, the caprolactam monomer is apparently intercalated, giving rise to a clay—polymer complex with an interlayer separation of 1.31 nm (Fujiwara and Sakamoto, 1976). Similarly, sheets of high flexural strength can be made from acrylic polymers grafted onto clays. This may be achieved by heating methyl methacrylate in the presence of acidic clays and an organic radical initiator (Ochirimi et al., 1976).

The addition of clays, such as kaolinites, to polyester and epoxy resins with which they mix well, gives rise to stable dispersions. Although the finer grades of clays tend to increase gel time (due to adsorption of the accelerator), they have the advantage of improving the thixotropic properties of the system (Cook, 1966). More importantly, clays prevent overheating during curing, making the mouldings resistant to cracking and shrinking. Seger and Sharpe (1975) have described a method of preparing an adhesive composition for joining ferrites. It consists of an epoxy resin, a thixotropic clay which, in this instance, is montmorillonite saturated with long chain alkylammonium ions (Bentone 38; see Jordan, 1963), and a curing agent. The product has a high breaking load as well as good heat and moisture resisting properties, making for strong bonding at the interface. General adhesives are usually prepared from emulsions of such polymers as polyvinyl acetate. Here again, however, clays may be added to good effect in that they markedly increase the viscosity of the emulsion without adversely affecting its adhesive properties (Kobayashi, 1975).

Clays as such, or after surface modification, have also been successfully incorporated into polyurethane foams with the principal objective of lowering the cost of the product. Since a low-density, open-celled material is desired, conventional inorganic additives, such as silica, are not satisfactory. This is so because they adversely affect the density and cell structure of the foam when used at a rate of 10% or more by weight. On the other hand, open-structured palygorskite, coated with water-soluble polymers (e.g. polyacrylamide, polyvinyl alcohol, polyvinyl pyrrolidone) and dispersed in the polyisocyanate—polyol system prior to its foaming, does not appear to cause an increase in density or a deterioration in texture (Ferrigno, 1962). An extension and modification of the above process to include a wide range of modified clays and related alumino-silicates, have subsequently been described by Showalter (1966).

More recently, Bengtson and Wooler (1975) have given a method for preparing porous, low-density polyurethane aggregates containing clay by mixing expanded clay particles with a dilute aqueous solution of sodium carboxymethyl cellulose and the polyurethane prepolymer. Although improved strength of the foam was claimed, the chief benefit of using clays lies in the attendant increase in foam volume for a given weight of reaction mixture. Clays, in this application, may therefore be considered as true extenders.

REFERENCES

Allison, F.E., 1968. Soil aggregation — some facts and fallacies as seen by a microbiologist. Soil Science, 106: 136—143.

Antonova, L.A., Khrustalev, Yu.A., Baramboim, N.K. and Krotova, N.A., 1976. Initiation of polymerization with the mechanical dispersion of crystals. Kolloidnyi Zhurnal, 8: 535—536.

Armbrust, D.V. and Dickerson, J.D., 1971. Temporary wind erosion control: Cost and effectiveness of 34 commercial materials. Journal of Soil Water Conservation, 26: 154—157.

Armbrust, D.V. and Lyles, L., 1975. Soil stabilizers to control wind erosion. In: Soil Conditioners; Soil Science Society of America Special Publication No. 7: 77—82.

Barrot, R.J., 1968. Interest of lamellar minerals in rubber compounding. Rubber Meeting of the Swedish Institution of Rubber Technology, Falsterbo, 14 pp.

Baum, B.O., 1969. Peroxide cured polyethylene compositions containing coated clays. United States Patent 3, 425, 980 (to Union Carbide Corp.).

Bear, F.E. (Editor), 1952. Synthetic soil conditioners. Soil Science, 73: 419—492.

Bengtson, O. and Wooler, A.M., 1975. Low-density aggregates bound with polyurethane foam. German Patent 2, 451, 393 (to Imperial Chemical Industries Ltd.).

Bixler, H.J. and Fallick, G.J., 1969. Reinforcing filler for polymers. United States Patent 3, 471, 439 (to Amicon Corp.).

Blavia, F.J., Moldenhauer, W.C. and Law, D.E., 1971. Materials for stabilizing surface clods of cropped soil. Soil Science Society of America Proceedings, 35: 119—122.

Blumstein, A., 1965. Polymerization of adsorbed monolayers. I. Preparation of the clay—polymer complex. Journal of Polymer Science, A, 3: 2653—2664.

Blumstein, R., Blumstein, A. and Parikh, K.K., 1974. Polymerization of monomolecular layers adsorbed on montmorillonite: cyclization in polyacrylonitrile and polymethacrylonitrile. Applied Polymer Symposium No. 25, 81—88.

Blumstein, A., Parikh, K.K., Malhotra, S.L. and Blumstein, R., 1971. Polymerization of monolayers. VI. Influence of the nature of the exchangeable ion on the tacticity of insertion poly(methyl methacrylate). Journal of Polymer Science A-2, 9: 1681—1691.

Bostwick, R. and Carey, R.H., 1950. Filled polyethylene compounds. Industrial and Engineering Chemistry, 42: 848—849.

Brandt, G.H., 1972. Soil physical property modifiers. In: C.A.I. Goring and J.W. Hamaker (Editors), Organic Chemicals in the Soil Environment, Vol. 2. Marcel Dekker, New York, N.Y., pp. 691—729.

Brockman, F.J. and Allenby, O.C.W., 1955. Factors affecting the laboratory evaluation of soil conditioners. Canadian Journal of Agricultural Science, 35: 27—34.

Canard, P. and Levy, A., 1975. Aqueous coating compositions for paper and carton. German Patent 2, 453, 910 (to Rhone Progil).

Carr, C.E. and Greenland, D.J., 1972. Preliminary results of an empirical study of the movement of polymers through soil and their effect on dispersion of clay from aggregates. Mededelingen Fakulteit Landbouwwetenschappen Rijksuniversiteit Gent, 37: 982—992.

Carr, C.E. and Greenland, D.J., 1975. Soil conditioners. Reports on the Progress of Applied Chemistry 1974, 59: 269—290.

Carter, L.W. Hendricks, J.G. and Bolley, D.S., 1950. Elastomer reinforced with modified clay. United States Patent 2, 531, 396 (to National Lead Co.).

Cook, L.E., 1966. China clays in the polymer industry. Rubber and Plastics Age, 47: 284—288.

Cook, L.E. and Pickering, F.G., 1968. Kaolinite fillers in rubber. Rubber Meeting of the Swedish Institution of Rubber Technology, Falsterbo, 14 pp.

Dannenberg, E.M., 1975. The effects of surface chemical interactions on the properties of filler-reinforced rubbers. Rubber Chemistry and Technology, 48: 410—444.

Davidson, R.R., 1975. Filler for paper manufacture. French Patent 2, 234 421 (to Associated Portland Cement Manufacturers Ltd.).

Davis, R.A., Tefertiller, B.A. and Savage, A.B., 1975. Aqueous pigment-containing binders for coating paper. German Patent 2, 450, 039 (to Dow Chemical Co.).

De Boodt, M. (Editor), 1972. Proceedings Symposium on the Fundamentals of Soil Conditioning. Mededelingen Fakulteit Landbouwwetenschappen Rijksuniversiteit Gent, 37: 897—1167.

De Boodt, M., 1975. Use of soil conditioners around the world. In: Soil Conditioners. Soil Science Society of America Special Publication No. 7: 1—12.

De Boodt, M., 1976. Soil conditioning: its roots and its meaning for the future. Mededelingen Fakulteit Landbouwwetenschappen Rijksuniversiteit Gent, 41: 19—22.

De Boodt, M. and Gabriels, D., 1973. The effect of synthetic soil conditioners on the physical fertility of loamy soils and in particular on the germination of sugar beets. Bulletin des Recherches Agronomiques de Gembloux, pp. 106—116.

De Boodt, M. and Gabriels, D. (Editors), 1976. 3rd International Symposium on Soil Conditioning. Mededelingen Fakulteit Landbouwwetenschappen Rijksuniversiteit Gent, 41: 1—464.

Dekking, H.G.G., 1965. Propagation of vinyl polymers on clay surfaces. I. Preparation, structure, and decomposition of clay initiators. Journal of Applied Polymer Science, 9: 1641—1651.

Dekking, H.G.G., 1967. Propagation of vinyl polymers on clay surfaces. II. Polymerization of monomers initiated by free radicals attached to clay. Journal of Applied Polymer Science, 11: 23—36.

De Nunzio, P.J. and Held, M.D., 1966. A new look at clay in SBR elastomers. American Chemical Society, Division of Rubber Chemistry, Spring Meeting, 15 pp.

Dowdy, R.H., 1972. Effects of hydroxyl-containing organics on the strength—energy characteristics of montmorillonite. Soil Science Society of America Proceedings, 36: 162—166.

Dowdy, R.H., 1975. The effect of organic polymers and hydrous iron oxides on the tensile strength of clays. In: Soil Conditioners. Soil Science Society of America Special Publication No. 7: 25—33.

Emerson, W.W., 1956. Synthetic soil conditioners. Journal or Agricultural Science (Cambridge), 47: 117—121.

Epshtein, S.M., 1976. Formation of synthetic soil aggregates. Pochvovedenie, No. 12: 117—124.

Fallick, G.J., Bixler, H.J., Marsella, R.A., Garner, F.R. and Fettes, E.M., 1968. Mineral-reinforced thermoplastics. Modern Plastics, 45, No. 5, 143, 144, 147, 148, 149, 209, 214.

Ferrigno, T.H., 1962. Polyurethane foams containing inorganic pigment coated with polymeric material. United States Patent 3, 024, 209 (to Minerals and Chemicals Philipp Corp.).

Friedlander, H.Z., 1964. Homopolymer forms by contact polymerization on oxides or with Ziegler catalyst on clay. Chemical and Engineering News, 42.

Fujiwara, S. and Sakamoto, T., 1976. Polyamide—organoclay composites. Japanese Patent 76, 109, 998 (to Unitika Ltd.).

Furter, W.F., 1964. Filled linear polyethylene compositions. The Canadian Journal of Chemical Engineering, pp. 77—81.

Gabriels, D., 1972. Response of different soil conditioners to soils. Mededelingen Fakulteit Landbouwwetenschappen Rijksuniversiteit Gent, 37: 1014—1034.

Gabriels, D. and De Boodt, M., 1972. Theoretical and practical approach for determining the optimal moisture content at the moment of soil conditioning. Mitteilungen der Deutschen Bodenkundlichen Gesellschaft, 15: 185—203.

Gabriels, D. and De Boodt, M., 1975. Erosion reduction for chemically treated soils: A laboratory experiment. In: Soil Conditioners. Soil Science Society of America Special Publication No. 7: 95—102.

Gabriels, D.M., Moldenhauer, W.C. and Kirkham, D., 1973. Infiltration, hydraulic conductivity, and resistance to water-drop impact of clod beds as affected by chemical treatment. Soil Science Society of America Proceedings, 37: 634—637.

Gardner, W.H., 1972. Use of synthetic soil conditioners in the 1950's and some implications to their further development. Mededelingen Fakulteit Landbouwwetenschappen Rijksuniversiteit Gent, 37: 1046—1061.

Gluesenkamp, E.W., 1957. Organic liquids and solids containing adducts. United States Patent 2, 795, 545 (to Monsanto Chemical Co.).

Goor, G., Bernaert, E., Sadones, M., Callebaut, F. and Schamp, N., 1976. The influence of chemical and physical properties of synthetic polymer emulsions on soil conditioning. Mededelingen Fakulteit Landbouwwetenschappen Rijksuniversiteit Gent, 41: 187—191.

Greenland, D.J., 1963. Adsorption of polyvinyl alcohols by montmorillonite. Journal of Colloid Science, 18: 647—664.

Haden, W.L., Jr, 1963. Attapulgite: properties and uses. Clays and Clay Minerals, 10: 284—290.

Harris, R.F., Chesters, G. and Allen, O.N., 1966. Dynamics of soil aggregation. Advances in Agronomy, 18: 107—169.

Hawthorne, D.G. and Solomon, D.H., 1974. Reinforcement of polyethylene by surface-modified kaolins. Journal of Macromolecular Science, Chemistry, A, 8: 659—671.

Hawthorne, D.G., Hodgkin, J.H., Loft, B.C. and Solomon, D.H., 1974. Polymerization of vinyl monomers on mineral surfaces; a novel method of preparing reinforcing fillers. Journal of Macromolecular Science, Chemistry, A, 8: 649—657.

Hedrick, R.M. and Mowry, D.T., 1952. The improvement of soil structure by synthetic polyelectrolytes. Chemistry and Industry, 652—656.

Heitz, E., 1975. Füllstoffe als qualitätsverbessernde Modifikatoren. Gummi, Asbest, Kunststoffe, 28: 286, 288, 290, 292, 294.

Helmer, B.M., Prescott, P.I. and Sennett, P., 1976. Surface-modified kaolin in plastic. 31st Annual Technical Conference, Reinforced Plastics/Composites Institute, The Society of the Plastics Industry, Inc., Section 8-G: 1—4.

Hofman, G., 1976. The influence of drying and storing soil samples on aggregate instability. Mededelingen Fakulteit Landbouwwetenschappen Rijksuniversiteit Gent, 41: 101—108.

Howard, E.G., Jr, 1975. Filled polyolefin compositions. German Patent 2, 459, 118 (to du Pont de Nemours, E.I., and Co.).

Ingles, O.G., 1968. Advances in soil stabilization, 1961—1967. Reviews of Pure and Applied Chemistry, 18: 291—310.

Jacovic, M., 1964. Initiation of polymerisation by means of dispersions of mineral solids. Double Liaison, pp. 83—91.

Jones, M.B. and Martin, W.P., 1957. Methods of evaluating aggregate stabilization by HPAN as it is affected by various inorganic salts. Soil Science, 83: 475—479.

Jordan, J.W., 1963. Organophilic clay-base thickeners. Clays and Clay Minerals, 10: 299—308.

Kaliski, A., 1972. Performance of some commercial clays in starch-containing paper-coating films. Part II. Gloss of coatings on a lightweight publication paper base sheet. Journal of the Technical Association of the Pulp and Paper Industry (Tappi), 55: 102—110.

Kargin, V.A. and Platé, N.A., 1959. Chemical grafting on crystal surfaces. Vysokomolekulyarnye Soedineniya, 1: 330—331.

Kargin, V.A. and Platé, N.A., 1961. Polymerization and grafting processes on freshly formed surfaces. Journal of Polymer Science, 52: 155—158.

Kobayashi, K., 1975. Kaolin-containing carboxylated poly(vinyl acatate) emulsions for adhesives. Japanese Patent 75 102, 646.

Koch, J.T., Kay, B.D. and Lange, G.L., 1974. A postulated mechanism for increasing the efficiency of organic soil conditioners. Soil Science Society of America Proceedings, 38: 586—590.

Kukharskaya, E.V. and Skorik, Yu.I., 1966. Modification of kaolin surface by organic and organosilicon polymers by means of ultrasonics. Kauchuk i Rezina, 25: 23—24.

Kullmann, A., 1974. Synthetische Bodenverbesserungsmittel. Deutsche Landwirtschaftsverlag, Leipzig, 144 pp.

Lambe, T.W. and Michaels, A.S., 1954. Altering soil properties with chemicals. Chemical and Engineering News, 32: 488—492.

Libby, P.W., Iannicelli, J. and McGill, C.R., 1967. Elastomer reinforcement with amino silane grafted kaolin. American Chemical Society, Division of Rubber Chemistry, Spring Meeting, 18 pp.

Low, A.J., 1954. The study of soil structure in the field and the laboratory. The Journal of Soil Science, 5: 57—74.

Martin, J.P., Martin, W.P., Page, J.B., Raney, W.A. and De Ment, J.D., 1955. Soil aggregation. Advances in Agronomy, 7: 1—37.

Matsumoto, E. and Ishida, T., 1975. Strippable wallpaper. Japanese Patent 75 14, 806 (to San Rokku Kogyo).

Mausbach, M.J. and Shrader, W.D., 1975. Influence of surface treatment of selected subsoil materials on infiltration and erosion. In: Soil Conditioners. Soil Science Society of America Special Publication No. 7: 83—93.

Moldenhauer, W.C. and Gabriels, D.M., 1972. Some uses of soil stabilizers in the U.S.A. Mededelingen Fakulteit Landbouwwetenschappen Rijksuniversiteit Gent, 37: 1076—1085.

Morris, H.H. and Brooks, L.E., 1974. Pigmentation of paper goods. In: T.C. Patton (Editor), Pigment Handbook, Vol. 2. John Wiley and Sons, Inc., New York, N.Y., pp. 205—213.

Muljadi, D., De Boodt, M., Vandevelde, R. and Lenvain, J., 1972. Soil conditioning of volcanic debris. Pedologie, Ghent, 22: 100—122.

Murphy, T.P., 1966. Reinforced and filled thermoplastics. Industrial and Engineering Chemistry, 58: 41—49.

Murray, H.H., 1963. Industrial applications of kaolin. Clays and Clay Minerals, 10: 291—298.

Nahin, P.G., 1963. Perspectives in applied organo-clay chemistry. Clays and Clay Minerals, 10: 257—271.

Nahin, P.G., 1965. Organo-clays bonded to polyethylene by ionizing radiation. Clays and Clay Minerals, 13: 317—330.

Nahin, P.G., 1966. Incorporating clays into thermoplastic vinyl polymers. United States Patent 3, 248, 314 (to Union Oil Co. of California).

Nahin, P.G. and Backlund, P.S., 1963. Organoclay—polyolefin compositions. United States Patent 3, 084, 117 (to Union Oil Co. of California).

Nakada, S. and Tagami, K., 1973. Thermoplastic resin composition. Japanese Patent 73 23, 331 (to Sekisui Chemical Co., Ltd).

Nicholls, R.L. and Davidson, D.T., 1957. Soil stabilization with large organic cations and polyacids. Proceedings of the Iowa Academy of Science, 64: 349—381.

Oades, J.M., 1976. Prevention of crust formation in soils by poly(vinyl alcohol). Australian Journal of Soil Research, 14: 139—148.

Ochirimi, T., Matsumoto, T. and Okubo, M., 1976. Compositions containing acrylic polymers and clays. Japanese Patent 76 148, 738 (to Nitto Electric Industrial Co., Ltd).

Pla, I., 1975. Effect of bitumen emulsion and polyacrylamide on some physical properties of Venezuelan soils. In: Soil Conditioners. Soil Science Society of America Special Publication No. 7: 35—46.

Plueddemann, E.P., Clark, H.A., Nelson, L.E. and Hoffman, K.R., 1962. Silane coupling agents for reinforced plastics. Modern Plastics, 39, No. 12, 135, 137, 138, 140, 144, 146, 187, 189, 191, 193.

Poletukha, V.V., Solomko, V.P., Vilenskaya, M.R., Uskov, I.A. and Yurzhenko, T.I., 1966. Grafting of polymethyl methacrylate and polystyrene on kaolin modified by organic peroxides. Dopovidi Akademii Nauk Ukrainskoi RSR., 1031—1033.

Polushkin, V.A., Bondarev, A.I., Lapin, V.V. and Tsetlin, B.L., 1976. Special characteristics of the rheological properties of aqueous solutions of kaolin modified by the grafting of polymeric acids. Kolloidnyi Zhurnal, 38: 496—501.

Quirk, J.P. and Panabokke, C.R., 1962. Pore volume-size distribution and swelling of natural soil aggregates. The Journal of Soil Science, 13: 71—81.

Quirk, J.P. and Williams, B.G., 1974. Disposition of organic materials in relation to stable aggregation. Transactions of the 10th International Congress of Soil Science, Moscow, 1: 165—173.

Raff, R.A.V., 1968. The grafting of organic polymers onto inorganic substances. Report to College of Engineering Research Division, Washington State University, 106 pp.

Ranney, M., Cameron, G.M. and Lipinski, B.W., 1973. Improvement of the static and dynamic properties of filled elastomers by silane coupling agents. Kautschuk und Gummi Kunststoffe 26: 409—416.

Rehner, J., Wiese, H.K. and Gessler, A.M., 1960. Cyclodienylhalosilane-treated mineral pigments. United States Patent 2, 928, 802 (to Esso Research and Engineering Co.).

Richard, F., 1959. Die Wirkung von "Krilium" in neutralen und saueren Ton- und Staubböden. Proceedings of the International Symposium on Soil Structure 1958, pp. 97—102.

Richard, F. and Chausson, J.S., 1957. Über die kunstliche Veränderung physikalischer Bodenfaktoren. Mitteilungen der Schweizerischen Anstalt für das forstliche Versuchswesen, 33, No. 1, 31 pp.

Rigole, W. and De Bisschop, F., 1972. The formation of adhesive links between granular particles by means of emulsions. Mededelingen Fakulteit Landbouwwetenschappen Rijksuniversiteit Gent, 37: 938—954.

Ruehrwein, R.A., 1957. Hydrophobic clays and related materials. United States Patent 2, 795, 567 (to Monsanto Chemical Co.).

Sadones, M., Bernaert, E. and Schamp, N., 1975. The effect of synthetic polymers on the water stability of soil aggregates. Zeitschrift für Pflanzenernährung und Bodenkunde, 6: 641—647.

Schamp, N., Huylebroeck, J. and Sadones, M., 1975. Adhesion and adsorption phenomena in soil conditioning. In: Soil Conditioners. Soil Science Society of America Special Publication No. 7: 13—23.

Schroeder, G. and Brehmer, E., 1976. Coatings for surface improvement of paper. German Patent 2, 442, 704 (to Chemische Werke Huels A.G.).

Schwarz, E.C., 1977. Polymer-modified hydrophilic inorganic fillers for thermoplastic polymeric materials. United States Patent 4, 017, 452 (to Presto Products, Inc.).

Seger, S.G., Jr and Sharpe, L.H., 1975. A technology for achieving durable joining of ferrites. Polymer Science and Technology, 9B: 577—595.

Showalter, W.E., 1966. Polyurethane foams containing clay fillers. United States Patent 3, 227, 666 (to Union Oil Co. of California).

Slater, R.W. and Kitchener, J.A., 1966. Characteristics of flocculation of mineral suspensions by polymers. Discussions of the Faraday Society No. 42: 267—284.

Smith, D.A., 1968. Compounding elastomers with fillers. Rubber Journal, 38—44.

Smith, H.E., Schwartz, S.M., Gugliemelli, L.A., Freeman, P.G. and Russell, C.R., 1958. Soil conditioning properties of modified agricultural residues and related materials: I. Aggregate stabilization as a function of type and extent of chemical modification. Soil Science Society of America Proceedings, 22: 405—409.

Soil Science Society of America, 1975. Special Publication No. 7. Soil Conditioners. Proceedings of a Symposium on Experimental Methods and Uses of Soil Conditioners, 186 pp.

Solomko, V.P., Molokoyedova, T.A. and Uskov, I.A., 1966. Effect of fillers on the supramolecular structures and mechanical properties of crystalline polymers. Vysokomolekulyarnye Soedineniya, 8: 110—115.

Solomko, V.P., Molokoyedova, T.A., Uskov, I.A. and Polichkovskaya, T.V., 1967. The influence of modified and unmodified fillers on the morphology and dimensions of spherulites and mechanical properties of polycaproamides and polyethylene films. Vysokomolekulyarnye Soedineniya, 9: 40—44.

Solomon, D.H., 1968. Clay minerals as electron acceptors and/or electron donors in organic reactions. Clays and Clay Minerals, 16: 31—39.

Solomon, D.H., 1977. Minerals, macromolecules, and man. Search, 10: 363—369.

Solomon, D.H. and Loft, B.C., 1968. Reactions catalyzed by minerals. III. The mechanism of spontaneous interlamellar polymerizations in aluminosilicates. Journal of Applied Polymer Science, 12: 1253—1262.

Solomon, D.H. and Rosser, M.J., 1965. Reactions catalyzed by minerals. I. Polymerization of styrene. Journal of Applied Polymer Science, 9: 1261—1271.

Stefanson, R.C., 1973. Polyvinyl alcohol as a stabilizer of surface soils. Soil Science, 115: 420—428.

Stepanova, G.A. and Daulbayev, E.A., 1970. Use of polymers in agriculture. Agrokhimiya, 156—157.

Sugimoto, S., Ohnishi, S. and Nitta, I., 1975. Radiation-induced polymerization of 2-hydroxyethyl acrylate adsorbed on sodium montmorillonite. Nippon Genshiryoku Kenkyusho Nempo, 5030: 43—47.

Suito, E. and Arakawa, M., 1964. Mechanism of reinforcement of rubber with hard clay. Bulletin of the Institute for Chemical Research, Kyoto University, 42: 105—114.

Tarasenko, Yu.G., Uskov, I.A. and Solomko, V.P., 1964. The effect of kaolin on the properties of polymethyl methacrylate and polystyrene. Ukrainskii Khimicheskii Zhurnal, 30: 86—90.

Theng, B.K.G., 1970. Interactions of clay minerals with organic polymers. Some practical applications. Clays and Clay Minerals, 18: 357—362.

Theng, B.K.G., 1974. The Chemistry of Clay—Organic Reactions. Adam Hilger Ltd, London, 343 pp.

Theng, B.K.G. and Walker, G.F., 1970. Interactions of clay minerals with organic monomers. Israel Journal of Chemistry, 8: 417—424.

Theng, B.K.G., Greenland, D.J. and Quirk, J.P., 1967. The swelling in water of complexes of montmorillonite with polyvinyl alcohol. Australian Journal of Soil Research, 5: 69—75.

Uskov, I.A., 1960a. Filled polymers. I. The filling of polymethyl methacrylate with aminated bentonite added directly to the monomer. Vysokomolekulyarnye Soedineniya, 2: 200—204.

Uskov, I.A., 1960b. Filled polymers. III. Polymerization of methylmethacrylate during the dispersion of sodium bentonite. Vysokomolekulyarnye Soedineniya, 2: 926—930.

Uskov, I.A. and Kusnitsyna, T.A., 1960. Filled polymers. II. The effect of dispersing aminated bentonite in the monomer upon the reinforcement of polymethyl methacrylate. Vysokomolekulyarnye Soedineniya, 2: 728—740.

Uskov, I.A., Tarasenko, Yu.G. and Kusnitsyna, T.A., 1961. Compounding of dissolved polymethyl methacrylates with fillers. International Chemical Engineering, 1: 132—134.

Uskov, I.A., Tarasenko, Yu.G., Surovtsev, V.I. and Solomko, V.P., 1963. Investigations of bentonite-filled polymethyl methacrylate and polystyrene. Vysokomolekulyarnye Soedineniya, Adgeziya Polimerov, 31—34.

Uskov, I.A., Tarasenko, Yu.G. and Solomko, V.P., 1964. Effect of the degree of dispersion of clay fillers on the properties of amorphous polymers. Vysokomolekulyarnye Soedineniya, 6: 1768—1772.

Uskov, I.A., Solomko, V.P. and Chemeris, N.P., 1965. Dispersion acceleration of radical polymerization. Ukrainskii Khimicheskii Zhurnal, 31: 1071—1073.

Uskov, I.A., Galinskaya, V.I. and Artyukh, L.V., 1966a. Calorimetric method of determining the interaction energy of polymers with fillers. Vysokomolekulyarnye Soedineniya, 8: 1853.

Uskov, I.A., Tertykh, L.I., Solomko, V.P. and Polishchuk, Yu.N., 1966b. Radiation polymerization of methyl methacrylate and styrene in the presence of mineral fillers. Vysokomolekulyarnye Soedineniya, 8: 26—30.

Vandevelde, R. and De Boodt, M., 1972. Reaction of soil textures on soil conditioning with an emulsion of bitumen and a solution of adhesive (PAM). Mededelingen Fakulteit Landbouwwetenschappen Rijksuniversiteit Gent, 37: 993—1013.

Walker, G.F. and Garrett, W.G., 1967. Silicate layer minerals and products made therefrom. United States Patent 3, 356, 611 (to Commonwealth Scientific and Industrial Research Organization, Australia).

Wilcox, J.R., 1961. Modified kaolin clay. United States Patent 2, 982, 665 (to Minerals and Chemicals Philipp Corp.).

Williams, B.G., Greenland, D.J. and Quirk, J.P., 1966. The adsorption of poly(vinyl alcohol) by natural soil aggregates. Australian Journal of Soil Research, 4: 131—143.

Williams, B.G., Greenland, D.J. and Quirk, J.P. 1967a. The tensile strength of soil cores containing polyvinyl alcohol. Australian Journal of Soil Research, 5: 85—92.

Williams, B.G., Greenland, D.J. and Quirk, J.P., 1967b. The effect of polyvinyl alcohol on the nitrogen surface area and pore structure of soils. Australian Journal of Soil Research, 5: 77—83.

Williams, B.G., Greenland, D.J. and Quirk, J.P., 1968. The water stability of natural clay aggregates containing polyvinyl alcohol. Australian Journal of Soil Research, 6: 59—66.

Zisman, W.A., 1965. Improving the performance of reinforced plastics. Industrial and Engineering Chemistry, 57: 27—34.

Part III

INTERACTIONS OF CLAY MINERALS WITH NATURALLY OCCURRING POLYMERS

Chapter 7

PROTEINS AND ENZYMES

7.1. INTRODUCTION

In view of its importance in many analytical, industrial, and medical applications, the behaviour of proteins and enzymes at interfaces has attracted the attention of workers from various fields and with diverse interests. This is attested by the numerous reports and reviews which have appeared over the past two decades (Fraser, 1957; Loeb, 1965; McLaren and Packer, 1970; Brash and Lyman, 1971; Fenstermaker et al., 1974). A classic example is the separation and purification of protein and enzyme systems by adsorption onto inorganic and organic supports (Zittle, 1953). Of more recent development is the immobilisation of enzymes by attachment to and occlusion in solid and gel matrices (Zaborsky, 1973; De Boeck, 1975; Messing, 1975a; Mosbach, 1976; Thomas and Kernevez, 1976). By this means both the activity and stability of enzymes can be controlled and modified to meet specific requirements, providing the basis for "enzyme engineering" (Wingard, 1972; Weetall and Suzuki, 1975). Equally relevant in this context is the ability of adsorbed proteins to alter the surface properties of the adsorbent. In such seemingly unrelated phenomena as blood clotting, dental plaque formation, and marine fouling, for example, the adhesion of arriving cells to the solid substrate is markedly influenced by the presence of a pre-adsorbed layer of protein at the surface (Baier, 1975).

Here we are specifically concerned with the interactions between clay minerals and proteins. Apart from its intrinsic value, this subject is of utmost interest and relevance to soil scientists and agronomists alike in that the clay—protein interaction exerts a profound influence on the fate and biological stability of proteinaceous materials when they are added to or enter the soil system (e.g. Bremner, 1965, 1967; McLaren and Peterson, 1965). Similarly, research into the behaviour of catalytically active proteins (i.e. enzymes) at clay mineral surfaces has been stimulated by the finding that most, if not all, enzyme reactions in the soil, as in the living cell, occur at solid/liquid or solid/gel interfaces rather than in homogeneous, aqueous solutions (Fraser, 1957; McLaren and Packer, 1970). This being so, enzymes in these systems express their activity within an organised structural framework where surface effects may be as important as or even take precedence over ordinary mass—action relationships (McLaren and Babcock, 1959; McLaren, 1960; Katchalski et al., 1971). In the soil, both clays and humic substances provide such a framework in that they can form stable complexes with various enzymes (Skujinš, 1967; Ladd and Butler, 1975). Indeed, in a sense, soil may be regarded as "a system of humus and clay immobilised enzymes" (McLaren, 1975).

We ought perhaps to interpolate at this point that, besides serving as an adsorbing and concentrating surface, clays are capable of catalysing the transformation and polymerisation of various organic compounds, including amino acids and their derivatives (Solomon, 1968; Theng and Walker, 1970; Theng, 1974a, b). Because of this dual function, clays have been implicated in chemical evolution and the abiogenic origin of life on earth (Bernal, 1951; Paecht-Horowitz, 1973; Fripiat and Cruz-Cumplido, 1974; Fripiat and Poncelet, 1974).

Although clay minerals undoubtedly play a part in degradative and synthetic reactions involving organic substances in the soil, the available evidence suggests that decomposition processes are largely catalysed by enzymes, a great proportion of which are presumed to be of microbial origin acting outside living cells (Hofmann and Hoffmann, 1966; Voets and Dedeken, 1966; Skujinš, 1967; Sequi, 1974; Aliev et al., 1975; Kiss et al., 1975). Accordingly, the overall level of enzyme activity may often serve to characterise soils in terms of their fertility, origin (location), and type (Galstyan and Tatevosyan, 1964; Durand, 1965; Hofmann and Hoffmann, 1966; Kuprevich and Shcherbakova, 1971; Thornton et al., 1975). Because soil enzymes, as indeed the microorganisms themselves, are apparently closely associated with the mineral colloids, their activity is related to the type of clay which is dominantly present in the system (Estermann and McLaren, 1959; Durand, 1965; Skujinš, 1967; Stotzky, 1972; Marshall, 1975). Similarly, the processes of humification and mineralisation of organic materials in the soil, being partially biochemical in nature, are influenced by clays (Kononova, 1966; Allison, 1973; Haider et al., 1975; Lespinat et al., 1976).

It is instructive to recall that nearly all of the nitrogen in most soils is organically bound (Bremner, 1965, 1967; Parsons and Tinsley, 1975). A large proportion (up to 50%) of this nitrogen can usually be released as α-amino acids on acid hydrolysis of the whole soil or of its organic matter extracts. Although the form in which these amino acids occur is still a matter of some debate, evidence is accumulating that they are primarily derived from peptides and proteins (e.g. Paul, 1970; Ladd and Butler, 1975). Failure on the part of many workers to extract and isolate amino acid polymers from soils or detect their presence in soil by conventional biochemical techniques only lends further support for the view that the amount of free proteins in the soil solution is small as compared with that which is attached to soil colloids or incorporated into humus. This condition is again largely ascribable to microbial (i.e. enzymic) activity by which proteins and other nitrogenous polymeric compounds, such as chitin and nucleic acids, are fairly rapidly broken down when they enter the soil environment.

Being proteins, free or unattached enzymes present in the soil solution are likewise susceptible to rapid proteolysis (e.g. Aliev and Zvyagintsev, 1974). Unlike that of catalytically inactive proteins, however, the occurrence of a wide variety of enzymes in soil has been amply documented (Durand, 1965; Hofmann and Hoffmann, 1966; Skujinš, 1967). Indeed, a number of (active) enzymes, either as such or in combination with humus, have been successfully extracted from soils (Sequi, 1974; Ladd and Butler, 1975), leaving little doubt that complex formation with soil colloids does not generally lead to enzyme inactivation. Rather, through this process, soil enzymes and proteins tend to be protected from microbial attack. What is readily demonstrable and measurable, of course, is the activity of a given enzyme acting on its substrate under some specified conditions rather than its (absolute) quantity.

The inherent resistance to microbial decomposition of soil organic nitrogen is further indicated by the observation that only a small proportion (<5%) of this nitrogen can be mineralised, that is, converted into inorganic forms during the growing season (Bartholomew, 1965; Bremner, 1967). Another indicator of biological stability is that amino acid polymers as well as humic substances, in general can persist in the soil over long periods of time (Skujinš and McLaren, 1968, 1969; Paul, 1970; Scharpenseel, 1971, 1972; Goh, 1972). This property of soil organic nitrogen is clearly of great practical importance in that it influences the availability of nitrogen to plants and hence, the fertility status of the soil.

Two seemingly divergent hypotheses have been advanced to account for the persistence of amino compounds, notably proteins, in soils. One is that such compounds are stabilised by combining with other organic substances through a variety of sequential reactions which ultimately yield a mixture of polymeric materials. By its nature, this concept is inseparably bound to views on the nature and origin or soil humic compounds. Alternatively, stability is thought to be achieved by complex formation with the mineral constituents of soil among which the clay fraction, by virtue of its large and reactive surface, is by far the most important. Although we are primarily concerned with the experimental basis and implications of the clay—protein interaction, the salient features of the first hypothesis will be mentioned for the sake of completeness.

According to Russell (1956, p. 266) the idea that humic substances were essentially a stable complex between lignin and protein was suggested by Hebert and Dehérain as early as 1892. It was not until the early nineteen-thirties, however, that the so-called ligno—protein theory of humus formation attracted wide interest due to the work of Hobson and Page (1932) and more particularly, of Waksman and Iyer (1932) who envisaged complex formation to occur by a Schiff-base type reaction between amino groups of the protein and carbonyl groups of the lignin. The appeal of this theory was largely based on the fact that ligno—protein complexes, prepared in the laboratory, displayed properties resembling those of soil humic acids. In addition, these complexes are very resistant to microbial decomposition, more so in fact, than some clay—protein assemblages (Estermann et al., 1959). However, attempts to isolate ligno—proteins from soil have not met with much success, indicating that an appreciable proportion of amino acid nitrogen in soils is not associated with lignin. A variant of Waksman and Iyer's concept is the "ligno—ammonia" theory of Mattson and Koutler-Andersson (1942, 1943), support for which has also come from the observation that the nitrogen, combined in this way, is biologically and chemically stable. This theory, however, does not adequately account for the amino nitrogen which can be recovered from humic substances. A more viable alternative is the concept that amino compounds can react with phenolic and quinonoid substances and in so doing, give rise to dark-coloured, biologically stable complexes resembling humic compounds. For references and more detail than can reasonably be included here, the reader is referred to the appropriate review articles by various authors contained in the monographs edited by Bartholomew and Clark (1965), McLaren and Peterson (1967), and Gieseking (1975).

Besides lignin and its derivatives, tannins have been implicated as being capable of reacting with proteins to form complexes which resist microbial degradation. In particular, this process may contribute to the formation of a mor type humus in some acid soils, such as those which develop under a coniferous or heather cover (Handley, 1961). The tanning process probably involves hydrogen bond formation between amide groups of the protein and phenolic groups of the tannin, supplemented by covalent bonding between the two parent components (Gustavson, 1966; Ladd and Butler, 1975). Starkey and co-workers (Basaraba and Starkey, 1966; Benoit et al., 1968; Lewis and Starkey, 1968) who have paid particular attention to the chemical and biological stability of various tannin—protein combinations, found that condensed tannins generally yielded more stable complexes as compared with their hydrolysable counterparts. The stability of tannin—protein complexes was also influenced by the pH of the system, declining under alkaline conditions which presumably caused the complex to dissociate. More recently, Yavich et al. (1975) have reported similarly for complexes between some leaf tannins and casein.

Although a unified theory on the formation of humic substances in soils is yet to be formulated, the view that they arise essentially from the sequential condensation of aromatic, phenolic, and amino compounds with the possible participation of carbohydrates, has gained widespread acceptance (cf. Chapter 12 and Fig. 12.1). A detailed discussion of the chemistry and biochemistry of humification is clearly beyond the scope of this monograph. The vast volume of literature on this topic has been reviewed from time to time by a number of authors, notably Scheffer and Ulrich (1960), Kononova (1966), Stevenson and Butler (1969), Paul (1970), Felbeck (1971), Flaig et al. (1975), Haider et al. (1975). The "heteropolycondensation" concept is also consistent with what is known about the nature, composition, and properties of humic substances which comprise a mixture of acidic, high molecular weight, dark-coloured, heteropolymeric compounds (Dubach and Mehta, 1963; Felbeck, 1965; Hurst and Burges, 1967; De Borger, 1972;

Schnitzer and Khan, 1972). It is this heterogeneous, polydisperse nature of humic substances made up of non-repeating subunits which, to a large extent, accounts for their resistance to microbial (enzymic) degradation. When these materials adsorb onto clay mineral surfaces, as they would do in the soil, this resistance is further enhanced.

7.2. GENERAL ASPECTS OF THE CLAY—PROTEIN INTERACTION

The protective effect of clays on organic nitrogen compounds in particular is indicated by the finding that clay minerals, notably montmorillonite, can appreciably reduce the decomposition rate of crop residues, nucleic acids, and proteins (Bremner, 1965, 1967; Allison, 1973). In addition, fine-textured soils tend to contain more nitrogen than do their coarse-textured counterparts and a positive correlation exists between the organic matter and the clay contents of many soils. There is, however, little direct evidence for the occurrence in soil of clay—protein complexes. Failure to detect their presence is only partly ascribable to practical difficulties involved in isolating such complexes from the soil system. More important is the fact that proteins may largely be incorporated into humic substances which, in turn, are attached to clays so that the association between proteins and clay minerals is, to this extent, indirect.

On the other hand, complexes of clays with a variety of proteins and enzymes can readily be prepared under controlled conditions in the laboratory. Accordingly, what information we have on the clay—protein interaction has mostly come from a systematic examination of the factors affecting the formation of such "synthetic" complexes. The data obtained in this way are nevertheless of great value in that they provide an insight into the behaviour of proteins when they come into contact with inorganic soil colloids. By the same token, such information should not be directly and unreservedly extrapolated to field situations.

Since the early work by Mattson (1932) and especially by Ensminger and Gieseking (1939, 1941, 1942), a considerable volume of literature has accumulated on the chemistry of clay—protein reactions. Although the different sets of results are broadly concordant, they do not always agree in detail. On reflection, this is hardly surprising because of the many variables which are known to influence these reactions. Thus, besides being sensitive to the pH of the system, the adsorption of proteins by clays is dependent on clay pretreatment and solution conditions. Even the way in which the reactants are brought together may affect the properties of the resultant complexes. These variables, either singly or together, determine protein adsorption by influencing the charge characteristics of both the clay and the protein, the solubility and conformation of the protein, and the relative accessibility of the clay surface to the solute. For those reasons, the reactions of clays with proteins and with biopolymers, in general, can at best be described in semi-quantitative terms.

As indicated in Chapter 1, the ambient suspension pH influences the charge characteristics of clays. This effect is pronounced with the kaolinite-type minerals (cf. Fig. 1.3) and more so with allophanes whose surface charge is almost entirely pH-dependent. On the other hand, the charge characteristics of the smectite group of minerals are relatively insensitive to variations in suspension pH because of the high amount of isomorphous substitution within their structure. The pH of the medium, of course, has a profound effect on the acid—base properties of proteins. Being essentially polyampholytes, proteins may either be positively or negatively charged depending on whether the ambient solution pH lies on the acid or the alkaline side of the isoelectric — or more correctly — the isoionic point (e.g. Fruton and Simmonds, 1961). At this point, the sum of positive charges on the protein molecule resulting from protonation of its bases, is equal to that of negative charges due to dissociation of its acidic residues. The average net charge is then zero, that is, the molecule behaves as a dipolar ion.

Since crystalline layer silicates acting as adsorbents are normally negatively charged, protein uptake by minerals of this class would be expected to increase as the pH of the system falls away from the isoelectric point (p*I*). By the same token, little, if any, adsorption might be predicted at pH values in excess of p*I*. These expectations, however, do not accord with practice; instead, adsorption tends to attain a maximum at or near the isoelectric point of the respective protein. The possible reasons for this behaviour will be given later. Clearly, electrostatic interactions between solute and surface cannot fully account for the adsorption of proteins by clays. Among the other factors involved, Van der Waals (dispersion) interactions operative at all pH values are important. Being additive, their contribution to the overall adsorption energy may be quite large, particularly when on adsorption the molecule can adopt an extended conformation and so establish numerous points of contact with the silicate surface. However, this type of conformation at solid surfaces, frequently encountered with synthetic uncharged linear macromolecules (Greenland, 1972), does not appear to obtain generally with proteins and enzymes. This is so because unlike the former class of polymers, most proteins are globular and as such do not exist as randomly coiled strands in aqueous solutions. Rather, the molecules are organised into several levels of structural complexity.

Conventionally, we may distinguish a primary structure denoting the sequence of amino acids linked through peptide bonds in a polypeptide chain; a secondary structure arising from the hydrogen bond-induced coiling of this chain into a more stable conformation, such as the α-helix or the random coil; a tertiary structure describing the folding of this helix or coil into a compact, globular molecule stabilised by interfold hydrogen bonding, Van der Waals and hydrophobic interactions. Disulphide links in both the secondary and tertiary structures may further contribute to holding the molecule in a particular conformation (e.g. Martin, 1964). Such a structural organisation imposes severe internal constraints limiting the extent to which the molecule can re-arrange itself at the clay surface.

There is evidence to suggest that adsorption onto clay mineral surfaces can and does induce alterations in the (solution) conformation of proteins. Such changes normally affect only the most labile tertiary structure of the protein. However, under certain conditions some proteins can undergo extensive conformational modification giving rise, in the extreme situation, to a more or less two-dimensional collapse or extension of the polypeptide chain ("β-structure") at the interface.

Expanding 2 : 1 type layer silicate minerals of which montmorillonite is the supreme example, have the obvious advantage of being able to intercalate a large variety of proteins. This process enables the thickness and probable conformation of the adsorbed species to be deduced from adsorption and X-ray diffraction data. In the absence of intercalation as with kaolinite, or when interlayer penetration is precluded for steric reasons, these parameters can only be estimated indirectly using such techniques as ellipsometry and viscometry. Although ellipsometry has been successfully applied to solid—protein systems (e.g. Vroman and Adams, 1969; Fenstermaker et al., 1974), no attempt has so far been made to extend the method to clay—protein complexes. Similarly, viscometry and related techniques for estimating a hydrodynamic thickness of adsorbed polymer layers (e.g. Garvey et al., 1976) have not been developed to full potential in studying the properties of clay-adsorbed proteins (Srivastava and Chauhan, 1975). Surprisingly little use has also been made of infrared spectroscopy in view of the insight it has provided into the mechanisms underlying the interaction of clays with organic micromolecules (Mortland, 1970; Theng, 1974b) as well as into the mode of bonding and conformation of proteins and synthetic polymers at solid/liquid interfaces (Fontana and Thomas, 1961; Thies 1965; Morrissey and Stromberg, 1974).

As alluded to above, the interaction of proteins with clays under acidic conditions partly involves a cation exchange mechanism. It seems appropriate therefore to recall some features of exchange reactions in which the replacing species consist of large and bulky organic cations. One is that the reaction may be non-stoichiometric in that less than the equivalent amount of inorganic cations, originally present at the clay surface, may be replaceable by the organic cations. This "cover-up" effect (Hendricks, 1941) is acribable to the surface requirement of the organic species which may occupy or "cover" an area exceeding the area per exchange site ($\sim$0.8 nm^2 for montmorillonite). However, for polymers carrying a large number of positively charged groups along their chain or coil surface, this effect may be less pronounced than for mono- or divalent cationic organics because a single polycation may simultaneously replace a number of exchangeable inorganic cations from the silicate surface. On the other hand, the cations so replaced may be sterically hindered from diffusing out into the bulk solution, giving rise to an apparent cover-up effect.

Another feature of these systems and one which is of greater importance and relevance here than the cover-up effect, is the phenomenon usually referred to as layer segregation or demixing. This process gives rise to organic cation-rich and inorganic cation-rich layers which are randomly interstratified within a single montmorillonite crystal as the exchange reaction proceeds. Perhaps because the occurrence of layer segregation has only relatively recently been recognised (Barrer and Brummer, 1963; Theng et al., 1967; McBride and Mortland, 1973), it has often been overlooked in interpreting basal spacing data on montmorillonite—protein complexes. Since the basal spacing of the fully exchanged organic clay is normally greater than that of the parent inorganic clay, the spacings of the partially exchanged (up to ~55%) samples lie intermediate between the two end-members. However, layer segregation is most readily characterised by the increase in basal spacing of partially exchanged complexes as intercalation progresses. In addition, the $d(001)$ peaks are rather diffuse and there is a non-integral or irrational series of basal reflections. To complicate matters further, the layer charge of most expanding 2 : 1 type layer silicates is by no means homogeneous, that is, the cation density varies from one interlayer space to another (Lagaly and Weiss, 1975).

It is against this background that we shall attempt to rationalise the scattered data on clay—protein systems and, in so doing, hopefully present them as a more or less coherent body of information.

7.3. FORMATION AND PROPERTIES OF COMPLEXES

As remarked on earlier, Ensminger and Gieseking (1939, 1941) pioneered the systematic investigation into the reactions of clays with proteins. To this end, they prepared complexes of montmorillonite with gelatin and albumin by acidifying an alkaline (pH 10) suspension of the sodium clay containing difference amounts of protein. Examination of the dry complexes using X-ray diffractometry showed that interlayer uptake had occurred and, for a given clay/protein ratio, the extent of intercalation increased with a fall in suspension pH (from 7 to 2.7). Since this treatment would have increased the net positive charge on the protein, they inferred that complex formation primarily involved an exchange between the cationic (-NH_3^+) groups on the amino acid side chains of the protein and the Na^+ ions occupying exchange sites at the montmorillonite surface. Further support for the cation exchange mechanism was provided by the finding that prior treatment of the protein with nitrous acid, formaldehyde, or lignin, all of which would have deactivated the amino groups on the molecule, resulted in a marked decrease in the extent of intercalation. In addition, complex formation reduced the cation exchange capacity of the clay sample. They also noted that on raising the amount of protein in the system the interlayer space progressively expanded, with basal spacings up to ~5 nm being recorded. This effect may be explained in terms of layer segregation, enhanced by the probable polydispersity of the protein samples. Similar results were obtained with other proteins such as casein, haemoglobin, pancreatin, pepsin, and protamine.

In examining the uptake of casein and gelatin by montmorillonite, Chiang and Chao (1964) found that the extent of adsorption was related to the exchange capacity of the clay. In addition, the ease with which these proteins could replace the inorganic ions from the montmorillonite surface was influenced by cation valency, again suggesting that a cation exchange process was involved. More recently, Morgan and Corke (1976) have proposed similarly for glucose oxidase.

A more detailed study into the intercalation of proteins by montmorillonite was made by Talibudeen (1950, 1955) using X-ray diffraction techniques. The respective complexes were formed by immersing air-dry oriented flakes of Na^+ montmorillonite in aqueous solutions (usually 2% w/v) and at a pH below the isoelectric point, of the protein. Using this approach or some variations of it, interlayer complexes of a different kind were obtained. In the case of gelatin (pI ~ 4.5) at pH 2.5, for example, the basal spacings recorded were indicative of the presence of a discrete number of extended molecular layers in the interlayer space. What was interpreted as a two-layer complex with a basal spacing of ~1.77 nm and several higher orders of basal reflection, appeared to be the most stable arrangement. A one-layer complex with a basal spacing of ~1.4 nm was generally obtained when the amount of protein in solution was relatively low (<0.5 w/v) and the pH was greater than 3.5. Under special conditions and then only rarely, a spacing of 2.63 nm was recorded (in addition to the 1.77 nm line) due possibly to the formation of a four-layer complex.

Subtracting the assumed thickness of an individual silicate layer (0.95 nm) from the observed basal spacings gave Δ-values or interlayer separations of 0.45 and 0.82 for the one-layer and two-layer complex, respectively. These values, after allowing for the normal amount of "keying" (Theng, 1974b), were comparable to those obtained for single- and double-layer complexes with amino acids and small peptides (Greenland et al., 1962, 1965; Weiss, 1969). By analogy, the protein layers therefore appeared to adopt an extended, slightly buckled, β-type conformation with the amino acid side chains lying more or less flat on the interlayer surface. Assuming that the gelatin molecules consisted of single chain, monodisperse, random coils in solution (Finer et al., 1975), such a conformation at the clay surface would imply that the overall adsorption energy was sufficiently large to offset the loss of entropy accompanying the coil → extended chain transition. The combination of cation exchange, entropy effects associated with the displacement of surface-adsorbed water, and Van der Waals attractive forces, could conceivably meet this requirement.

The adopted β-keratin type structure on interlayer surfaces of montmorillonite, however, is apparently not confined to fibrous proteins such as gelatin. Under similar conditions, edestin and pepsin, both of which are globular, may also yield stable complexes with a basal spacing of ~1.8 nm. This would again suggest the presence of two layers of extended polypeptide chains in the interlayer space. In this instance, adsorption onto montmorillonite induced a conformational change affecting both the tertiary and secondary structure of the proteins. In the absence of adsorption and kinetic data we can only speculate on the manner in which the proteins entered the interlayer space. It seems reasonable to suppose that the molecules change shape just prior to or immediately after coming into contact with the accessible external crystal surface of the clay. Still possessing some mobility, the partially unfolded protein could then slowly penetrate the montmorillonite interlayers and here attain its final extended form. It is difficult to conceive how edestin and pepsin could otherwise form an interlayer complex of the type suggested for gelatin. Besides being able to account for the observed cover-up effect, such a mechanism is in keeping with the kinetic data of McLaren et al. (1958) who noted that although the initial uptake of proteins by montmorillonite was very rapid, many hours were required for the system to attain equilibrium.

Interestingly, Weiss (1969) also found that the basal spacing of montmorillonite complexes with a number of proteins (gelatin, gliadin, human albumin, egg albumin, pepsin, serum albumin, salmin, and zein) never exceeded 1.8 nm despite the fact that the amount adsorbed varied over a fairly wide range. He described the complexes as "co-precipitates" of anionic and cationic polyelectrolytes implying that protein uptake involved a cation exchange mechanism, but did not comment on the X-ray diffraction data. However, by analogy with the expansion characteristics of his own complexes with alkylammonium ions and amino acids, the observed spacings would indicate the intercalation of up to two layers of extended polypeptide chains as Talibudeen (1950) has proposed. Weiss further noted that for a given protein and at pH values in which the molecule carried a net positive charge, adsorption was greater for Na^+- than for Ca^{2+}-montmorillonite. Being more dispersed, that is, showing more extensive interlayer swelling to begin with, the sodium clay system would have offered a larger extent of accessible surface to the protein than did the divalent ion sample. In addition, the basal spacing of the complexes (e.g. with salmin) tended to increase with the amount adsorbed which, in turn, rose as the solution pH was increased from 2.7 to near the isoelectric point (~12.1) of the protein. Other examples of this behaviour will be discussed more fully below.

By combining X-ray diffraction measurements with infrared spectroscopy, Tarasevich et al. (1975) have been able to provide more direct evidence for the probability that proteins may adopt an extended β-type conformation in the interlayer space of montmorillonite. The adsorption isotherms were typically of the H-type, uptake increasing with a fall in pH from 8 to 3.6. Again, considerably larger amounts of protein were adsorbed by the sodium clay as compared with the calcium saturated sample. The complex which they examined in some detail was one between Na^+-montmorillonite and serum albumin (molecular weight ~ 65 000), formed at pH 3.5 and containing 250 mg protein per g clay. The infrared spectrum of the air-dry complex (basal spacing 2.3 nm) showed NH stretching bands at frequencies close to those given by polypeptides and albumin in the solid state. This observation was interpreted in terms of the formation of N—H···O = C hydrogen bonding between adjacent, more or less extended, polypeptide chains in the adsorbed albumin. Heating the complex to 423—573 K disrupted these interpeptide links, inducing the formation of N—H···O—Si bonding between the interlayer protein (or its decomposition products) and the silicate surface. The occurrence of certain absorption bands in the frequency range of 1800—1200 cm^{-1} provided further support for the view that adsorption onto montmorillonite led to the development (in the interlayer species) of a more pronounced β-type conformation at the expense of the α-helix structure, as compared with the protein in the solid state. This was particularly indicated by the presence in the spectrum of the complex of a component to the amide P band with a maximum at 1524 cm^{-1}, characteristic of albumin in a β-conformation. However, as Tarasevich et al. (1975) have stressed, adsorption merely caused a diminution in the proportion of albumin molecules having an α-helix character rather than a complete disappearance of such species.

In an attempt to test the biological stability of clay-adsorbed proteins, Pinck and co-workers (Pinck and Allison, 1951; Pinck et al., 1954; Pinck, 1962) prepared montmorillonite complexes with some proteins by a method similar to that used by Ensminger and Gieseking (1939). The complexes so formed were then incubated in sand or soil and the rate of protein decomposition was followed by measuring the amount of carbon dioxide evolved over a period of some weeks. By concurrently examining the complexes using X-ray diffractometry, the extent of decomposition could be related to the changes in basal spacing which took place during incubation. In keeping with Ensminger and Gieseking but unlike Talibudeen (1950), Pinck et al. (1954) noted that both gelatin and ovalbumin gave complexes which showed a gradual rather than a step-wise rise in basal spacing as the amount adsorbed increased. Thus, in the range of ~10 to ~55% (w/w) of protein present in the complex at pH ~ p*I*, basal spacings of ~1.5 to ~3 nm were recorded (Table 7.1). Some of the spacings (not included) were close in value to those reported by Talibudeen for his one-, two-, and four-layer complexes and were interpreted in similar terms. They further assumed that the complex with a basal spacing of ~1.5 nm and containing ~10% (w/w) protein,

TABLE 7.1

Basal spacing of some montmorillonite—protein complexes before and after four weeks' incubation with soil (from Pinck, 1962)

Protein	Amount in complex (%)	Amount evolved as carbon dioxide (%)	Basal spacing (nm)	
			Initial	After incubation
Gelatin	10.6	18.4	1.50	1.46
Ovalbumin	9.4	17.4	1.54	1.50
Gelatin	54.8	82.0	2.92	1.27
Ovalbumin	53.8	86.4	3.00	1.18

was one in which a single layer of protein molecules was present in the interlayer space. Since the basal spacing of this complex remained sensibly constant over the incubation period, the interlayer material was seemingly resistant to enzymic degradation (proteolysis). On this basis, the 18% or so of protein carbon (in the "single-layer" complex) evolved as CO_2 was assigned to the portion adsorbed into external crystal surfaces of the clay since the material in this position would have been readily accessible to extracellular enzymes. On the other hand, the interlayer protein in complexes assumed to contain two or more layers of molecules (35—54% w/w) was decomposed to about the same extent (~80%) as the protein which was added as such to the soil. At the same time, the basal spacing fell from an initial value of ~3 nm to ~1.2 nm at the end of the incubation period. Pinck (1962) suggested that the latter value might be due to the presence of interlayer NH_4^+ ions produced during the decomposition process. These observations led to the conclusion that extracellular enzymes in soils were incapable of penetrating the montmorillonite interlayers containing a single layer of adsorbed protein molecules. In other words, when the distance separating individual silicate layers in a crystal was of the order of 0.45 nm, enzymes were prevented from entering the interlayer space. Since the interlayer protein material in complexes with two or more intercalated protein layers was apparently degradable, a basal spacing of 1.8 nm or greater would be required if enzymes were to gain access to the interlayer space.

Earlier, Ensminger and Gieseking (1942) noted that albumin and haemoglobin adsorbed onto montmorillonite were largely resistant to hydrolysis by pepsin (acidic pH) and pancreatin (alkaline pH). By contrast, extensive protein decomposition occurred for the corresponding complexes with kaolinite. Similar results were reported by Birch and Friend (1956) for gelatin. Although the basal spacing of the (montmorillonite) samples were not recorded, it seemed probable that the protective effect of montmorillonite was related to its ability to intercalate the respective proteins, something that kaolinite did not possess.

What should be borne in mind, therefore, is that interlayer sorption per se does not invariably or necessarily impart immunity to the intercalated molecules against enzymic attack (Estermann and McLaren, 1959; Estermann et al., 1959), although the rate at which such species can be decomposed might be appreciably lower as compared with the material adsorbed on external crystal surfaces. Rather, steric factors related to the extent of interlayer expansion have a determining influence on the biological stability of intercalated proteins in montmorillonite complexes. This point will be taken up again below in connection with complexes involving lysozyme.

The interactions of clays with proteins and enzymes have been extensively studied by McLaren and his colleagues, and reported in a series of papers (McLaren, 1954a,b; McLaren and Estermann, 1956, 1957; McLaren et al., 1958; McLaren and Peterson, 1965; McLaren and Packer, 1970; Skujinš et al., 1959; Benesi and McLaren, 1975; McLaren, 1975). A good proportion of their work deals with the formation and properties of complexes with kaolinite. Besides being incapable of intercalating proteins and enzymes, kaolinite has a relatively low exchange capacity so that the extent to which proteins may be adsorbed is rather limited. This is not to say that the interaction process is any less complicated. However, to preserve the thread of our discussion, we shall direct our attention firstly to that part of their work for which montmorillonite has been used as the adsorbent.

One of the early papers on this topic was that by McLaren et al. (1958) who described the reactions of some proteins (chymotrypsinogen, gelatin, haemoglobin, β-lactoglobulin, lysozyme, and pepsin), dissolved in buffer solutions of different pH and ionic strength, with Na^{+}-montmorillonite suspensions. The process was characterised by determining the adsorption isotherms for the respective protein in function of pH and relating the basal spacing of the resulting complexes to the amount adsorbed. Like lysozyme, for which the isotherms are depicted in Fig. 7.1, the other proteins examined gave H-type curves showing a steep initial rise and levelling off to a long plateau at fairly low solute concentrations. Such isotherms are characteristic of high molecular weight compounds and macromolecules and indicate that the solute has a high affinity for the adsorbing surface (Giles, 1970). Besides being dependent on the ionic strength of the system, the maximum or plateau adsorption (A_{max}) at a given pH value is also influenced by the kind of buffer used to control the pH. Another feature which typifies the clay—protein and, indeed, the solid—protein interaction in general, is that the curve relating A_{max} to pH shows a maximum close to the isoelectric point of the solute (Kragh, 1964; Brash and Lyman, 1971; Samsonov, 1974). Fig. 7.2 illustrates this point for a number of proteins adsorbed onto montmorillonite.

The decrease in adsorption at pH > pI, represented by the "right branch" of the A_{max}—pH curve, is commonly ascribed to electrostatic repulsion between the protein anion and the negatively charged montmorillonite surface.

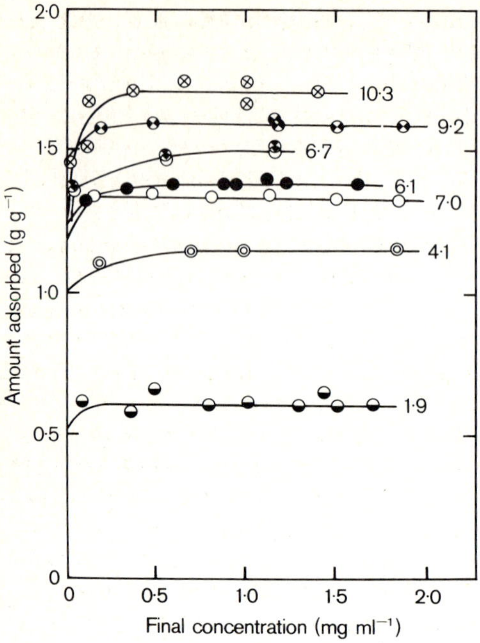

Fig. 7.1. Isotherms at 298 K for the adsorption of lysozyme by montmorillonite at the indicated pH values. All curves were determined at an ionic strength of 0.0375 except for the curve at pH 1.9 (ionic strength 0.053). The data for pH 1.9, 4.1, 6.1, and 6.7 were obtained in sodium citrate buffer; those at pH 7.0, 9.2 and 10.3 in sodium borate buffer (after McLaren et al., 1958).

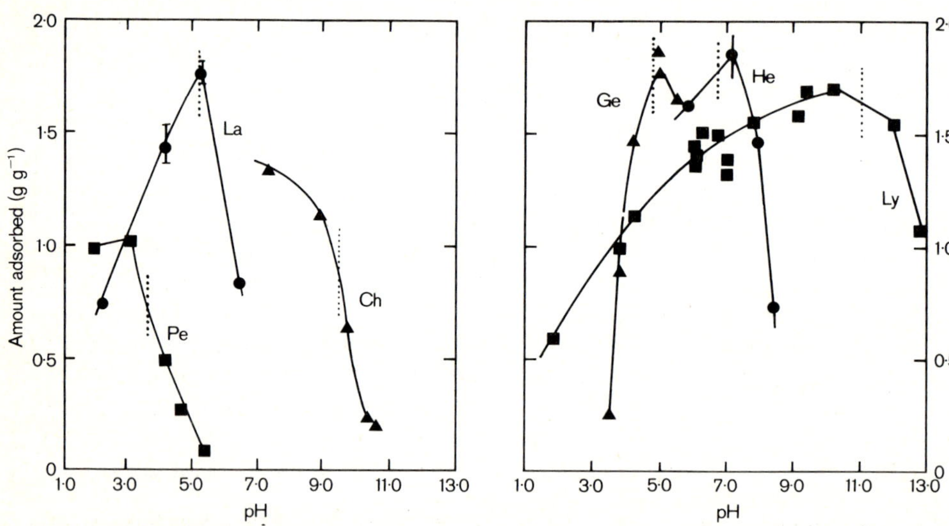

Fig. 7.2. Relationship between the maximum extent of adsorption and suspension pH for various proteins, adsorbed by montmorillonite. The ionic strength was 0.0375. Vertical dotted lines indicate the isoelectric point of the respective proteins. Ch, chymotrypsinogen; Ge, gelatin; He, haemoglobin; La, β-lactoglobulin; Ly, lysozyme; Pe, pepsin (after McLaren et al., 1958).

This effect is further enhanced by the cations in solution (e.g. Na^+) competing with the protein for the surface. Several explanations have been put forward to account for the "left branch" of the curve, that is, the apparent anomaly that adsorption decreases rather than increases with falling pH below the isoelectric point. One is that the protons in solution compete with the protein cation for exchange sites at the clay surface (McLaren et al., 1958). Alternatively, less protein is required to satisfy the negative charges on the clay as the total number of positively charged groups on the protein increases with falling pH (Armstrong and Chesters, 1964). In addition, the solubility of most proteins is lowest near their isoelectric point since intramolecular electrostatic interactions would be maximal at this point, causing the (dipolar) molecule to adopt the most compact conformation. Accordingly, protein adsorption reaches a peak at pH ~ pI and falls away at either side of the isoelectric pH. It is interesting to note that chymotrypsinogen does not show this behaviour (Fig. 7.2) in accord with the finding that this protein is not precipitated — as are the others — at its generally accepted isoelectric pH (~9.5) but rather near pH 5.4 (Fruton and Simmonds, 1961).

The variation in basal spacing of the lysozyme complexes in function of the amount adsorbed is reproduced fully in Fig. 7.3 because its interpretation has become the subject of a polemic between Laura (1975, 1976) on

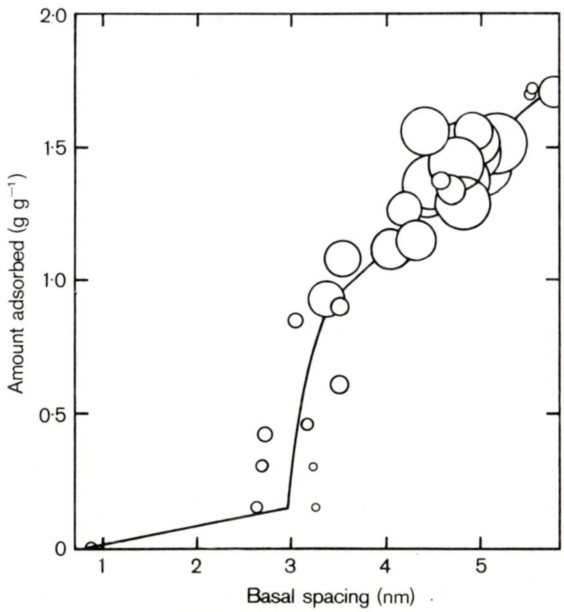

Fig. 7.3. Variation in basal spacing with the amount of lysozyme adsorbed by montmorillonite. Note that the pH varied from 1.9 to 10.3 and the ionic strength from 0.0375 to 0.1375 with most points obtained at pH 6.6 and an ionic strength of 0.0375 (after McLaren et al., 1958).

the one hand, and McLaren and Barshad (1976) on the other. The explanation offered by McLaren et al. (1958) was that interlayer expansion occurred in two steps of approximately equal magnitude as intercalation progressed. The first one took the spacing from ~1 nm for the parent (untreated) clay to ~2.7 nm (corresponding to an amount adsorbed of 0.15—0.8 g protein per g clay). This was followed by a second expansion step which increased the spacing from ~2.7 to ~4.4 nm (for complexes containing 1.0—1.6 g g^{-1}). The 2.7 nm spacing was thought to arise from random interstratification within a crystal of fully expanded layers ($d(001) = 4.4$ nm) and fully collapsed layers ($d(001) = 1.0$ nm) in a 1 : 1 proportion, that is, $(4.4 + 1.0)/2 = 2.7$. In other words, they ascribed the 4.4 nm spacing to a complex in which a single layer of (globular) lysozyme molecules was present in the interlayer space.

It is this interpretation which has been called in question by Laura (1975). Basing his arguments on the data of Talibudeen (1955) and Weiss (1969), discussed earlier, Laura suggested that a spacing of this magnitude could only be caused by the intercalation of multilayers of lysozyme molecules. This suggestion was made in the context of the hypothesis that the decomposition of soil organic matter could be explained in chemical terms, more particularly in terms of the protolytic action of water (Laura, 1974). Being mobile, multilayers of clay-adsorbed organics are thought to be susceptible to protolysis (by water) whereas the thermodynamically stable monolayer arrangement is not. This view runs counter to the finding of Estermann et al. (1959) who reported that interlayer lysozyme, in what they believed to be a one-layer complex with montmorillonite, was to a large extent enzymically degradable. For this reason, the basal spacing of montmorillonite—protein systems, in terms of the shape and thickness of the intercalated species, has become a point of issue (Laura, 1976; McLaren and Barshad, 1976).

Whilst we do not wish to enter into the polemic, we are obliged to make a few comments on this point, if only to put matters in perspective. That the decomposition of soil organic matter may be chemically induced and proceed without the participation of microorganisms, is, of course, quite feasible. We have already referred to the ability of clays to activate a variety of organic reactions. Nevertheless, it seems incontestible that the decomposition of organic compounds, at least in unsterilised soil systems, is essentially an enzymically mediated process. As Stotzky and Kunc (1975) have pointed out in their comment on Laura's hypothesis, the microbial or enzymic concept of degradation is supported by a large, solid body of experimental evidence. Admittedly, the data concerning the protective effect of clays on adsorbed organics have not always been concordant. Such inconsistencies, however, are probably inevitable because of the inherent complexity of the system and the interplay of many independent variables. The protolytic action of water is but one of these variables, however important a part it may play in the overall process.

Our second comment relates to the conformation of proteins adsorbed at interlayer surfaces of montmorillonite. A one-layer complex with a basal spacing exceeding 4 nm would imply that the interlayer protein retains its gross native globular shape, that is, little molecular uncoiling does not occur on adsorption. In view of the tendency of many proteins to unfold and spread out as a monolayer at the air/water interface (e.g. Fraser, 1957), such an interlayer conformation may, at first sight, seem unrealistic. However, there is good evidence to indicate that adsorption onto solid surfaces does not commonly result in extensive uncoiling of the molecule (e.g. Brash and Lyman, 1971). Indeed, the immobilisation of enzymes by attachment to solid substrates or their chromatographic separation and fractionation using solid supports would be impractical if the adsorbed enzymes were to unfold and stretch at the surface since this would lead to a loss in activity. Similarly, the fact that enzymes in soil can express their specific activity would suggest that no drastic structural alterations occur when they interact with soil colloids. Lysozyme, in particular, is likely to retain its native globular shape at the clay surface since the molecule is held together by four intramolecular disulphide bridges. Moreover, the eluted protein showed little or no reduction in enzyme activity (Fevold, 1951; McLaren, 1954b), again suggesting that it does not unfold appreciably on adsorption by clays.

On these grounds, an interlayer complex with montmorillonite containing a monolayer of (globular) lysozyme molecules can acceptably have a relatively high basal spacing. If so, the degradability of the interlayer material ceases to be problematical since it would be accessible to extracellular enzymes in accordance with Pinck's (1962) observation. Nevertheless, the X-ray diffraction data of McLaren et al., depicted in Fig. 7.3, are amenable to a different explanation. Further, the choice of a 4.4 nm spacing for such a complex corresponding to a monolayer capacity of 1.25 g g^{-1} seems somewhat arbitrary. Inspection of Fig. 7.3 shows that the interlayer expansion is more gradual than step-wise as adsorption progresses although the presence of two expansion stages is indicated. The first stage is more or less discernible and appears to end at a spacing of ~3.3 nm but the second one is less distinct, possibly terminating near or slightly below 6 nm. The gradual nature of the expansion with rising amounts adsorbed is particularly noticeable as the second stage develops beyond a spacing of ~3.3 nm. It is also apparent that the basal peaks in this region were rather broad, indicating an irrational series of basal reflections. On the basis of this analysis we may presume that the formation of a one-layer complex was completed at a basal spacing of ~3.3 nm giving a monolayer capacity of ~0.85 g g^{-1}. As uptake continued beyond this level, a second layer of protein formed in the interlayer space, attaining completion at ~1.7 g g^{-1}. Basal spacings recorded in the range of 1—3.3 nm and 3.3—6 nm probably arose from segregation of 0 and 1 and of 1 and 2 molecular layers within a single crystal, respectively. If so, the 4.4 nm spacing did not so much represent a monolayer complex as one in which single and double layers of molecules co-existed. The observed spread of basal spacing for a given amount present also reflected the effect on molecular conformation of pH and ionic strength, neither of which was kept invariant. This only added to the difficulty in selecting the point which marked the completion of a monolayer coverage.

That montmorillonite may intercalate more than one layer of lysozyme molecules depending on the solution concentration and pH, is further indicated by the data of Armstrong and Chesters (1964). Using a sample of Mg^{2+}-montmorillonite as the parent material, they were able to prepare interlayer complexes with lysozyme giving basal spacings of 3.5, 5.9, and 7.4 nm. In keeping with the finding of McLaren et al. (1958), they noted that the initial reaction was rapid, over 90% of the added lysozyme being adsorbed within the first hour, and uptake continuing up to 12 h after mixing the protein with the clay suspension. The isotherms were again of the H-type, plateau adsorption (A_{max}) increasing with rising pH and reaching a maximum near the isoelectric point ($pI \sim 10.5$). The basal spacings quoted above were for complexes formed at pH 10.5 at three (increasing) levels of protein/clay ratio, corresponding to A_{max} values of 0.93, 1.91, and 2.45 g g^{-1}, respectively. In this instance, however, the basal peaks were sharp and at least one higher (rational) order of basal reflection was recorded.

The apparent absence of layer segregation and the persistence of the 3.5 nm line in the X-ray diffractograms strongly suggest that the first or lowest level of expansion ($d(001) \sim 3.5$ nm) represents the formation of a single-layer complex for which the monolayer capacity is 0.93 g g^{-1}. This suggestion is in line with our interpretation of the data reported by McLaren et al. (1958), shown in Fig. 7.3. The observed monolayer thickness (Δ-value) of ~2.5 nm may be compared with the value of ~3.3 nm for the diameter of an equivalent sphere, derived by McLaren and co-workers (McLaren et al., 1958; McLaren and Peterson, 1961), indicating that lysozyme adsorbed in the montmorillonite interlayers is more ellipsoidal than spherical in shape. Of the many possibilities examined by Armstrong and Chesters (1964), a 2.5 nm × 3.2 nm ellipsoid with the shorter axis perpendicular to the silicate layer, appears to satisfy both the X-ray diffraction and adsorption data. By the same token, the complex with a basal spacing of 5.9 nm (Δ-value ~ 4.9 nm) and containing 1.91 g g^{-1} can be identified with the presence of a double layer of intercalated lysozyme molecules adopting essentially the same orientation in the interlayer space. However, the 7.4 nm complex (Δ-value ~ 6.4 nm) containing 2.45 g g^{-1} cannot be explained in terms of the intercalation of three molecular layers of lysozyme since the extent of interlayer expansion and adsorption for such an arrangement would be three times that for the monolayer complex. Rather, a basal spacing of 7.4 nm at the indicated capacity can reasonably be ascribed to the presence in the interlayer space of a double layer in which the molecules are arranged with their shorter axis parallel to the silicate surface. The three probable modes of orientation of interlayer lysozyme consistent with basal spacings of 3.5, 5.9, and 7.4 nm or Δ-values of 2,5, 4.9, and 6.4 nm, respectively, are shown in Fig. 7.4. A summary of the calculations taking $750/2 = 375\ m^2\ g^{-1}$ as the available interlayer surface of montmorillonite and assuming no overlap between close-packed molecules, is given in Table 7.2.

It would therefore appear that lysozyme adsorbed in the interlayer space of montmorillonite, although not as asymmetrical in shape as its counterpart in aqueous solutions (Young, 1963), is by no means spherical. It should be borne in mind, however, that the molecular dimensions as set out in Table 7.2, refer to solution conditions near the isoelectric pH at which, as indicated earlier, the molecules would attain their maximum compactness. The change in orientation from one in which the shorter axis of the molecule is perpendicular (Fig. 7.4B) to one in which this axis is parallel to the surface (Fig. 7.4C) as the amount adsorbed increases, is not uncommon. Such a reorientation from a "side-on" to an "end-on" arrangement resulting in more surface being exposed to the solute, has been shown to occur in many complexes with small, well-defined organic species (Theng, 1974b).

Using the published values for partial specific volume and molecular weight of the respective proteins (Edsall, 1953) and taking a slightly larger interlayer surface area for montmorillonite, McLaren and Peterson (1961)

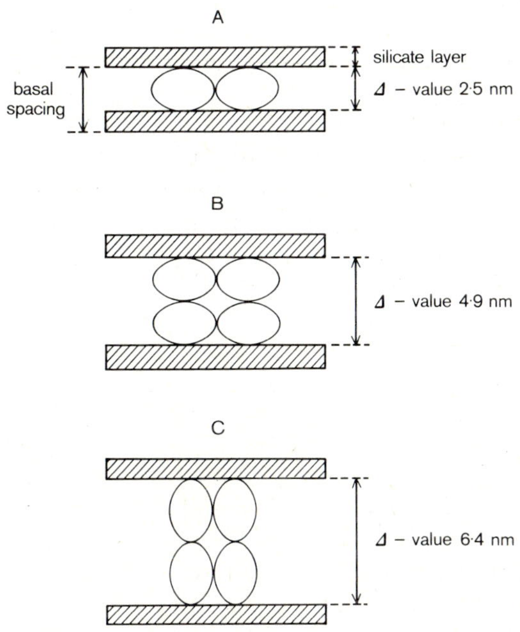

Fig. 7.4. Possible orientations of lysozyme molecules in the interlayer space of montmorillonite to give the observed interlayer spacings or Δ-values (Redrawn from Armstrong and Chesters, 1964).

TABLE 7.2

Adsorption and X-ray diffraction data on montmorillonite—lysozyme complexes formed at pH 10.5 in the presence of 0.02 M borate buffer. The probable organisation of lysozyme molecules in the interlayer space, deduced from these data, is also indicated (from Armstrong and Chesters, 1964)

Amount added [a]	Amount adsorbed [a]		Basal spacing (nm)	Δ-value [b] (nm)	Axial lengths oriented parallel to silicate layer (nm)	Nature of complex
	Observed	Calculated				
1.0	0.93	0.92	3.5	2.5	3.2 × 3.2	Single layer
2.0	1.91	1.84	5.9	4.9	3.2 × 3.2	Double layer
4.0	2.45	2.36	7.4	6.4	2.5 × 3.2	Double layer

[a] g protein per g clay.
[b] These values are only approximate, being obtained by subtracting the thickness of an individual montmorillonite layer, taken as 1 nm, from the observed basal spacing.

have made similar calculations for a number of proteins, as set out in Table 7.2 for lysoszyme. Assuming that the intercalated proteins were essentially spherical, covering an area equal to their cross-section, the measured interlayer separation (Δ-value) of the complexes was found to agree reasonably well with the corresponding (calculated) molecular diameter. This led them to propose that montmorillonite might be used as a caliper for the size of protein molecules. However, in view of the ability of montmorillonite to intercalate more than one layer of protein and the uncertainties as to the shape and orientation the molecules may adopt in the interlayer space, the use of montmorillonite for this purpose is only of limited applicability. Moreover, as they themselves pointed out, some proteins like pepsin failed to intercalate without undergoing extensive unfolding whereas haemoglobin and β-lactoglobulin seemed to be intercalated as a monomer and a dimer, respectively.

What is important to appreciate at this point is that adsorption onto clay mineral surfaces does not invariably lead to a drastic change in the solution conformation of protein molecules. Indeed, extensive molecular unfolding or uncoiling may be more the exception than the rule. Accordingly, unless the adsorption data would indicate otherwise, basal spacings in excess of 3 nm might be expected to obtain for montmorillonite complexes with the majority of proteins, containing a single layer or at most a double layer, of molecules in the interlayer space.

The essential preservation of molecular conformation when proteins adsorb from solutions onto solid surfaces has also been found in systems other than clays. Using monodisperse polystyrene latex particles as the adsorbent, Deželić et al. (1971) obtained a value of about —9 kcal mol^{-1} for the free energy of binding of human serum albumin to the latex surface. Despite this relatively high affinity of the solute for the solid substrate, some expansion of the tertiary structure of the protein appeared to occur only in that part of the molecule which was in actual contact with the surface. On the other hand, the portion which was oriented towards the solution phase seemed to retain its native conformation since the adsorbed protein remained immunochemically active. More direct evidence for the conformational stability of adsorbed proteins has been provided by the elegant studies of Stromberg and his associates (Fenstermaker et al., 1974; Morrissey and Stromberg, 1974). Using infrared difference spectroscopy and ellipsometry, these workers obtained a bound fraction (p) of ~0.11 for the carbonyl groups of bovine serum albumin and prothrombin at the silica surface irrespective of the amount adsorbed. This value is only 1/5—1/3 of that predicted and commonly observed for linear uncharged polymers which exist as random coils in solution but tend to unfold and collapse at the surface (Silberberg, 1962a,b, 1967; Greenland, 1972; cf. Chapter 2). The relatively low value of p together with molecular extension data derived from ellipsometric measurements indicate that these blood proteins retain their essential native conformation in the adsorbed state. Apparently, their intramolecular bonding is sufficiently strong to maintain the (globular) tertiary structure of the molecule. The results for albumin and prothrombin adsorption onto chrome, platinum, carbon, and polyethylene could be explained in similar terms, that is, that conformational changes occurring on adsorption are relatively small. This conclusion is further supported by computer simulation studies using lysozyme. Other examples illustrating this point have been described by Brash and Lyman (1971).

The question arises as to why some proteins, notably gelatin and pepsin, can apparently unfold when they adsorb at the interlayer surface of montmorillonite. The reason for this behaviour may be sought in the solution properties of the proteins concerned as much as in the experimental approach used in forming the clay—protein complex. Being fibrous, gelatin would tend to adsorb side-on, that is, with its shorter axis perpendicular to the silicate layer (cf. Fig. 7.4 A and B). This enables many contacts to be established with the surface, inducing the molecule to unfold and stretch. Likewise, pepsin tends to unfold and become a more or less linear polyanion in solution whose ambient pH lies below the isoelectric point of the protein (McLaren and Lewis, 1950). This behaviour may be ascribed to a high electrostatic repulsion between cationic groups on the molecule. Similarly, the ability of bovine serum albumin to unfold at the solid/liquid interface is influenced by the solution pH and protein concentration (Bull, 1956; Chattoraj and Bull, 1959; MacRitchie, 1972). In the final analysis, the mode of conformation which a protein would adopt at the interface is determined by the balance between internal forces tending to maintain the structural integrity of the molecule and the affinity the molecule has for the adsorbing surface.

Leaving aside the effect of adsorption on protein conformation and stability, the question may also be asked as to what determines the relative layer ordering in montmorillonite—protein complexes. The examples which we have cited from the literature seem to indicate that layer segregation, giving rise to randomly interstratified structures within a crystal, occurs when the protein solution is mixed with a suspension of montmorillonite in the sodium form (Ensminger and Gieseking, 1939, 1941; Pinck et al., 1954; McLaren et al., 1958). On the other hand, little or no segregation appears to take place when air-dry, oriented flakes of the Na^+-montmorillonite are used (Talibudeen, 1950, 1955) or when the protein is added to a divalent ion clay suspension (Armstrong and Chesters, 1964).

Montmorillonite samples saturated with small inorganic cations (H^+, Li^+, Na^+) normally show extensive interlayer (i.e. intracrystalline) expansion in water and in dilute (<0.3 M) aqueous solutions of the respective cation salt (cf. section 1.2). Their X-ray diffraction patterns are characterised by the presence of a broad diffuse basal peak, corresponding to an average interlayer separation (Δ-value) of 2 nm or greater. On the other hand, samples containing polyvalent exchangeable cations do not swell beyond a Δ-value of ~1 nm, even in distilled water and their basal lines are generally sharp and well-defined (Norrish, 1954). The answer to our question must therefore lie in the stacking mode of the silicate layers making up a crystal. In an aqueous low-salt or salt-free suspension of Na^+-montmorillonite, these layers tend to be displaced relative to each other and separated by distances which are relatively large but by no means uniform. Although protein molecules would therefore not meet with much steric hindrance as they penetrate the interlayer space, the mutual alignment of the silicate layers in the (final) complex would remain somewhat disordered. At a concentration in excess of what is required for monolayer coverage, some interlayers may contain a single layer, and others a double layer, of protein molecules. The variation in basal spacing as a function of the amount adsorbed might therefore be expected to show the behaviour depicted in Fig. 7.3. On the other hand, for air-dry oriented Na^+-montmorillonite flakes and polyvalent ion clay systems, the stacking of the silicate layers in a crystal is relatively ordered. Whilst this arrangement would not preclude layer segregation from occurring during protein intercalation, more ordered single- and double-layer complexes would be expected to form by comparison. However, because the interlayer swelling of the parent clays is limited, little intercalation would occur unless the protein can uncoil as it apparently did for Talibudeen (1950, 1955) and Weiss (1969). The fact that lysozyme was capable of intercalating into a Mg^{2+}-montmorillonite as a globular molecule (Armstrong and Chesters, 1964) would indicate that some interlayer swelling of the sample preceded intercalation of the protein. It seems reasonable to suppose that this pre-expansion was induced by the replacement of interlayer Mg^{2+} ions for the Na^{+} ions supplied by the (borate) buffer solution. The X-ray diffraction data of Harter and Stotzky (1973) who used unbuffered systems tend to support our interpretation.

It ought perhaps to be interpolated here that layer segregation may not occur until the complexes are dried prior to X-ray diffraction analysis, as Vansant and Uytterhoeven (1972) have argued. There is good evidence to suggest, however, that segregation would obtain in both wet and dry montmorillonite systems, unless the wet clay is more fully expanded in suspension, in which case cation size is less important than the extent of interlayer expansion (McBride and Mortland, 1973). This proviso would apply to the aqueous Na^{+}-montmorillonite system at low protein concentration and pH. Homogeneity might thus be expected to obtain until drying restricts the uniform distribution of protein cations in every layer throughout the montmorillonite crystal. In view of all these considerations, some caution should be exercised in interpreting basal spacing data of clay—organic complexes in terms of the thickness and disposition of the interlayer molecule, especially for systems involving macromolecular compounds.

Nevertheless, there is little doubt that, despite their bulky size and globular shape, proteins and enzymes can penetrate the interlayer space of 2 : 1 type layer silicates under appropriate (laboratory) conditions. Adsorption is confined to external crystal surfaces in situations where interlayer expansion of the parent clay is either absent, as in kaolonite, or limited, as in halloysite, and in montmorillonite containing polyvalent exchangeable cations. However, there is as yet no unequivocal evidence for the occurrence in soil of interlayer montmorillonite—protein complexes. Even if such complexes were to form in montmorillonitic soils, the interlayer protein would generally be accessible to and hence fairly rapidly degraded by soil enzymes (Estermann and McLaren, 1959; Estermann et al., 1959). Only if the basal spacing of the complex is limited to ~1.4 nm because the intercalated molecule can unfold, or has fallen to this value after partial decomposition of the interlayer protein (cf. Table 7.1), might the adsorbed species be protected against enzymic attack (Pinck, 1962) or protolysis by water (Laura, 1975).

We have already seen how the exchangeable inorganic cations at the montmorillonite surface may influence protein uptake. Besides being directly involved in exchange reactions with the protein, the nature of these cations influences the extent of interlayer swelling of the clay and hence, the ease with which the protein may be intercalated. These and related aspects have been more closely examined by Harter and Stotzky (1971, 1973). Their approach was to react montmorillonite samples containing different exchangeable cations (H^{+}, Na^{+}, Ca^{2+}, Al^{3+}, La^{3+}, Th^{4+}) with a number of proteins, such as α-casein, catalase, α-chymotrypsin, edestin, β-lactoglobulin, lysozyme, ovalbumin, ovomucoid, and pepsin, covering a wide spectrum of molecular weight (14 000—310 000) and isoelectric point (2—11). Adsorption and retention ("binding") isotherms were determined, the latter being obtained by measuring the amount of protein remaining in the respective complexes after repeated washing with distilled water. The use of buffers was avoided so that the initial and final pH of the system varied somewhat with the clay sample and with the protein species. However, the pH of an aqueous suspension of the various clay—protein complexes was generally close to neutrality.

Positive uptake was recorded for all protein—clay combinations, even when the net charge on the solute molecule was negative. The binding curves indicated that uptake depended on the valency of the cations initially present at the clay surface, decreasing in the order $H^+ > Na^+ > Ca^{2+} > Al^{3+} > La^{3+} > Th^{4+}$. This behaviour reflects the decreasing ability of a given protein to replace the inorganic cations from their exchange sites as well as the relative accessibility of the interlayer surface to the solute. The position of Na^+ relative to Ca^{2+} in the series is consistent with the results of Weiss (1969), Tarasevich et al. (1975), and Morgan and Corke (1976). X-ray diffraction analysis indicated that, with the exception of catalase, there was little or no steric hindrance to intercalation (for casein, chymotrypsin, lysozyme, and ovomucoid) at clay/protein ratios >1/5 with H^+- and Na^+-montmorillonite. Failure of catalase to intercalate was undoubtedly related to its relatively high molecular weight and hence its large (native) molecular dimension. Interestingly, the lysozyme complex with the monovalent ion clay samples gave a basal spacing of ~3.5 nm, close to the value reported by Armstrong and Chesters (1964) for their single-layer complex with Mg^{2+}-montmorillonite. On the other hand, the lysozyme complex with Ca^{2+}-montmorillonite of Harter and Stotzky (1973) failed to expand beyond a $d(001)$ spacing of ~1.9 nm. This observation lends further support to our suggestion that the Mg^{2+} ions in Armstrong and Chesters' example, exchanged for the Na^+ ions supplied by the buffer solution, enabling the interlayers to expand and accommodate the lysozyme. Thus, in the absence of buffers none of the proteins used (except possibly for casein with La^{3+}-montmorillonite) appeared to be intercalated into the polyvalent ion clays, at all protein/clay ratios, a behaviour which may be ascribed to steric effects associated with the limited interlayer expansion of the parent clay samples.

The effect of molecular weight and size was also reflected in the extent of adsorption. If the amount adsorbed was expressed on a w/w basis, adsorption rose with molecular weight although the magnitude of the increase was not proportional to molecular weight. However, on a mol g^{-1} basis, uptake tended to increase with falling molecular weight. This finding was explained in terms of the facility with which the protein could establish contact with the clay surface which, in turn, was influenced by steric factors and by the number of interacting groups on the molecule. It seems likely, however, that the smaller, more compact molecules can effect a more intimate contact with and closer packing at the surface (without necessarily being more unfolded) and so can occupy a greater area of the mineral as compared with their larger counterparts. Such a mechanism has been suggested by Chernyakhovskaya and Runova (1975) who observed a similar relationship between uptake and molecular weight for a number of proteins adsorbed by colloidal suspensions of $Al(OH)_3$, $Ca_3(PO_4)_2$, and $Al_2(PO_4)_3$.

In another paper, Albert and Harter (1973) described the interactions of lysozyme and ovalbumin with different layer silicates, such as biotite-vermiculite, illite, kaolinite as well as montmorillonite. Again, no buffers were used, the suspension pH being varied by titrating the acid-washed clays with NaOH. In agreement with previous workers (McLaren et al., 1958; Armstrong and Chesters, 1964; Weiss, 1969), adsorption increased as the ambient pH rose to the isoelectric point of the respective protein. Following Armstrong and Chesters, they explained this finding in terms of a decrease in the net positive charge on the protein as the pH was increased, resulting in more solute being required to satisfy the negative charge on the mineral. As pointed out earlier, the decrease in protein solubility and the adoption of a more compact conformation of the molecule in solution must also contribute to the observed increase in adsorption as the pH approaches the isoelectric point. With the possible exception of kaolinite, an ion exchange mechanism was involved since the amount of sodium ions released into solution at pH $\sim$7 was proportional to the quantity of protein retained by the clay samples. Less sodium, however, was released from biotite-vermiculite and illite as a proportion of the respective exchange capacity, as compared with montmorillonite. This indicated a cover-up effect which, in montmorillonite, with its lower charge density would be relatively unimportant.

Harter (1975) has extended this work to montmorillonite saturated with different cations, viz. the first five of the alkali and alkaline-earth series as well as H^+, Al^{3+}, and La^{3+}. In an attempt to relate the extent of adsorption (A) to ionic properties, a multiple correlation technique was applied to the data, which for lysozyme yielded the following relationship,

$$A = 290.2 - 5.5 \text{ (valency)} - 96.8 \text{ (ionic radius)} - 114.3 \text{ (Pauling electronegativity)} - 6.5 \text{ (ionisation potential)}$$

Values of A were derived from the inverse slope of the straight lines obtained when the data were plotted according to the linear transform of the Langmuir equation. It should be noted that the isotherms (for lysozyme) were more of the H-type than of the Langmuir form. Moreover, the values for both electronegativity and ionisation potential strictly refer to the corresponding atoms or elements rather than to ions. Even if the above relationship were generally valid, it tells little of the mechanisms underlying the adsorption process.

The finding that adsorption decreased with an increase in ionic radius (for cations of equal valency) could partly be ascribed to variations in the extent of the accessible surface. Thus the sodium clay adsorbed nearly five times as much lysozyme as did the cesium sample. The importance of accessibility is further illustrated by the following observations. Firstly, uptake by H^+-montmorillonite depended on the length of time during which the clay sample was stored ("aged") prior to being reacted with the protein. Prolonged ageing generally led to a decline in sorptive capacity. This is due to the slow conversion of hydrogen clays to H^+/Al^{3+} forms since acid-treated montmorillonite is essentially unstable, the protons readily displacing octahedrally-coordinated Al^{3+} ions from the interior of the silicate layers to interlayer exchange sites (Coleman and Craig, 1961; Eeckman and Laudelout, 1961). Secondly, adsorption was progressively reduced by adding increasing amounts of a neutral electrolyte to the system. Thus, in the presence of 2 M NaCl, uptake by Na^+-montmorillonite was only half as much as in NaCl solutions of $\leqslant 0.1$ M. In the former situation, the diffuse double layers on interlayer surface were presumably so compressed as to make them inaccessible to lysozyme. In addition, the solubility of the protein would, if anything, be enhanced as the NaCl concentration is raised. This salting-in effect of 1 : 1 electrolytes extending to high ionic strengths (e.g. Mahler and Cordes, 1968) would contribute to the observed decrease in adsorption. By contrast, Armstrong and Chesters (1964) found that addition of KCl or $NaNO_3$ solutions (<0.02 M) to the system at $pH < pI$ increased rather than decreased the uptake of lysozyme by Mg^{2+}-montmorillonite. They suggested that the added anions might partially screen the positive charges on the protein. However, the exchange of Mg^{2+} ions initially present at the clay surface by K^+ or Na^+ ions, as suggested before, seems to be a more plausible explanation.

Unlike lysozyme, ovalbumin yielded linear isotherms and its adsorption was generally insensitive to variations in the nature of the exchangeable cation. Thus, with the possible exception of the H^+- and Be^{2+}-saturated samples, the slope of the curves was similar in all instances, an observation for which no satisfactory explanation was given. It does not seem improbable, however, that ovalbumin can unfold reversibly, as the X-ray diffraction data of Weiss (1969) would indicate. If this were the case, the unfolded species could wedge its way into the montmorillonite interlayers and in so doing expose more sites to the following molecules. The nature of the saturating cation would then be relatively unimportant in determining the extent of adsorption. Linear isotherms, however, can obtain without being accompanied by interlayer sorption. The chief requirement for linearity which is that fresh sites are being created as adsorption progresses, could equally be met by penetration of the intercrystalline pores by the unfolded (globular) ovalbumin molecules. This process presupposes that the clay mass is made up of "domains" which are regions of oriented clay crystals (Aylmore and Quirk, 1960). Intercrystalline or intradomain expansion induced by solute uptake has been proposed to account for the behaviour of humic and fulvic acids at montmorillonite surfaces (Theng and Scharpenseel, 1975; Theng, 1976).

Desorption studies by Harter (1975) indicated that only a small proportion (<10%) of adsorbed lysozyme and ovalbumin could be eluted from their complexes with water or moderately concentrated (up to 2 M) solutions of sodium chloride, sodium acetate, and ammonium acetate (pH 4.8 and 7). This finding further supports the view that interactions other than ion exchange are involved in the adsorption process. Similarly, Brash and Lyman (1969) have reported that the adsorption of some plasma proteins by a number of hydrophobic, uncharged polymers is generally irreversible. The retention of proteins by clays, however, is sensitive to pH in that treatment of the complex with the same solution but at a higher pH than that at which it is formed may lead to appreciable desorption (McLaren, 1954a; Armstrong and Chesters, 1964; Morgan and Corke, 1976).

So far, our attention has been focused on the interactions of proteins with montmorillonite. We now wish to summarise the available information on the behaviour of proteins at the surface of other clay minerals.

As might be expected, reactions involving minerals of the kaolinite group have many features in common with those of the expanding 2 : 1 type layer silicates. Thus, more than 90% of the maximum adsorbable amount is usually taken up within minutes of contacting the clay with the protein solution, after which there is a slow increase in uptake until equilibrium is attained. Again, the isotherms tend to be of the H-type, adsorption levelling off to a plateau whose height depends on the ambient solution pH. Unlike that for the montmorillonite system, however, the pH-uptake curve for kaolinite usually shows a broad, less well-defined maximum which may be slightly displaced towards the acid side of the isoelectric point of the protein in question (McLaren, 1954a; McLaren et al., 1958). Fig. 7.5 gives such a plot for lysozyme indicating that adsorption, measured at an equilibrium concentration of 0.5 mg ml^{-1}, increased sharply from pH ~ 3 to 6, followed by a gentle decline in uptake between pH 7 and 11. The isotherm at pH 11.8 levelled off to a higher plateau than at any pH value between 7 and 12. This is indicated in Fig. 7.5 by the occurrence of a shallow but distinct maximum at pH 11.8, beyond which value there was an abrupt fall in uptake. These observations may largely be explained in terms of electrostatic interactions between solute and surface and the pH-dependent charge characteristics of the reactants.

The recent work by Bolland et al. (1976) indicates that the positive surface charge of kaolinite decreases as the suspension pH increases from 3 to 7, a decline which may continue up to pH 9—10. On the other hand, the permanent negative charge due to isomorphous replacement within the framework of kaolinite remains more or less constant over this pH range.

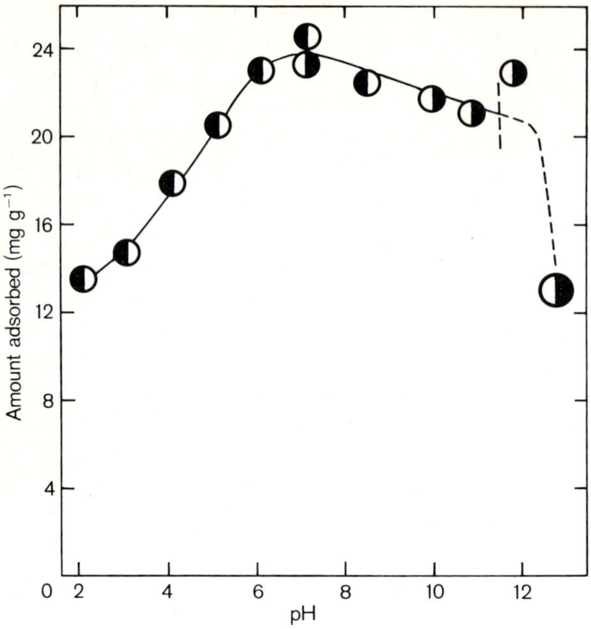

Fig. 7.5. Relationship between the maximum extent of adsorption and suspension pH for lysozyme adsorbed by kaolinite from universal buffer. The vertical dashed line indicates the isoelectric point of lysozyme. The shaded points refer to two different clay samples (after McLaren, 1954a).

If so, the net negative charge would effectively increase and this accounts for the gradual rise in plateau adsorption with increasing pH from 3 to 7. At the lower end of this range, adsorption might be enhanced by hydrogen bonding of COOH and NH_3^+ groups on the protein to surface silanols on the clay, as McLaren (1954a) has suggested. The contribution to uptake from this source, however, would not be large if, as seems likely, the adsorbed protein molecule retained its globular conformation. At pH 7—8 the amount adsorbed together with the estimated net positive charge on lysozyme agreed well with the cation exchange capacity of the kaolinite used. This was further evidence for the involvement of an ion exchange mechanism, although a strong cover-up effect was observed. Above pH 7 the kaolinite surface progressively acquired negative charges due to dissociation of exposed silanol groups (e.g. Rand and Melton, 1975) whereas the positive charge on lysozyme (p$I \sim$ 11.5) continued to diminish. A decline in adsorption might therefore be expected as the suspension pH was raised from 7 to ~11.5 and this accorded with the experimental data. Only at pH 12 and beyond when both the clay and the protein were negatively charged did uptake fall off sharply (Fig. 7.5). The small maximum in the pH—uptake curve at pH 11.8 was ascribed by McLaren to protein denaturation by the strongly alkaline medium. However, by analogy with the corresponding montmorillonite—lysozyme system, this behaviour was probably due to the attainment of minimum solubility and hence maximum compactness of the molecule at this point.

That the adsorption of proteins by kaolinite, expressed on a w/w basis, is quantitatively small as compared with montmorillonite (cf. Fig. 7.2) is due more to the limited extent of the accessible surface in the former case than to its low surface charge density. Proteins and organic macromolecular compounds, in general, are incapable of penetrating the kaolinite interlayers so that adsorption is confined to the external crystal surface. McLaren noted that the maximum extent of (lysozyme) adsorption ranged from 26 to 35 mg g^{-1} for the various kaolinites used. These values correspond reasonably well to the magnitude of the (external) surface area of the respective clay samples on the assumption that 1 mg of protein occupies 1 m^2 of accessible surface.

As with montmorillonite, the isotherms for the adsorption of proteins by kaolinite do not invariably conform to the H-type. A case in point is catalase which, at an ambient solution pH close to its isoelectric point (~5.7), gives a linear isotherm (Aliev and Zvyagintsev, 1974). Since the possibility of interlayer uptake does not arise in this instance, an intradomain type of solute penetration, as suggested above for ovalbumin in montmorillonite systems, may well occur.

The reactions of proteins and enzymes with kaolinite will be further described in the following section in connection with the way in which complex formation may influence protein stability and enzyme activity.

As compared with kaolinite, less information is available on the reactions of proteins with halloysite. Since halloysite can intercalate a single layer of water molecules, there is scope for some proteins to effect interlayer entry under suitable solution conditions. Whether interlayer complex formation is practically feasible, however, has yet to be demonstrated.

The interactions of halloysite with whole casein and its α_{s1}-, β-, γ- and κ-components, have been examined by Mills and Creamer (1971). They noted that the rate of uptake, initially rapid, fell off after the first hour of contacting the protein with the clay suspension. Although adsorption isotherms were not determined, the data indicated that casein has a high affinity for the halloysite surface.

As might be expected, the ambient pH of the system strongly influenced the extent of adsorption which declined rather sharply with rising pH from 6 to 8.5 and then levelled off as the pH was increased further to 10. This effect is shown in Fig. 7.6 for whole casein at various protein/clay ratios. The pH range of 6 to 10 for which uptake was measured, was selected on the basis of the limited solubility of casein below pH 6 and its susceptibility to hydrolysis above pH 10. Since the isoelectric point of casein and its components is close to pH 5, the curves in Fig. 7.6 represent the "right-hand"

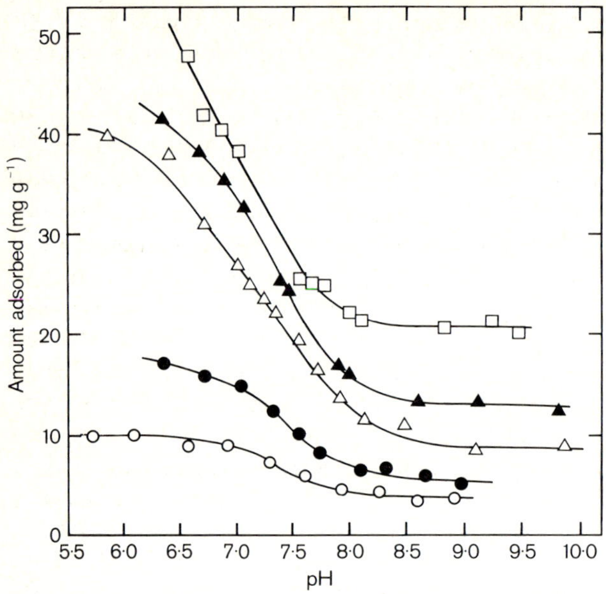

Fig. 7.6. Adsorption of whole casein by halloysite as a function of pH and at different casein to clay ratios. A 6% (w/v) clay suspension was used. The ratios (g protein per g clay) were: (○), 0.01; (●), 0.02; (△), 0.04; (▲), 0.06 and (□), 0.10. Adsorption was measured after shaking for 20 h at 293 K (after Mills and Creamer, 1971).

branch only of the pH—uptake relationship. (cf. Fig. 7.2); the maximum, normally occurring at pH ~ p*I*, is therefore not observed in this instance. Between pH 6 and 10 the protein carries a net negative charge whose amount increases with rising pH. This effect together with increasing competition by Na^+ ions for exchange sites at the clay surface would partly account for the observed decrease in uptake. That casein was at all times positively adsorbed rather than repelled by the negatively charged surface would indicate the involvement of secondary interactions (Van der Waals, hydrophobic and possibly hydrogen bonding) in the adsorption process. These interactions may further be augmented by electrostatic attraction between positively charged regions on the casein molecule and the halloysite layers. Thus, the apparent order of attraction γ-casein > β-casein > α-casein is also the sequence of increasing net negative charge on the protein.

With the possible exception of κ-casein, whose monomers are held together by disulphide links forming a high molecular weight aggregate (polymer), casein molecules are aggregated by hydrophobic forces and exist in equilibrium with their corresponding monomers. Although the basal spacing of individual complexes was not recorded, it seems conceivable that the monomeric forms of caseins can penetrate the interlayer space of (hydrated) halloysite.

The apparent preferential uptake among the protein components and the relatively slow rate of adsorption are at least consistent with this concept. Incidentally, the particular halloysite sample used by Mills and Creamer (from Northland, New Zealand) has been shown to readily intercalate small organic compounds (G.J. Churchman and B.K.G. Theng, unpublished data, 1976).

In view of the ability of allophane-rich soils to retain and stabilise large amounts (up to 20%) of organic matter (Aomine and Kobayashi, 1964a,b; Hétier et al., 1974; Duchaufour, 1976), surprisingly little is known about the reactions of allophane with proteins, enzymes, and organic compounds in general. This lack of information is a reflection partly of our still incomplete understanding of the structural and surface properties of allophane, and partly of the relatively restricted geographical distribution of allophanic soils, derived from volcanic ash. It should be recalled (cf. section 1.1) that allophane is a generic term denoting a series of X-ray amorphous, hydrous soil aluminosilicates. Allophane is also rather unique in that its surface charge, both in sign and quantity, is almost wholly pH-dependent. In other words, it may be regarded as an inorganic polyampholyte with an isoelectric point or zero point of charge between pH 4.5 to 6. This property of allophane has an important bearing on its reactivity towards charged species (Fieldes and Claridge, 1975).

Aomine and Kodama (1956) and Harada (1959) reported that the rate at which albumin was degraded on incubation with allophanic soils was much lower as compared with soils containing layer silicate minerals as their principal clay component. Earlier, Jackman (1955) suggested that enzymes might be partially inactivated when they adsorbed onto allophane surfaces and this was a factor contributing to the marked stabilisation of organic materials by allophane. Experimental evidence for this concept will be presented later.

It is only relatively recently, however, that the quantitative aspects of the allophane—protein interaction have been examined when Milestone (1971) measured the adsorption of a number of proteins (bovine serum albumin, gelatin, ovalbumin) by some (soil) allophanes. As compared with the crystalline layer silicates, discussed earlier, the rate of uptake by allophane was low, a half-time of 90 min or longer being commonly observed. This was interpreted in terms of a slow conformational change which the protein underwent at the interface. In keeping with this view, the adsorbed protein was found to resist elution by a 10^{-2} M solution of NaOH or Na_2CO_3. Alternatively, as seems probable, the small adsorption rate is due to the slow diffusion of solute molecules into the pores of the aluminosilicate (gel) network. Steric effects of this type have been suggested by Theng (1972) to account for the interaction of alkylammonium salts with allophane, and by Messing (1975b) for the behaviour of proteins at porous glass surfaces.

The adsorption isotherms, some of which are reproduced in Fig. 7.7, generally conform to the L-type. The indicated plateau values, however, are

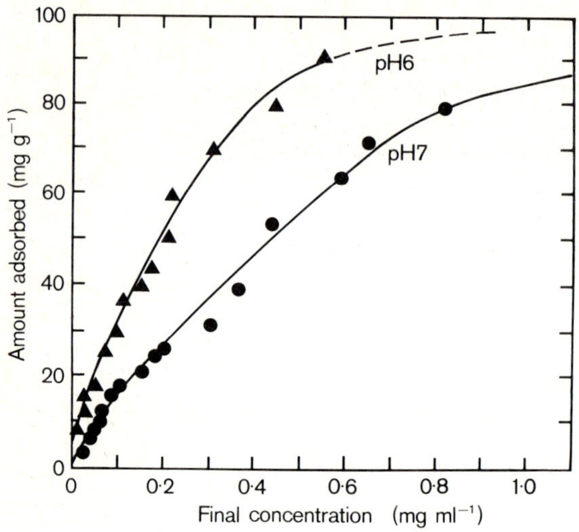

Fig. 7.7. Isotherms for the adsorption of bovine serum albumin by allophane at two pH values (after Milestone, 1971).

more apparent than real in that a second rise frequently occurs when the solute concentration is raised beyond the stated range. If it is again assumed that 1 mg protein covers 1 m^2 of surface, the apparent extent of adsorption would correspond to a surface coverage of only 25—35%, a value which probably represents the surface contained in the more readily accessible (interparticle) pores. For a comparable concentration of protein in solution, adsorption generally attains its highest level between pH 5 and 6. In this pH range, the net charge on the allophane is slightly positive or zero whereas that on the proteins (p*I* 4.6—4.9) is slightly negative. Above pH 6 both solute and surface are negatively charged causing adsorption to decline markedly. Similarly, charge—charge interactions would partly account for the reduction in uptake below pH 5 since then the net charge on both the protein and the mineral is positive. That positive adsorption still occurs at pH 7 may again be ascribed to the operation of secondary attractive interactions, such as H-bonding (Messing, 1975b). This suggestion would also explain the observed resistance of adsorbed proteins to elution by dilute electrolyte solutions.

7.4. COMPLEX FORMATION AND ENZYME ACTIVITY

As McLaren has pointed out in several places (McLaren and Babcock, 1959; McLaren, 1960; McLaren and Skujinš, 1968; McLaren and Packer,

1970), enzymes attached to or complexed with clay minerals, operate in a structurally restricted system. Accordingly, the behaviour of such enzymes towards their respective substrates is different from that obtained in a homogeneous, aqueous solution.

A generally observable effect, imposed by the negatively charged (polyanionic) environment of the solid matrix, is the shift to higher pH values of the pH—activity profile for adsorbed enzymes. The same holds true in the reverse situation where free enzymes (in solution) act on their adsorbed substrates. This displacement in pH at which the enzyme shows optimal activity has been well documented for a number of enzymes in clay and synthetic polyanion systems (McLaren and Packer, 1970; Katchalski et al., 1971). As might be expected polycationic derivatives of the same enzymes show the opposite effect, that is, their pH—activity profile is shifted towards more acidic pH values. The extent of this displacement is of course dependent on the surface charge density of the solid matrix. It is also influenced by the ionic strength (I) of the suspending medium, becoming effectively zero for $I \geqslant 1$. Another effect is that the apparent Michaelis constant for the rate reaction of the polyanionic—enzyme derivative is normally lower as compared with the unadsorbed enzyme. Again, the magnitude of this depression becomes vanishingly small at high ionic strengths.

The pH effect may be explained in terms of the influence that the polyanionic surface exerts on the distribution of protons and positively charged substrates in its vicinity. Since cationic species, in general, tend to distribute themselves in a diffuse ionic double layer (cf. Fig. 1.13), their concentration at the solid/solution interface is greater than that in the surrounding (bulk) solution (see Van Olphen, 1963).

In terms of pH the distribution of protons at the interface may be written as

$$\mathrm{pH_s} = \mathrm{pH_b} + 0.43\, ze\psi/kT \qquad (7.1)$$

where the subscripts s and b refer to the surface and bulk phases, respectively; z is the valency of the cationic species which, in this case, equals unity; e is the electronic charge; ψ is the local electric potential, k is the Boltzmann constant, and T the absolute temperature. For practical purposes, the following approximation can be applied (McLaren and Packer, 1970)

$$-\Delta\mathrm{pH} = \mathrm{pH_s} - \mathrm{pH_b} = \zeta/60 \qquad (7.2)$$

where ζ is the electrokinetic or zeta potential which can be estimated from the electrophoretic mobility (μ) of the suspended particle, using the Smoluchowski equation

$$\mu = \zeta\epsilon/4\pi\eta \qquad (7.3)$$

μ is considered to be negative for particles which move towards the anode; ϵ and η are the dielectric constant and viscosity of the suspending medium, respectively.

For particles with an equivalent spherical diameter (e.s.d.) equal to or greater than 6×10^{-4} m, suspended in distilled water at 298 K, combination of equations (7.2) and (7.3) after inserting the appropriate numerical values, yields the relationship

$$-\Delta pH = pH_s - pH_b = 0.217\mu \tag{7.4}$$

For small particles (e.s.d. $\leqslant 10^{-6}$ m) of which clays usually consist, the equation proposed by Hartley and Roe (1940)

$$-\Delta pH = pH_s - pH_b = 0.325\mu \tag{7.5}$$

is more appropriate.

Marshall (1964) has listed values of μ for some representative 2 : 1 and 1 : 1 type layer silicates in the sodium form. They range from −3.15 to −4.34 microns s^{-1}/Volt cm^{-1} which correspond to ΔpH of between 1.02 and 1.41. This is to say that the mineral "surface" is at least 10 times more acid than the bulk solution (suspension), "surface" being the plane of shear which is 0.5 nm away from the oxygen or hydroxyl plane of the silicate layer.

Using infrared spectroscopy, Harter and Ahlrichs (1967) derived a value of $\Delta pH \sim 2$ for montmorillonite, suspended in a solution at $pH_b \sim 7$. However, as Mortland (1970) has pointed out, the magnitude of the pH shift is influenced by such factors as the water content and the nature of the saturating (exchangeable) cation at the clay surface. The ionic strength of the system also affects ΔpH since the zeta potential is reduced by increasing the electrolyte concentration of the ambient solution (cf. Fig. 1.13C; equation 7.2). In practice, it is difficult to compare enzyme activity at different values of ionic strength because addition of electrolytes to the system, as we have seen in the preceding section, would change the adsorption behaviour of proteins. A useful approach is to estimate pH_s at a given value of pH_b and ionic strength (I) by means of equation (7.5). For lysozyme and kaolinite in a medium at $pH_b = 8.05$, this yields pH_s values of ~9.4 and ~7 or ΔpH of approximately −1.35 and 1.05, respectively (McLaren, 1960). This may be compared with the value of $\Delta pH = 0.54$ for lysozyme adsorbed onto kaolinite at the same ionic strength ($I = 0.05$). Thus, the pH at the "surface" of lysozyme in solution is about two units higher than that of the protein at the kaolinite surface.

Since enzyme activity is controlled by the pH environment in the vicinity of the substrate surface, the pH optimum for an enzyme acting on lysozyme in solution would be expected to be two pH units less than that in a system where lysozyme (substrate) is bound to kaolinite. This expectation is borne out by the measurements of McLaren and Estermann (1957) on the digestion of heat-denatured lysozyme, adsorbed on kaolinite, by chymotrypsin (Fig. 7.8). An enzyme acting on its substrate at solid/solution interfaces may therefore serve, in McLaren's phrase, as a "molecular pH-meter". Table 7.3

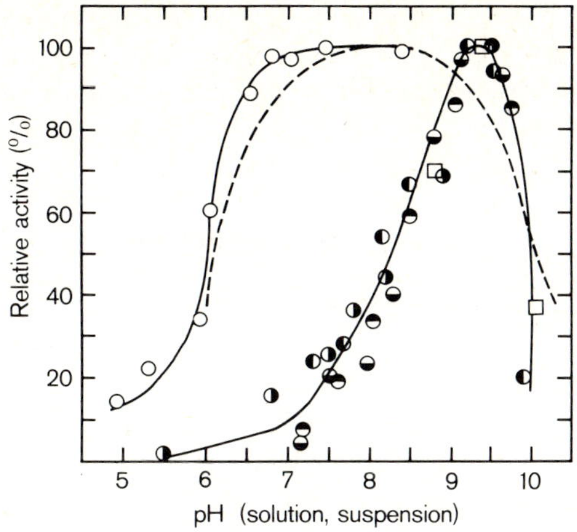

Fig. 7.8. Digestion of denatured lysozyme in solution (open circles) and when adsorbed onto kaolinite (half-shaded circles for universal buffer, squares for ethylammonium buffer) by chymotrypsin as a function of pH. Dashed curve is from Northrop's study with casein (in solution) (after McLaren and Estermann, 1957).

lists some examples illustrating this effect for various enzymes acting on their respective substrates in the presence of clays and synthetic polymers. In accord with prediction, ΔpH is positive in those instances where clay minerals act as a carrier. For systems involving polycationic supports, such as the DEAE—cellulose—invertase—sucrose assemblage, ΔpH is negative. On the other hand, this parameter is sensibly zero for the ribonuclease—cation exchange resin—RNA system when both carrier and substrate are negatively charged.

The dependence of enzyme activity on pH is commonly ascribed to the dissociation of ionising groups on the enzyme which are actively involved in catalysing the reaction. The chemical identity of such groups or "active sites" may often be inferred from the apparent dissociation constant (pK_a') as indicated by the pH—activity profile of the enzyme in question. In the case of chymotrypsin ($pK_a' \sim 7$), for example, one of the reactive amino acid side chain is the imidazole residue, which chemical and crystallographic studies have independently identified as histidine-57 (Hartley, 1964; Sigler et al., 1968). Thus, the observed displacement in the pH-optimum of bound enzymes may alternatively be explained in terms of a shift in the pK_a' value of the ionising active group as induced by the polyelectrolyte environment.

It should be noted that many of the examples listed in Table 7.3 refer to systems in which carrier-bound enzymes act on their substrates in solution. Such systems have received a considerable amount of attention in view of their importance and relevance to biochemical processes occurring in the

TABLE 7.3

Observed ΔpH values for some carrier-bound enzymes acting on their respective substrates

Enzyme	Carrier	Substrate	ΔpH	References
Catalase	Kaolinite	Hydrogen peroxide	+0.65	Aliev et al. (1976)
Catalase	Montmorillonite	Hydrogen peroxide	+1.0	Aliev et al. (1976)
Chitinase	Kaolinite	Chitin	+1.1	Skujinš et al. (1974)
Chymotrypsin	Kaolinite	Lysozyme	+2.0	McLaren and Estermann (1957)
Chymotrypsin	Montmorillonite	Lysozyme	+2.0	McLaren and Peterson (1965)
Chymotrypsin	Carboxymethyl cellulose	Acetyltyrosin	0	Mitz and Summaria (1961)
Ficin	Carboxymethyl cellulose	Benzoylarginine ethyl ester	+0.3	Hornby et al. (1966)
Invertase	DEAE-cellulose	Sucrose	−2.0	Suzuki et al. (1966)
Papain	Kaolinite	Benzoylarginine ethyl ester	+0.9	Benesi and McLaren (1975)
Phosphatase	Kaolinite	β-Naphtylphosphate	+0.7	Ramirez-Martinez and McLaren (1966)
Ribonuclease	Cation exchange resin	Ribonucleic acid	0	Barnett and Bull (1959)
Urease	Montmorillonite	Urea	+0.6	Durand (1964a)

living cell (McLaren and Packer, 1970; Katchalski et al., 1971). In soils, however, both substrates and enzymes are likely to be closely associated with the mineral (carrier) surface and it is difficult to distinguish between enzyme- and substrate-immobilised systems. This is especially true when the substrate on which the enzyme acts, is itself a protein. The kaolinite—lysozyme—chymotrypsin assemblage is a good example of a system in which the added enzyme (chymotrypsin) is adsorbed by the clay—lysozyme complex. Indeed, the formation of a clay—substrate—enzyme complex appears to be a prerequisite for proteolysis to occur (McLaren and Estermann, 1956). The rate of (enzyme) adsorption is generally more rapid than that of (substrate) digestion which, in turn, is about 2/3 of the rate obtained in a homogeneous aqueous solution. The lysozyme used here again refers to the material which has previously been inactivated or denatured by heat treatment since the native protein, either in solution or attached to the clay surface, is only minimally hydrolysable by chymotrypsin (McLaren, 1954b). As might be expected, the amount of chymotrypsin adsorbed by either the kaolinite or montmorillonite complex with lysozyme (denoted by KHLC and MLHC, respectively) was less than by the parent, lysozyme-free mineral. The rapid decrease in enzyme activity at $pH > 9.6$ (Fig. 7.8) coincided with the sharp decline in the amount of chymotrypsin adsorbed by KHLC since this pH range was far above the isoelectric point ($pI \sim 8.6$) of the enzyme. At pH 9.0, for example, about 95% of the added enzyme was taken by KHLC, increasing to 100% at $pH \sim 8.3$ (in universal buffer). The close correspondence between adsorption and activity of chymotrypsin lends further support to the notion that a clay—enzyme—substrate complex is formed before substrate decomposition can occur.

Following on this work, Estermann et al. (1959) examined the relative stability of lysozyme in complexes with kaolinite, lignin, montmorillonite, and silica gel, towards digestion by chymotrypsin, mixed soil cultures, and pure cultures of *B. subtilis*, *B. mycoides*, and *Pseudomonas* sp. The rate of hydrolysis was estimated from the increase in optical density at 280 nm (D_{280}) of the solution due to the release of the digestion products. Fig. 7.9 shows that this rate fell in the order kaolinite > montmorillonite > lignin. The resistance of lysozyme, associated with silica gel, to digestion by *Pseudomonas* sp. was comparable to that of the protein in the lignin complex.

The smaller susceptibility to enzymic attack of organic substrates attached to montmorillonite as compared with kaolinite is of general applicability to these systems (Ensminger and Gieseking, 1942; Mortland and Gieseking, 1952; Lynch and Cotnoir, 1956). This finding has been ascribed to the fact that most of the substrate in montmorillonite is present in the interlayer space whereas adsorption in kaolinite is confined to external crystal surfaces. Although the interlayer material would largely be accessible to extracellular enzymes, as emphasised earlier, substrate intercalation would impose a measure of steric constraint on the formation of an enzyme—substrate complex.

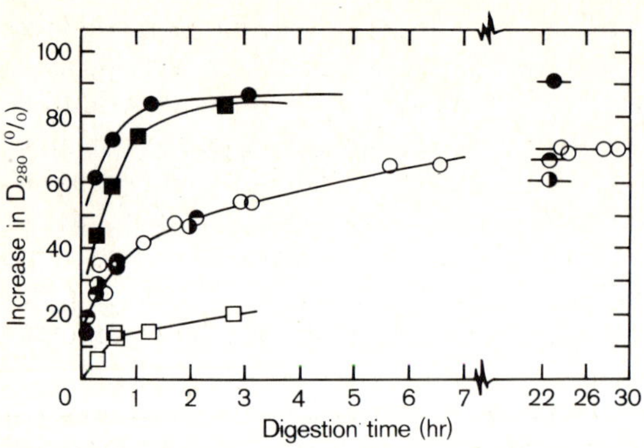

Fig. 7.9. Digestion of lysozyme in complexes with various solids. (□) lignin—lysozyme complex with 143 μg chymotrypsin in 0.133 M ethylamine hydrochloride, 0.08 M $CaCl_2$, pH 8.9; (○), 1/1 monolayer of lysozyme on montmorillonite with 71 μg chymotrypsin in 0.13 M ethylamine hydrochloride, pH 9.1; (◓), 1/2 monolayer of lysozyme on montmorillonite with 69 μg chymotrypsin in 0.13 M ethylamine hydrochloride, pH 8.9; (◑), 1/3 monolayer of lysozyme on montmorillonite with 69 μg chymotrypsin in 0.133 M ethylamine hydrochloride, pH 9.0; (●), 1/1 monolayer of lysozyme on kaolinite with 70 μg chymotrypsin in 0.133 M ethylamine hydrochloride, pH 9.1; (■), 1/1 monolayer of lysozyme on kaolinite with 70 units Nagarse proteinase in 0.133 M ethylamine hydrochloride, pH 8.9 (after Estermann et al., 1959).

In addition, the diffusion of enzyme into, and of reaction products out of, the interlayer space may be retarded.

This interpretation is borne out by experiment, at least for the system being considered here. Thus the (initial) basal spacing of the montmorillonite—lysozyme complex was ~4.6 nm, corresponding to an interlayer separation of ~3.6 nm which would be sufficiently wide for the enzyme to gain entry into the clay interlayers. As digestion progressed, this spacing decreased until at the end of the incubation period (10 days) a spacing of ~1.7 nm was recorded. When this stage was reached, little further decomposition would have occurred. Secondly, the hydrolysable amount of adsorbed lysozyme, but not its proportion, was dependent on the initial surface coverage of the clay by the protein (Fig. 7.9).

Interestingly, proteins associated with lignin were even more resistant to enzymic hydrolysis, confirming the earlier observation by Lynch and Lynch (1958). Similarly, Burns et al. (1972a) have reported that whereas the activity of jackbean urease, complexed with montmorillonite, was markedly reduced on incubation with pronase (a proteolytic enzyme), little or no suppression of activity occurred on adding lignin to the system. To a lesser degree, proteins are also stabilised when they enter into complex formation with silica gel.

These observations were ascribed by Estermann et al. (1959) to pore penetration by which the protein became enmeshed in the gel structure of the solid matrix. Although a Schiff base type condensation as proposed by Waksman and Iyer (1932) is possible with lignin acting as the matrix, such a reaction would not occur with silica gel. The suggested mechanism implies that proteins in lignin and silica gel systems are relatively inaccessible to, and hence protected against the action of, extracellular enzymes. An analogous situation and one which is perhaps more relevant to soils, is encountered with allophane which, like silica gel, has a porous network structure. This may explain why soil allophane tends to be more effective in suppressing the activity of adsorbed enzymes as compared with crystalline layer silicates (Aomine and Kobayashi, 1964a,b, 1966).

Lysozyme adsorbed onto kaolinite and montmorillonite surfaces was equally susceptible to being hydrolysed by cultures of microorganisms and by soil (Estermann and McLaren, 1959). Apparently, the agents responsible in this instance were a variety of exoenzymes secreted by the cultures or present in the soil. The rate at which the adsorbed protein is broken down can, in fact, be enhanced as compared with the clay-free system. The stimulating effect occurs whether or not the microorganisms are attached to the clay. Thus, treatments which would dislodge the bacteria from KHLC particles, such as shaking the system, do not sensibly change the rate of hydrolysis as measured by the amount of ammonia produced. On the other hand, this effect is absent with a non-adsorbed substrate, such as (denatured) lysozyme which has previously been digested by chymotrypsin. Clays can therefore act as a concentrating surface for both substrate and enzyme and so accelerate substrate decomposition.

In line with Pinck's (1962) observation (cf. Table 7.1), digestion of adsorbed lysozyme in MHLC led to a marked contraction in basal spacing, from an initial value of ~4.6 nm to as low as 1.4—1.7 nm at the end of the incubation period. The magnitude of this contraction is influenced by the type of microorganisms used. This would suggest that although bacterial enzymes can, on the whole, enter the interlayer space of MHLC, a certain amount of species-dependent steric hindrance is encountered. As the legend of Fig. 7.9 indicates, the parent complexes (i.e. before incubation) are presumed to contain a single layer of (globular) lysozyme molecules in the interlayer space. However, as remarked on earlier, a basal spacing in the order of 4.6 nm seems more compatible with the presence of a double layer of molecules in the majority of the montmorillonite interlayers.

In an attempt to simulate soil systems, McLaren and Estermann (1956) have examined the behaviour of chymotrypsin towards KHLC in paste form. Some of the results are shown in Fig. 7.10 from which we may draw the following inferences. As in suspension form, adsorbed lysozyme in a paste of KHLC is susceptible to digestion by chymotrypsin at a rate which is smaller than that observed for the unadsorbed protein. For a given period of digestion, however, the amount of lysozyme hydrolysed is less in the paste than in suspension but this difference in rate tends to vanish when the initial

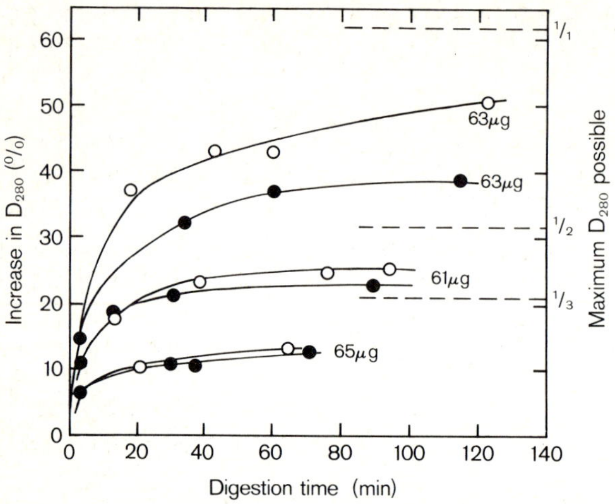

Fig. 7.10. Digestion of KHLC: (○), in suspension and (●) in a paste by chymotrypsin, corrected for digestion of eluted substrate. Horizontal dashed lines indicate maximum possible D_{280} for digestion of lysozyme on kaolinite at 1/3, 1/2 and 1/1 coverage by substrate. The amount of chymotrypsin (μg) used is indicated on the curves (after McLaren and Estermann, 1956).

surface coverage by the protein substrate is small in comparison with the "monolayer" capacity.

These observations may be explained in terms of the relative mobility of, and surface coverage by, the enzyme. It seems probable that the mobility of chymotrypsin over the KHLC surface in the paste is more restricted than that in suspension. In addition, the diffusion of hydrolysis products out of the complex is likely to be faster for the suspension system. Accordingly, there is a faster build-up of surface vacancies on which the enzyme can adsorb as compared with the system in paste form. Since the rate of hydrolysis is primarily controlled by the amount of adsorbed enzyme, suspension conditions and a low surface coverage by the substrate would favour a more rapid breakdown of lysozyme in the complex. By the same token, the difference in hydrolysis rate between KHLC in suspension and that in paste form would diminish as the initial surface coverage by lysozyme decreases. Irrespective of coverage, however, the rate at which adsorbed proteins are digested by exoenzymes tends to reach a limiting or plateau value which is appreciably lower than the maximum attainable. This would suggest that after a given period of contact the enzyme may "strike" a strongly adsorbing site at the clay surface. At this point the enzyme becomes more or less immobilised leaving patches of undigested substrate.

An analogous situation has recently been described by Morgan and Corke

(1976) who noted that the specific activity of glucose oxidase in complexes with different clay minerals was influenced by the extent of adsorption. At low levels of uptake enzyme activity was markedly suppressed as compared with that in solution. As the amount (of enzyme) adsorbed increased, its specific activity rose until at maximum adsorption it approached that of the free enzyme.

Also relevant in this context is the work of Aliev and Zvyagintsev (1974) who have compared the activity of catalase and pronase in the presence of kaolinite, montmorillonite, and some soils. The activity of clay-adsorbed enzymes was about half of that in solution, montmorillonite being more effective than kaolinite in suppressing enzyme activity. Like jackbean urease, adsorbed catalase could be hydrolysed by pronase but at a much lower rate than the non-adsorbed form. Since the adsorption of catalase was presumably confined to external crystal surfaces (Harter and Stotzky, 1973), the observed reduction in hydrolysis rate was probably due more to conformational and surface (mobility, coverage) effects than to steric and accessibility factors.

Using chromatographic techniques, Skujinš et al. (1959) have compared the hydrolysis products of *B. subtilis* N′ growing in a paste of KHLC with those of a purified protease (Nagarse), extracted from these organisms, acting on denatured lysozyme in solution. Qualitative and quantitative differences were observed between the products arising from these two systems but the causative factors were not clearly understood. Apparently the protease was partially adsorbed by KHLC at pH 8, showing an activity comparable to that of chymotrypsin. Nagarse, acting on casein, had a broad pH optimum between 7 and 10 but on lysozyme in a KHLC paste this optimum was narrower, lying between pH 9.5. Although the pH near the paste surface was probably further depressed, enzyme activity was retained.

We should point out that diffuse ionic double layers would not obtain in a paste. However, the principle that the acidity near the surface is greater than that at some distance away from it, would still hold although the relationship between ionic strength and pH in a paste is clearly more complex as compared with that in dilute aqueous suspension. Indeed, in the extreme situation where the water content of the clay system falls below about 5% by weight, the clay "surface" becomes very acidic due to the polarisation of residual water molecules by the exchangeable cations (e.g. Mortland, 1968; Theng, 1974b).

An interesting feature of clay—enzyme—substrate systems is that the order in which the enzyme and the substrate are added to the clay can markedly influence the properties of the resultant complex. Thus, for the system comprising kaolinite, chymotrypsin, and lysozyme discussed above, the rate of substrate hydrolysis is much lower if the adsorption of chymotrypsin (the enzyme) precedes that of lysozyme (the substrate) than if the sequence of addition were reversed (McLaren and Estermann, 1956). This effect lends further support to the concept that the adsorbing sites on the clay surface are non-uniform in strength and affinity towards proteins. By the same token, chymotrypsin added to the substrate-free clay shows a relatively low activity towards, and is not displaced by, lysozyme presumably because the enzyme tends to occupy the most strongly adsorbing sites first. Skujinš et al. (1974) have reported similarly for the system consisting of kaolinite, chitinase, and chitin, as the data of Table 7.4 indicate.

It seems clear from what we have seen that the formation of an enzyme—substrate complex at the clay surface is an essential preliminary to substrate decomposition. This is to say that the enzyme-catalysed decomposition of clay-adsorbed substrates occurs at the mineral/solution interface rather than in the surrounding bulk phase. With reference to the kaolinite—chymotrypsin—lysozyme system of McLaren and Estermann (1956), for example, only about 10% of the total enzyme activity could be ascribed to the free enzyme acting on lysozyme which was released from KHLC into the bulk solution. On the other hand, Pinck and Allison (1961) have suggested that urease complexed with kaolinite and montmorillonite must first be eluted from the clay surface before it can act on urea in solution. It should be noted, however, that these workers used a commercial, impure sample of urease and furthermore, the ammonia produced by the reaction increased the pH of the system to levels where little urease would have been retained by the clay. Indeed, most of the enzyme initially adsorbed eventually became detached from its complex. The retention of essential, albeit modified activity by adsorbed enzymes is, of course, basic to the synthesis and application of immobilised enzyme systems.

The presence of clays in the system clearly interferes with enzyme behaviour in the sense that the rate of reaction is frequently, but by no means

TABLE 7.4

The effect of sequence of enzyme and substrate addition on the activity of chitinase, in the presence and absence of kaolinite (from Skujinš et al., 1974)

Sequence of addition	Relative activity (%)
(A) Chitin + chitinase + buffer (control)	100
(B) Chitin + kaolinite + chitinase + buffer	15.5
(C) Chitinase + kaolinite + chitin + buffer	5.5
(D) 50% of (B) + 50% of (C)	10.6

The measurements were made under the following conditions: time, 30 min; temperature, 310 K; chitinase, 14 μg ml^{-1}; buffer, 0.03 M sodium acetate at pH 4.5; kaolinite, 7 mg ml^{-1}; chitin, 1 mg ml^{-1}. After adding the clay, the mixture was shaken end-over-end for 10 min. The structural formula of chitin is shown in the Appendix.

invariably, reduced. However, apart from the fact that either the enzyme or the substrate may become immobilised at the clay surface, adsorption per se of either component should not materially influence the frequency of enzyme—substrate encounter. A decrease in reaction rate or enzyme activity, if observed, may therefore be ascribed at least partly to diminished accessibility of the active sites on the enzyme due to steric and/or conformational effects (Doonan, 1969). If, in addition, the active sites were involved in the adsorption process and if either the substrate or the enzyme is capable of penetrating the interlayer space of the clay, the rate of reaction may be further reduced. Because of their large surface area and exchange capacity, as well as their ability to intercalate organic compounds, montmorillonite-type minerals are generally more effective in suppressing enzyme activity than the 1 : 1 type layer silicates, as the following examples illustrate.

We have already referred to the work by Ensminger and Gieseking (1941) who found that the presence of montmorillonite markedly reduced the hydrolysis of albumin and haemoglobin by pepsin (pH 2.25) and pancreatin (pH 8.5) as compared with the clay-free system. Kaolinite, on the other hand, had but a minimal effect on enzymic proteolysis. In comparing the influence of various clay minerals on the hydrolysis of some small organic phosphorus compounds (fructose diphosphate, phytin, glycerophosphate, and lecithin) by phosphatase, Mortland and Gieseking (1952) observed that in all cases the presence of clays (at a constant enzyme/substrate ratio) reduced the rate of hydrolysis. The extent of inhibition, however, depended on the type of clay used, decreasing in the order montmorillonite > illite > kaolinite. Nearly complete inhibition could be achieved by raising the clay content of the system, the amount of inorganic phosphate released during hydrolysis being an inverse, linear function of clay concentration (Fig. 7.11).

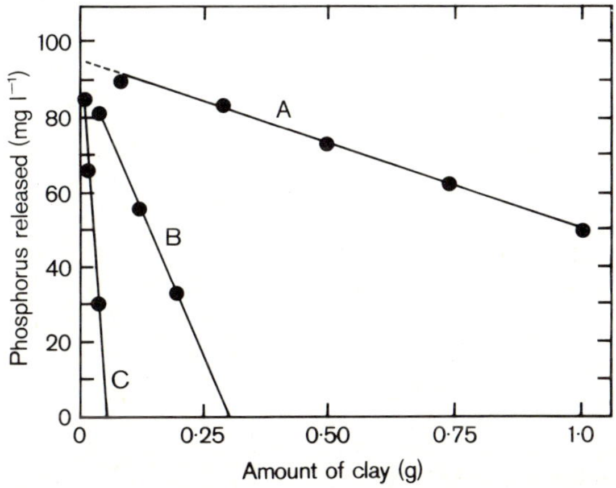

Fig. 7.11. The effect of the amount of different clay species in the system on the activity of phosphatase. A, kaolinite; B, illite; C, montmorillonite (after Mortland and Gieseking, 1952).

The extent of hydrolysis was correlated with the exchange capacity of the minerals, suggesting that the clay—enzyme interaction involved an ion exchange mechanism. The kinetics of the phosphatase reaction in clay and soil systems will be discussed more fully later but the formation of an enzyme—substrate complex at the mineral/solution interface is generally agreed upon. This would mean that the reaction rate is influenced by the concentration of adsorbed substrate, an effect which should be borne in mind in comparing the decomposition of various substrates. This point also emerged from Erickson's (1948) earlier work when he examined the hydrolysis of tyrosine, *p*-cresol, and phenol by tyrosinase in the presence of montmorillonite. By the same token, the reaction products may be adsorbed to different extents by various clays (and soils). Rate measurements based on the amount of, say, phosphorus released into solution (Fig. 7.11) may therefore give a less than accurate reflection of the relative efficiency of clays in inhibiting enzyme activity (Kroll and Kramer, 1955). Lastly, the pH—activity profile of an enzyme-catalysed reaction at the clay/solution interface is, as remarked on earlier, normally displaced towards more alkaline pH values. Because the extent of this pH shift may not be identical for all clays, comparisons of reaction rates between different mineral species should be viewed with the pH effect in mind.

Thus, Ramirez-Martinez and McLaren (1966) have reported that adsorption of an acid phosphatase on kaolinite led to a substantial (~75%) reduction in the rate of Na-β-naphtylphosphate hydrolysis as well as to a positive shift in the pH optimum (Fig. 7.12). Incidentally, the pH—activity profile for "phosphatase" acting on glycerophosphate in a sterilised soil displays a rather broad optimum between pH 6 and 7 (McLaren, 1963). This is not

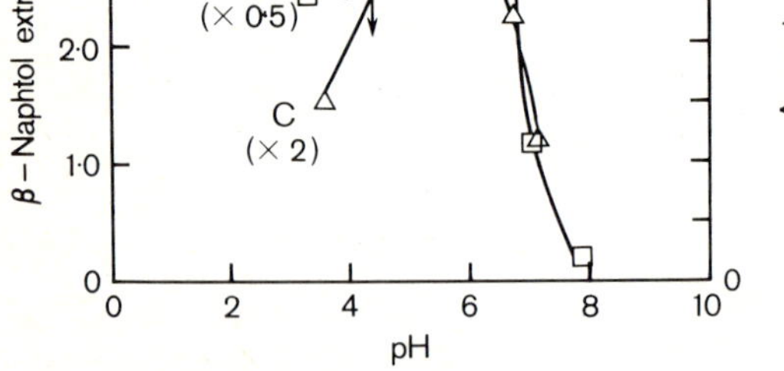

Fig. 7.12. Adsorption and activity of phosphatase in function of pH in solution or suspension (after Ramirez-Martinez and McLaren, 1966).

altogether unexpected since soil phosphatase comprises a mixture of acid and alkaline species, secreted by a variety of microorganisms, and acting in different micro-environments.

Similarly, Durand (1964a) observed a reduction in reaction rate and a positive ΔpH (Table 7.3) for the hydrolysis of urea by urease adsorbed onto montmorillonite. Furthermore, Cu^{2+} ions lost much of their inhibitory effect on urease activity in the presence of clay. These findings are consistent with the view that the reaction occurs at the clay surface rather than in the surrounding solution. Uricase (uric oxidase) attached to montmorillonite behaved similarly in that dissociation or detachment of the enzyme from its clay complex was not required for degrading uric acid. Complex formation with montmorillonite, however, again resulted in a reduction of enzyme activity (Durand, 1964b).

The effect of soil clays, consisting dominantly of either allophane, halloysite or montmorillonite, on the activity of some (adsorbed) enzymes, has been studied by Aomine and Kobayashi (1964a,b, 1966) and by Kobayashi and Aomine (1967). The usual approach was to mix the clay suspensions with a buffer solution containing the enzyme before adding the appropriate substrate to the mixture. Enzyme activity was estimated by measuring the concentration of either the hydrolysis products appearing or the (unhydrolysed) substrate remaining in the system after a specific reaction time. As might be expected, this activity, for a given clay/enzyme ratio, was related to the adsorption capacity of the clay for the enzyme (Table 7.5) and, with the exception of cellulase in halloysite and montmorillonite systems, was less than the activity in (clay-free) solution. Although the order of inhibition of allophane > montmorillonite > halloysite generally followed the order of enzyme adsorption by these clays, the extent of inactivation was not proportional to the amount adsorbed. This was probably a reflection of the relative accessibility of the active sites on the enzyme to the respective substrate. The discrepancy between the amount adsorbed and the extent of inhibition was particularly marked for β-amylase. The reason for this behaviour was not clear although the β-amylase preparation used was apparently polydisperse, the enzymically active fraction possibly being preferred by the clays. This may also partly explain the observation that the activity of β-amylase, unlike that of protease, was more strongly inhibited by montmorillonite than by allophane.

Using glucose oxidase, Ross and McNeilly (1972) have reported similarly in that the activity of this enzyme was much more inhibited by montmorillonite than by allophane. As already mentioned, Morgan and Corke (1976) have been able to show that the specific activity of glucose oxidase, attached to montmorillonite, increased with the amount adsorbed, approaching that of the free enzyme at maximum uptake. To the factors referred to above we may therefore add surface concentration of the enzyme as one which affects enzyme activity. This is perhaps hardly surprising since enzyme—substrate

TABLE 7.5

Influence of some soil clays on the relative activity of different enzymes, taking the activity of the clay-free system as 100% (from Aomine and Kobayashi, 1964a)

Enzyme	Substrate	Enzyme added (mg)	Clay added (mg)	Relative activity (%)		
				Allophane	Halloysite	Montmorillonite
Protease [a]	Haemoglobin	0.1	0.3	62.9	93.8	81.0
		0.1	0.6	51.7		
		0.1	0.9	44.3		
		0.1	1.2	42.0		
α-Amylase [b]	Amylose	0.05	0.08	86.2	88.1	87.7
		0.1	0.3	76.8	78.2	78.9
β-Amylase [b]	Soluble starch	0.1	0.3	27.3	64.6	15.5
		0.1	1.5	13.2	45.7	8.7
Cellulase [b]	Na^+ Carboxy-methylcellulose	0.1	0.3	85.6	99.3	97.2
		0.2	3.0	83.9	98.3	88.6

[a] pH 5.0; [b] pH 5.6.

complex formation must precede substrate decomposition. For all these reasons it is often difficult to compare published data on the part played by clays in enzyme-catalysed reactions, unless the conditions and methods used to assess them are the same. Ross and McNeilly (1972), for example, observed that kaolinite had a negligible effect on the activity of glucose oxidase whereas Zvyagintsev and Velikanov (1968) had reported differently for a similar system.

In common with the kaolinite—lysozyme—chymotrypsin and kaolinite—chitin—chitinase systems (cf. Table 7.4), the activity of β-amylase in the presence of clays relative to that of the enzyme in solution, depends on the order in which the enzyme and the substrate are added to the clay (Aomine and Kobayashi, 1966). In this instance, the activity is less inhibited when the substrate (starch) is added to the clay before the enzyme than when the sequence of addition is reversed. This is another illustration of the influence of enzyme and substrate affinity for the mineral surface and of their relative surface requirement or coverage on reaction rates. Incidentally, the starch and sodium carboxymethylcellulose on which β-amylase and cellulase act, respectively, were soluble in the aqueous (buffer) solution used by Aomine and Kobayashi (1964a, 1966) who analysed the rate reaction in terms of the classical Michaelis—Menten kinetics (cf. equation (7.7)). In many practical situations, however, such as during the breakdown of ingested food, these and related enzymes have to deal with particulate, insoluble substrates. The action of amylase on starch granules and of cellulase on cellulose are examples of such systems. By assuming that the adsorption of the enzyme onto its substrate surface conforms to either the Langmuir or Freundlich relationship, and that the digestion rate is proportional to the amount of enzyme adsorbed, the appropriate rate equations may be set up (e.g. McLaren and Packer, 1970). The kinetics may then differ from those of Michaelis and Menten in that the initial rate is not linearly related to the total enzyme concentration in the system. Rather, it is proportional to the surface area of the (solid) substrate which is in contact with the solvent.

For somewhat different reasons, a departure from Michaelis—Menten kinetics has recently been reported by Irving and Cosgrove (1976) for soil acid phosphatase acting on *p*-nitrophenylphosphate. In most instances, enzyme-catalysed reactions involving clays, soils and charged carriers, in general, appear to follow Michaelis—Menten kinetics, or at least, are assumed to do so. However, the magnitude of the apparent Michaelis constant (K_m) of the reaction will differ from that observed in the carrier-free system.

In its simplest form, the classical model for an enzyme-catalysed reaction may be presented by the following scheme

$$E + S \underset{k_2}{\overset{k_1}{\rightleftharpoons}} ES \overset{k_3}{\rightarrow} E + P \qquad (7.6)$$

where E, S, ES, and P refer to enzyme, substrate, enzyme—substrate complex, and reaction products, respectively; k_1, k_2, and k_3 are the corresponding velocity constants of the assumed processes. K_m is equated with the ratio $k_2 + k_3/k_1$; but for $k_2 >> k_3$, $K_m \simeq k_2/k_1$ and represents the thermodynamic dissociation constant of ES. The relationship between K_m, E, and S, expressed in terms of reaction velocities, is given by the Michaelis—Menten equation

$$v = V_{max}\,(S)/K_m + (S) \tag{7.7}$$

where v and V_{max} are the measured initial and maximal velocity, respectively; and (S) is the substrate concentration. Since for $v = \frac{1}{2}\,V_{max}$, K_m is numerically equal to (S), K_m has the dimension of concentration. In order to determine V_{max} and K_m more easily and accurately, a number of linear transformations of equation (7.7) have been suggested (Dowd and Riggs, 1965). Among these, the double reciprocal form due to Lineweaver and Burk

$$1/v = K_m/V_{max} \cdot 1/(S) + 1/V_{max} \tag{7.8}$$

and the single reciprocal form due to Eadie and Hofstee

$$v = V_{max} - K_m \cdot v/(S) \tag{7.9}$$

are frequently used.

The observed change in the apparent Michaelis constant of a carrier-bound enzyme, K'_m, acting on a charged substrate relative to that of the corresponding enzyme in a homogeneous solution, K_m, is brought about by the unequal distribution of substrate species between the polyelectrolyte surface and the bulk solution (e.g. Katchalski et al., 1971). When the charge on the carrier-enzyme derivative and on the substrate is of opposite sign, the concentration of substrate near the surface, $(S)_s$, is greater than that in the bulk solution, $(S)_b$. V_{max} is thus obtained at a lower $(S)_b$ value as compared with the carrier-free system and accordingly $K'_m < K_m$. If, on the other hand, the charge on the enzyme derivative and on the substrate are of the same sign, $(S)_s < (S)_b$ and hence $K'_m > K_m$.

The relationship between K'_m, K_m, and the electrostatic potential, ψ, is completely analogous to equation (7.1) relating pH to ψ, that is

$$\Delta pK_m = pK'_m - pK_m = 0.43\,ze\psi/kT \tag{7.10}$$

Clearly, equation (7.10) does not apply to a situation where either the carrier or the substrate is uncharged, in which case $K'_m \simeq K_m$, provided that diffusion effects are negligible. However, if the diffusion of substrate to the carrier-enzyme/solution interface is retarded so that $(S)_s$ is effectively less than $(S)_b$, the following empirical relationship is applicable

$$K'_m = K_m + V_{max} \cdot t/D \tag{7.11}$$

where t is the diffusion layer thickness and D the diffusion coefficient. Diffusion effects may be important with high molecular weight substrates. The validity of the above analyses have been substantiated by experimental measurements using enzymes attached to polymer matrices, acting on their respective substrates (Hornby et al., 1968; Ladd and Butler, 1975).

Since clay—enzyme derivatives are generally negatively charged, the direction in which K'_m will change with respect of K_m can, in principle, be predicted from the charge characteristics of the substrate. It must be borne in mind, however, that enzyme reactions at clay (and soil) surfaces are influenced by factors other than charge interactions. Steric, conformational, and diffusion effects touched on above may be of overriding importance. We have already seen that the hydrolysis of lysozyme, complexed with kaolinite, by bacterial exoenzymes was stimulated rather than retarded as compared with the clay-free system.

One of the few attempts at measuring reaction rates and apparent Michaelis constants of enzymes in the presence of clays is that of Kobayashi and Aomine (1967). These workers reacted allophane and montmorillonite suspended in a buffer solution with the enzyme before adding the appropriate substrate to the clay—enzyme "mixture" or the clay—enzyme "complex". The latter referred to the material after centrifuging the mixture and washing the precipitate with sodium acetate buffer (pH 5) to remove any unadsorbed and weakly attached enzyme. Fig. 7.13 shows the results for pronase (a protease extracted from *Streptomyces griseus*) acting on a dipeptide, carbobenzoxy-L-glutamyl-L-tyrosine or CBGT, in the absence (control) and presence of clay. The adsorbed enzyme clearly obtained a maximal velocity at a lower substrate concentration than did its free counterpart in solution, as might be expected on theoretical ground. The effect of montmorillonite on the reaction velocity was greater than that of allophane. The average V_{max}

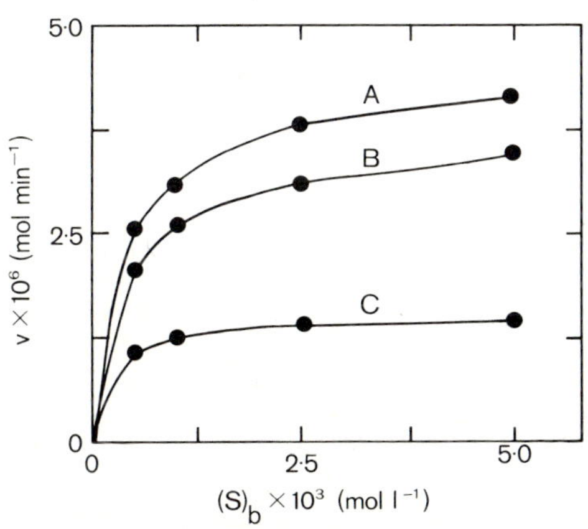

Fig. 7.13. Variation of initial reaction rate (v) with substrate concentration in the bulk phase (S_b) for the hydrolysis of carbobenzoxy-L-glutamyl-L-tyrosine by protease. A, control (in solution); B, in the presence of allophane; C, in the presence of montmorillonite (after Kobayashi and Aomine, 1967).

and K_m values derived from Fig. 7.13 are summarised in Table 7.6. These refer to the hydrolysis of CBGT by the clay—enzyme "mixture", and hence would include the contribution from any unadsorbed enzyme.

It is interesting to note that the montmorillonite—pronase "complex" showed almost no enzymic activity towards CBGT whereas the corresponding "complex" with allophane retained much of its initial capacity to hydrolyse the dipeptide. From this they inferred that the activity of the unwashed montmorillonite—enzyme system was due to pronase attached to external crystal surfaces and, to some extent, to the unbound enzyme both of which were presumably removed during washing. If a proportion of the adsorbed pronase were present in the interlayer space, the intercalated enzyme was evidently incapable of forming an enzyme—substrate complex. On the other hand, most of the allophane-bound pronase was apparently retained against washing, in keeping with Milestone's (1971) finding that proteins are tenaciously held by allophane.

Unlike CBGT, haemoglobin could be hydrolysed by pronase in the montmorillonite "complex" although the rate at which this occurred was lower than that shown by the enzyme in the corresponding "mixture". The reason why haemoglobin was apparently more susceptible to proteolysis, as Kobayashi and Aomine (1967) have suggested, was not obvious. The retention of enzymic activity by the "complex" (i.e. after washing) would imply, however, that some pronase was present in the montmorillonite interlayers. At pH 5 haemoglobin carries a net positive charge. This and its high molecular weight (~63 000) would favour its adsorption by the clay—enzyme "complex". Furthermore, haemoglobin is capable of being intercalated by montmorillonite (McLaren et al., 1958) and so can presumably combine with the interlayer enzyme prior to being hydrolysed. Similar results were obtained for α-amylase acting on amylose but, in this instance, the reduction in V_{max}

TABLE 7.6

Influence of allophane and montmorillonite on the apparent Michaelis constant (K_m) and maximum reaction velocity (V_{max}) for the hydrolysis of carbobenzoxy-L-glutamyl-L-tyrosine by a protease from *Streptomyces griseus* (from Kobayashi and Aomine, 1967)

	Treatment		
	Control (solution)	Allophane	Montmorillonite
K_m (mol l^{-1})	4.26×10^{-4}	4.10×10^{-4}	2.03×10^{-4}
V_{max} (amino-N, mol min^{-1})	4.52×10^{-6}	3.74×10^{-6}	1.54×10^{-6}

The values quoted are an average, derived from various linear transformations of the Michaelis—Menten rate equation (cf. equations 7.8 and 7.9).

and K_m, relative to the corresponding values in solution, was greater for allophane than for montmorillonite. On the other hand, K_m for β-amylase complexed with allophane was higher than for the mineral-free system, partly because the activity of this enzyme was considerably inhibited in the presence of allophane (Table 7.5). In addition, the β-amylase preparation was very far from being monodisperse and substrate diffusion to the interface might be retarded.

The influence of diffusion effects on K_m is illustrated by the measurements of Usami and Inoue (1974). Using glucoamylase attached to an acid clay acting on soluble starch, these workers found that K'_m was 5—6 times greater than K_m for a relatively high molecular weight (8×10^3) substrate. However, when lower molecular weight starch was used, the apparent Michaelis constant for the bound enzyme was similar in magnitude to that of its free counterpart. Steric effects may further reduce the rate of enzyme—substrate complex formation for high molecular weight (uncharged) substrates. Kinetic studies of enzyme action in the presence of clays (and soils) thus lend further support to the concept that substrate adsorption by the clay—enzyme derivative precedes its hydrolysis or digestion.

An example where the apparent Michaelis constant of an enzyme (soil acid phosphatase) acting on its substrate is measured in conjunction with substrate adsorption by the soil—enzyme system, has been described by Cervelli et al. (1973). To this end they took whole soils rather than the respective clay fractions and further assumed that only the fraction of the total substrate left in solution was available to the enzyme. The isotherm for the adsorption of substrate (*p*-nitrophenyl phosphate or PNP) obeys the Freundlich equation

$$x/m = K_f(S)_b^{1/n} \tag{7.12}$$

where x and m are the weight of PNP and soil, respectively; K_f and n are constants and $(S)_b$ is the equilibrium substrate concentration. Writing $(S)_b = (x/m)^n/K_f{}^n$ and substituting this function for (S) in equation (7.8) gives

$$\frac{1}{v} = K_f^n K_m/V_{max} \frac{1}{(x/m)^n} + 1/V_{max} \tag{7.13}$$

The value of K'_m for the process in which part of the added substrate was assumed to be immobilised by adsorption onto the soil and therefore did not participate in the reaction, was obtained by plotting $1/v$ against $1/(x/m)^n$. This yielded a series of straight lines for which $K'_m = (x/m)^n/K_f{}^n$ at $1/v = 0$. In another set of experiments the enzyme was assumed to act on the total amount of substrate present, yielding the corresponding K_m values. For all four soil samples used, $K'_m < K_m$. It should be noted, however, that the comparison made here is not between a soil system and a homogeneous solution. Rather, the study compares reaction rates in a soil system in which substrate adsorption is accounted for with one in which such a process is neglected. The results bear out the presupposition that more substrate is required if the reaction is to attain the same limiting rate as for the situation where no substrate immobilisation (by adsorption) occurs. Earlier, Durand (1964a) came to a similar conclusion for urease in that a higher urea concentration was required to reach V_{max} when the enzyme was attached to montmorillonite as compared with urease in solution.

At first sight, we might expect K_m' to be higher than K_m rather than the other way round since both the enzyme derivative and PNP are negatively charged. The fact that PNP, in this instance, was positively adsorbed by the soils indicated that factors other than charge—charge interactions were important. Alternatively, the negative surface charge of the adsorbing soil constituents may have been effectively masked by attachment of positively charged gel particles of hydrated aluminium and iron hydroxides. Leaving aside the involvement of microorganisms in the production of phosphatases in the soil, the above effects make it difficult to predict the direction towards which K_m' may be displaced relative to K_m. Thus, Parks (1974) has presented data to show that for an alkaline phosphatase (from *Escherichia coli*) acting on PNP, the apparent Michaelis constant for the clay—enzyme system was generally higher than for the corresponding (free) enzyme in solution. The clays used were unfractionated, raw samples of a kaolinite, an illite, a montmorillonite, and a palygorskite ("attapulgite"), reaction rates being measured in the presence of 1 M Tris buffer at pH 8 which coincided with the pH optimum of the enzyme. Under these conditions, $(S)_s$ was presumably less than $(S)_b$ so that the adsorbed enzyme reached a limiting or maximal velocity at a higher bulk substrate concentration than alkaline phosphatase in a homogeneous solution.

It is perhaps not altogether surprising that the apparent Michaelis constant of enzymes in soil systems does vary with soil type and condition, as Tabatabai and Bremner (1971) have reported for arylsulphatase and phosphatase and subsequently, Brams and McLaren (1974) did for phosphatase. The former workers noted, for example, that when the system was shaken prior to measuring reaction rates, both the magnitude and variation in K_m between soils were reduced. Shaking presumably promoted the adsorption of substrate onto the soil—enzyme complex, enhancing the probability of enzyme—substrate encounter. Alternatively, or in addition, this treatment would increase the rate of substrate diffusion from the bulk phase to the interface (cf. equation (7.11)).

In all the examples described above, it was implicitly assumed that the enzyme-catalysed process conformed to the classical kinetic model. This assumption, however, has recently been called into question by Irving and Cosgrove (1976). They found that at least for acid phosphatase acting on PNP in their (krasnozem) soil sample, the reaction did not follow Michaelis—Menten kinetics in that a plot of v against $v/(S)$ according to equation (7.9) was non-linear. To account for this behaviour, the following mechanism was suggested (cf. equation (7.6))

$$S_b \underset{k_{-1}}{\overset{k_1}{\rightleftharpoons}} S_s \underset{k_{-4}}{\overset{k_4(A^+)}{\rightleftharpoons}} AS$$

$$S_s \underset{k_{-2}}{\overset{k_2(\mathrm{E})}{\rightleftharpoons}} \mathrm{ES} \xrightarrow{k_3} \mathrm{E} + \mathrm{P}$$

(A^+) and AS being the concentration of adsorbing sites in the soil and of the adsorbed or relatively non-diffusible substrate, respectively. A departure from linearity in the Eadie—Hofstee plot would occur if the kinetics of step 1 and/or step 4 were non-linear with respect to S_b.

On the other hand, in situations where Michaelis—Menten kinetics are followed, considerable merit attaches to the determination of the apparent Michaelis constant for soil enzymes since K_m can usefully serve as an index of (enzyme) activity. However, caution should be exercised in interpreting K_m values and in making intra- and inter-soil comparisons. Besides being dependent on many variables already referred to, enzyme-catalysed reactions in soils are also influenced by the type and numbers of microorganisms present. The difficulty here lies in distinguishing between the activity due to extracellular enzymes (attached to clay or humus particles) and that attributable to intracellular (microbial) enzymes.

Paulson and Kurtz (1969) have suggested a useful approach to this problem. They were able to establish that the urease activity, at least of the soil they used, was linearly related to the number of ureolytic microorganisms. However, an appreciably high "background" activity remained as indicated by the positive Y-intercept when the regression line was extrapolated to zero population. This residual activity was ascribed to urease which was attached to soil colloids. By applying Michaelis—Menten kinetics and assigning the total urease activity to microbially-derived and adsorbed forms of the enzyme, they arrived at K_m values of 57×10^{-3} and 252×10^{-3} M, respectively.

Some extracellular enzymes in soil have been shown to be associated more with the organic matter or humus fraction than with the inorganic or mineral colloids (Chalvignac and Mayaudon, 1971; Ladd, 1972; Ladd and Butler, 1975; Cacco and Maggioni, 1976) but only in the case of urease has this been persuasively demonstrated (Burns et al., 1972a,b). However, the association between enzyme and humus is apparently non-uniform in that some fractions of the soil organic matter display a greater urease activity than do others (McLaren et al., 1975). Further, unlike its free counterpart, the humus-bound urease is resistant to proteolysis by pronase (McLaren and Pukite, 1975) presumably because the proteolytic enzyme is, on steric ground, prevented from entering the pores of the polyanionic (humus) matrix.

Finally, we wish to mention the recent data of Benesi and McLaren (1975) who have compared the relative activity of the SH-enzyme, papain, in solution and when attached to kaolinite. Requiring a free SH group, the activity of papain depends on the redox potential of the system and hence, on the ratio (R) of disulphide/thiol. By analogy with H^+ ions and positively charged species, in general, the relative distribution of charged disulphide and thiol compounds near the surface of the enzyme or the clay—enzyme complex would, to a first approximation, obey a Maxwell—Boltzmann relationship (cf. equation (1.1)). For a redox pair, such as $^+R{-}S{-}S{-}R^+$ and $^+R{-}SH$, the former would be more strongly attracted to the negatively charged clay surface. The approach used was to determine enzyme activity under nitrogen for papain in solution at pH_b = 6.1 and when adsorbed onto kaolinite at pH_b = 6.9 (cf. Table 7.5), in the presence of different disulphide—thiol mixtures at various values of R. One mixture (A) consisted, for example, of dithiodiglycol and β-mercaptoethanol, both of which were uncharged; a second mixture (B) of dithioglycol and β-aminoethylmercaptan (monovalent cation); a third (C) of cystine ethyl ester (divalent cation, disulphide) and cysteine ethyl ester (monovalent cation, thiol).

The results, depicted in Fig. 7.14 show that in the presence of pair (A), the curve for papain in solution coincided with that of the adsorbed enzyme.

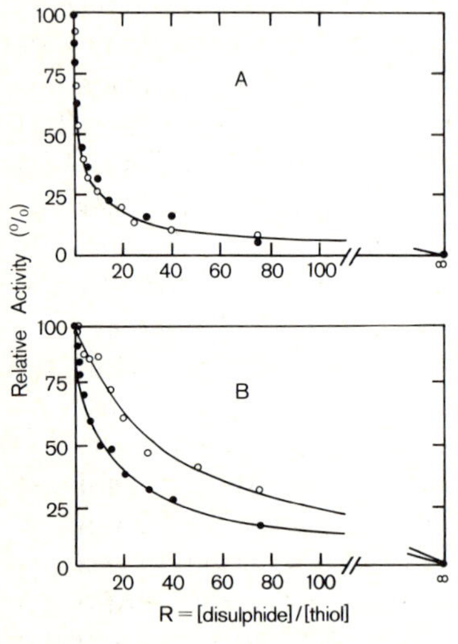

Fig. 7.14. Relative activity of papain with benzoyl arginine ethyl ester as a function of the ratio (oxidising disulphide)/reducing thiol). (○), soluble papain (pH_b 6.1, ionic strength ~ 0.07); (●), papain—kaolinite complex (pH_b 6.9, ionic strength ~ 0.07). A, dithiodiglycol and β-mercaptoethanol; B, dithiodiglycol and β-amino ethylmercaptan (after Benesi and McLaren, 1975).

This would indicate that although the amount of active, reduced SH-enzyme clearly depended on R, this ratio was apparently the same in solution as near the clay—enzyme complex surface. On the other hand, the relative activity of the clay-bound enzyme for pair (B) was markedly reduced as compared with the corresponding free enzyme. This difference was more pronounced for pair (B) because the concentration of $^{+}R{-}S{-}S{-}R^{+}$ was increased over and above that of $^{+}R{-}SH$ in the vicinity of the surface.

7.5. ENZYME BEHAVIOUR IN SOILS

It is not intended here to discuss at length the occurrence, activity, and biological significance of enzymes in soil. Besides lying beyond the scope of this monograph, these aspects of soil biology and biochemistry have been adequately dealt with in a number of reviews (Durand, 1965; Hofmann and Hoffmann, 1966; Voets and Dedeken, 1966; Skujinš, 1967; Kuprevich and Shcherbakova, 1971; Sequi, 1974; Kiss et al., 1975). Rather, to round off our discussion on clay—enzyme systems, we wish to insert a few remarks about the behaviour of enzymes in soil. In so doing, we hope to establish a link between the aseptic chemically definable system, emphasised in the preceding section, and the "living" system which soil essentially is.

It is generally accepted that microorganisms, both living (proliferating) and dead, are the primary source of enzymes in soil although the animal population as well as the vegetation must also make a sizeable contribution to the overall enzyme pool. However, conventional biochemical techniques used to measure enzyme activities in soil do not normally discriminate between extra- and intracellular species. Nonetheless, we may reasonably expect to find a positive correlation between enzyme activity and microbial numbers. Such relationships have indeed been reported for many soils, as illustrated by the work of Paulson and Kurtz (1969) to which we have already referred. Other data, however, tend to support the view that enzyme activity bears little or no relationship to either microbial population as a whole or to the number of microbes in each of the main groups of microorganisms. Similarly, experimental results on the extent to which seasonal factors affect the dynamics and accumulation of soil enzymes are not always concordant. This is also true for the influence of clay addition to soil on enzyme activity. These and other apparent inconsistencies in the literature on enzyme behaviour in soil are perhaps to be expected since microbial numbers are influenced by climatic and soil conditions and, in the case of cultivated soils, by management practices as well. In addition, enzyme activity depends on the synthesis of the enzyme in question for which not all of the microorganisms present in the soil may be responsible. As Ramirez-Martinez and McLaren (1966) have pointed out, the dilution plate technique of assessing microbial numbers does not accurately reflect the actual state of microbial population unless what is being isolated and counted refers to those species which are actively involved in breaking down the particular substrate. Furthermore, the addition of substrate to the soil may stimulate the production or activate the appropriate enzyme systems.

Kiss (1958a) found, for example, that although both kaolinite and montmorillonite were capable of inhibiting invertase activity "in vitro", they failed to show this effect when soil was added to the system. By the same token, the addition of these clay species to a soil did not affect its natural invertase activity; only when the minerals were added together with sucrose did a marked stimulation of enzyme activity occur (Kiss, 1958b). From this he inferred that sucrose induced the microbial synthesis of invertase which, outside the living cell, was stabilised by adsorption onto the (added) clay surface. The induction of enzyme synthesis in this way is by no means uncommon. Thus, the microbial production of amylase (Drobnik, 1955), dextranase (Drăgan-Bularda and Kiss, 1972), levan sucrase (Kiss, 1961), and xylanase (Sørensen, 1955) in soil has been shown to increase in the presence of their respective substrates.

Similarly, Sørensen (1969, 1972) has reported that when montmorillonite and carbohydrates were added together to a soil, there was a notable increase in the activity of the appropriate enzymes. Using ^{14}C-labelled substrates, he was able to recover an appreciable proportion (6—12%) of the carbohydrate carbon in the form of amino acids during the first 10—30 days of incubation. In the presence of montmorillonite (5% w/w), the amount of recoverable amino acids was substantially increased, indicating that part of the added carbon was rapidly transformed into enzyme proteins which were then protected from proteolysis by adsorption onto the clay.

If, as seems probable, living cells contribute but a small proportion to the overall enzyme pool, the biological activity of a soil (often regarded as an index to its fertility at the time of sampling) must be attributed more to the activity of extracellular enzymes than to microbial numbers per se. This hypothesis has stimulated a great deal of research into the detection and measurement of extracellular enzymes in soil systems (e.g. Hofmann and Hoffmann, 1966). A common approach to separating the activity of such enzymes from that due to the presence of proliferating microorganisms is to sterilise the soil in such a way that microbial activity is arrested (without causing cell lysis) while leaving the extracellular enzymes more or less intact. This is normally achieved by adding sterilising (biocidal) agents to the soil of which toluene is perhaps the most widely used. Alternatively, the soil is exposed to high-energy radiation, such as γ-rays. The merits and disadvantages of each method in terms of enzyme activity have been described by Skujinš (1967), McLaren (1969), Sequi (1974) and Kiss et al. (1975). The subject of soil irradiation and the effect of biocidal treatments on soil organisms and properties has been summarised by McLaren (1969), Cawse (1975), and Powlson (1975). Although expensive, radiation sterilisation is perhaps the more ideally suited for enzyme studies because no extraneous compounds or their residues are introduced into or remain in the system after treatment. An added advantage is that although enzymes do vary in their sensitivity to irradiation, they are largely resistant to a radiation dose which all microorganisms would fail to survive (~2.5 Mrad). In other words, most soil enzymes retain their activity, and in some instances may become more active, in irradiated soils. This activity, however, is partly due to intracellular enzymes released from cells which have ceased to proliferate because of damage to their nucleic acid content.

Clearly, the ideal method and one which does not lead to any ambiguity would be to extract and isolate the enzyme(s) concerned from the soil mass, followed by its (their) purification and characterisation. This has proved to be a rather difficult and laborious exercise. That the reward has not matched the effort can be ascribed to the intimate association of enzymes with the mineral and organic constituents of soil. Only relatively few enzymes have been isolated from soils and then mostly in combination with the soil colloids. Thus, Briggs and Segal (1963) separated a mixture of proteins with urease activity, and Martin-Smith (1963) one which was active towards uric acid. Subsequently, Bartha and Bordeleau (1969) obtained a preparation with oxidase and peroxidase activities towards aniline and *o*-anisidine, respectively, whilst Chalvignac and Mayaudon (1971) reported the extraction of an enzyme system which catalysed the transformation of tryptophan into indole acetic acid. Similarly, Ladd (1972) succeeded in extracting a peptidase from soil, and Satyanarayana and Getzin (1973) an esterase which decomposed malathion.

As previously indicated, the urease-active material extracted from soil by Burns et al. (1972a,b) and by McLaren et al. (1975) was identified as an enzyme—humus conjugate, representing 20—40% of the total activity of the soil. Nannipieri et al. (1974) reported similar yields from a soil which, incidentally, contained the same number of microorganisms after extraction. The association of extracellular urease with organic matter in an alpine humus soil has also been suggested by Lloyd (1975). Likewise, the diphenol oxidases of Mayaudon et al. (1973a,b) and Mayaudon and Sarkar (1974) were largely complexed with humus whereas the esterase of Satyanarayana and Getzin (1973) appeared to be a carbohydrate—protein derivative.

This is not to say, however, that the mineral or clay fraction of soils plays only a minor part or is not involved in complex formation with extracellular proteins and enzymes. Even if most of the enzymically active preparations extracted from soil were in the form of a humus derivative, they would be indirectly bound to the mineral constituents, since humic substances by themselves are, for the most part, attached to clay surfaces (cf. Chapter 12). In the instances quoted above, the clay—humus bond might be weak as compared with that between enzyme and humus. The methods used to extract enzymically active substances from soils could conceivably release the humus—enzyme conjugates from their respective mineral (clay) complexes, as Mayaudon et al. (1973a,b) have suggested for their system.

Another piece of evidence for the involvement of the inorganic components of soils, notably the clay fraction, in binding (soil) enzymes, is provided by soil fractionation studies. Quite early on Haig (1955) found that in a fine sandy loam esterase activity, as measured by its effect on phenylacetate, decreased in the order clay > silt >> sand. Indeed, the enzyme was so strongly attached to the (montmorillonitic) clay fraction that it effectively resisted elution by organic bases and phosphate; nor could it be separated from its clay complex by sonic treatment. Hoffmann (1959) has reported similarly for some carbohydrates and urease. The virtual absence of microorganisms from the clay fraction further indicated that the clay-bound enzymes were largely extracellular. Likewise, Galstyan and co-workers (Galstyan et al., 1968; Galstyan and Havoundjian, 1970) have presented data to show that the activity of various enzymes in the surface horizon of two genetically different soils decreased with an increase in mean particle size, being concentrated in the silt and clay fractions and absent from sand particles. They also noted that enzyme activity was positively correlated with the humus content of the respective particle-size fractions. In fractionating a series of soils by ultrasonic dispersion in a dense liquid, Ross (1975) and Speir (1977) observed that although the "light" fraction was enriched in a number of enzymes, most of the activity was in the residue. This was in accord with the results of Aliev et al. (1975) who found large amounts of (soil) particle-bound extracellular enzymes which could not be released into solution by ultrasonic treatment. The strong attachment of enzymes to the inorganic soil colloids was also inferred by Dalal (1975) and Khaziev (1975) from measurements of the enthalpies and entropies of activation.

There is evidence to show that the total enzyme activity of a soil is a characteristic index of soil type (Kuprevich and Shcherbakova, 1971). This concept has important implications for both soil classification and forensic purposes since the detection limits of enzymes are very much lower than any other organic compound in the soil. The forensic aspect of soil enzymology has recently been examined by Thornton et al. (1975) who found that not only were enzyme activity levels diagnostic of geographical locality but soils collected from adjacent sites within a given location could be characterised by their individual enzyme pattern. The forensic application of enzyme activity measurements may be considerably enhanced by determining the apparent Michaelis constant of the enzyme-catalysed reaction, a value which is independent of sample size. When the sensitivity of the assay is further increased by using radioactive substrates, K_m ratios for various substrates can usefully serve as soil markers.

We have already seen that clays either inhibit, stimulate, or exert no measurable influence on enzyme activity although the inhibitory effect is the more prevalently observed. This behaviour of enzymes in the presence of clays also extends to soils but the interpretation of experimental data for such systems is more difficult and equivocal because biochemical processes occurring "in vivo" are more of a dynamic and consecutive than a steady-state nature. During the infiltration of a soil column by a substrate solution, for example, the ratio of $(S)/(P)$ (cf. equation 7.6) depends on the position within and the time of passage through the column, that is, the process has a vectorial as well as a temporal component (McLaren, 1970). In addition, a particular compound present in or added to soil, may undergo a series of consecutive transformations not all of which may be mediated by extracellular enzymes, as exemplified by the sequence urea $\rightarrow NH_4^+ \rightarrow NO_2^- \rightarrow NO_3^- \rightarrow N_2O \rightarrow N_2$. The nitrification step, in particular, has received much attention in terms of both experimental and theoretical considerations (McLaren, 1969; McLaren and Ardakani, 1972; Ardakani et al., 1973, 1974, 1975). Often the value of K_m for an enzyme reaction in a dynamic column is appreciably higher than in the corresponding static batch situation, due probably to diffusion and surface charge effects (McLaren, 1972, 1975).

In keeping with the data on "pure" clay systems, montmorillonite in soil generally affords greater stability to adsorbed organic compounds or substrates as compared with either illite or kaolinite (Ensminger and Gieseking, 1942; Lynch and Cotnoir, 1956; Sørensen, 1972). The higher "protective" effect of montmorillonite relative to the non-expanding clays is undoubtedly related to its higher cation exchange capacity and larger accessible area. It seems also probable that montmorillonite in soil is capable of intercalating organic substrates which then become relatively inaccessible to extracellular enzymes. However, there is as yet little direct evidence for the occurrence in soil of interlayer montmorillonite complexes. Even if they were formed quite readily, the resulting (biological) stability of organic substances adsorbed in this way may not be permanent, particularly if the soil is subjected to repeated drying and wetting (Birch and Friend, 1956; Birch, 1958; Sørensen, 1974).

REFERENCES

Albert, J.T. and Harter, R.D., 1973. Adsorption of lysozyme and ovalbumin by clay: Effect of clay suspension pH and clay mineral type. Soil Science, 115: 130—136.

Aliev, R.A. and Zvyagintsev, D.G., 1974. Effect of protolytic enzymes on free and adsorbed catalase and on the catalase activity of soils. Vestnik Moskovskogo Universiteta. Pochvovedenie, 29: 97—104.

Aliev, R.A., Guzev, V.S. and Zvyagintsev, D.G., 1976. Influence of adsorbents on the optimum pH of catalase. Vestnik Moskovskogo Universiteta. Pochvovedenie, 31: 67—70.

Aliev, R.A., Zvyagintsev, D.G. and Koshevin, P., 1975. Determination of extracellular and intracellular enzymes in soil by treatment with ultrasound and subsequent differential centrifugation. Vestnik Moskovskogo Universiteta. Pochvovedenie, 30: 76—80.

Allison, F.E., 1973. Soil Organic Matter and Its Role in Crop Production. Elsevier, Amsterdam, 637 pp.

Aomine, S. and Kobayashi, Y., 1964a. Effect of allophane on the activity of some enzymes. Transactions of the 8th International Congress of Soil Science, Bucharest, 3: 697—703.

Aomine, S. and Kobayashi, Y., 1964b. Effects of allophane on the enzymatic activity of a protease. Soil Science and Plant Nutrition (Tokyo), 10: 28—32.

Aomine, S. and Kobayashi, Y., 1966. Effects of allophanic clays on the enzymatic activity of beta-amylase. Soil Science and Plant Nutrition (Tokyo), 12: 7—12.

Aomine, S. and Kodama, I., 1956. Clay minerals in some arable soils in Miyazaki prefecture. Journal Faculty of Agriculture, Kyushu University, 10: 325—344.

Ardakani, M.S., Rehbock, J.T. and McLaren, A.D., 1973. Oxidation of nitrite to nitrate in a soil column. Soil Science Society of America Proceedings, 37: 53—56.

Ardakani, M.S., Schulz, R.K. and McLaren, A.D., 1974. A kinetic study of ammonium and nitrite oxidation in a soil field plot. Soil Science Society of America Proceedings, 38: 273—277.

Ardakani, M.S., Volz, M.G. and McLaren, A.D., 1975. Consecutive steady state reactions of urea, ammonium and nitrate nitrogen in soil. Canadian Journal of Soil Science, 55: 83—91.

Armstrong, D.E. and Chesters, G., 1964. Properties of protein—bentonite complexes as influenced by equilibration conditions. Soil Science, 98: 39—52.

Aylmore, L.A.G. and Quirk, J.P., 1960. Domain or turbostratic structure of clays. Nature (London), 187: 1046—1048.

Baier, R.E. (Editor), 1975. Applied Chemistry at Protein Interfaces. Advances in Chemistry Series No. 145, 400 pp.

Barrer, R.M. and Brummer, K., 1963. Relations between partial ion exchange and interlamellar sorption in alkylammonium montmorillonites. Transactions of the Faraday Society, 59: 959—968.

Barnett, L.B. and Bull, H.B., 1959. Optimum pH of adsorbed ribonuclease. Biochimica et Biophysica Acta, 36: 244—246.

Bartha, R. and Bordeleau, L.M., 1969. Cell-free peroxidases in soil. Soil Biology and Biochemistry, 1: 139—143.

Bartholomew, W.V., 1965. Mineralization and immobilization of nitrogen in the decomposition of plant and animal residues. In: W.V. Bartholomew and F.E. Clark (Editors), Soil Nitrogen. American Society of Agronomy, Monograph 10, Madison, Wisconsin, pp. 285—306.

Bartholomew, W.V. and Clark, F.E. (Editors), 1965. Soil Nitrogen. American Society of Agronomy, Monograph 10, Madison, Wisconsin, 615 pp.

Basaraba, J. and Starkey, R.L., 1966. Effect of plant tannins on decomposition of organic substances. Soil Science, 101: 17—23.

Benesi, A. and McLaren, A.D., 1975. Radox dependence of papain enzyme activity on charged kaolinite surfaces and in solution. Soil Biology and Biochemistry, 7: 379—381.

Benoit, R.E., Starkey, R.L. and Basaraba, J., 1968. Effect of purified plant tannin on decomposition of some organic compounds and plant materials. Soil Science, 105: 153—158.

Bernal, J.D., 1951. The Physical Basis of Life. Routledge and Kegan Paul, London, 80 pp.

Birch, H.F., 1958. The effect of soil drying on humus decomposition and nitrogen availability. Plant and Soil, 10: 9—31.

Birch, H.F. and Friend, M.T., 1956. Humus decomposition in East African soils. Nature (London), 178: 500—501.

Bolland, M.D.A., Posner, A.M. and Quirk, J.P., 1976. Surface charge on kaolonites in aqueous suspension. Australian Journal of Soil Research, 14: 197—216.

Brams, W.H. and McLaren, A.D., 1974. Phosphatase reactions in columns of soil. Soil Biology and Biochemistry, 6: 183—189.

Brash, J.L. and Lyman, D.J., 1969. Adsorption of plasma proteins in solution to uncharged, hydrophobic polymer surfaces. Journal of Biomedical Materials Research, 3: 175—189.

Brash, J.L. and Lyman, D.J., 1971. Adsorption of proteins and lipids to nonbiological surfaces. In: M.L. Hair (Editor), The Chemistry of Biosurfaces, Vol. 1. Marcel Dekker, New York, N.Y., pp. 177—232.

Bremner, J.M., 1965. Organic nitrogen in soils. In: W.V. Bartholomew and F.E. Clark (Editors), Soil Nitrogen. American Society of Agronomy, Monograph 10, Madison, Wisconsin, pp. 93—149.

Bremner, J.M., 1967. Nitrogenous compounds. In: A.D. McLaren and G.H. Peterson (Editors), Soil Biochemistry, Vol. 1. Marcel Dekker, New York, N.Y., pp. 19—66.

Briggs, M.H. and Segal, L., 1963. Preparation and properties of a free soil enzyme. Life Sciences, 2: 69—72.

Bull, H.B., 1956. Adsorption of bovine serum albumin on glass. Biochimica et Biophysica Acta, 19: 464—471.

Burns, R.G., El-Sayed, M.H. and McLaren, A.D., 1972a. Extraction of a urease-active organo-complex from soil. Soil Biology and Biochemistry, 4: 107—108.

Burns, R.G., Pukite, A.H. and McLaren, A.D., 1972b. Concerning the location and persistence of soil urease. Soil Science Society of America Proceedings, 36: 308—311.

Cacco, G., Maggioni, A., 1976. Multiple forms of acetyl-naphthyl esterase activity in soil organic matter. Soil Biology and Biochemistry, 8: 321—325.

Cawse, P.A., 1975. Microbiology and biochemistry of irradiated soils. In: E.A. Paul and A.D. McLaren (Editors), Soil Biochemistry, Vol. 3. Marcel Dekker, New York, N.Y., pp. 213—267.

Cervelli, S., Nannipieri, P., Ceccanti, B. and Sequi, P., 1973. Michaelis constants of soil acid phosphatase. Soil Biology and Biochemistry, 5: 841—845.

Chalvignac, M.A. and Mayaudon, J., 1971. Extraction and study of soil enzymes metabolising tryptophan. Plant and Soil, 34: 25—31.

Chattoraj, D.K. and Bull, H.B., 1959. Electrophoresis of adsorbed protein. Journal of the American Chemical Society, 81: 5128—5133.

Chernyakhovskaya, V. and Runova, V.F., 1975. Protein sorption with mineral colloids. Zhurnal Mikrobiologii, Epidemiologii i. Immunobiologii, pp. 104—108. See: Chemical Abstracts, 83: 127748 (1975).

Chiang, C.M. and Chao, C.H., 1964. Adsorption of protein by soil colloids. T'u Jang Hsueh Pao, 12: 411—420. See: Chemical Abstracts, 65: 1429a (1966).

Coleman, N.T. and Craig, D., 1961. The spontaneous alteration of hydrogen clay. Soil Science, 91: 14—18.

Dalal, R.C., 1975. Effect of toluene on the energy barriers in urease activity of soils. Soil Science, 120: 256—260.

De Boeck, S., 1975. Geïmmobilizeerde enzymen. Natuurwetenschappelijk Tijdschrift (Gent), 57: 129—157.

De Borger, R., 1972. Huidige kennis omtrent de bouw der humusstoffen. Landbouwtijdschrift, 25: 37—50.

Deželić, G., Deželić, N. and Telišman, Z., 1971. The binding of human serum albumin by monodisperse polystyrene latex particles. European Journal of Biochemistry, 23: 575—581.

Doonan, S., 1969. The chemistry and physics of enzyme catalysis. Royal Institute of Chemistry Reviews, 2: 117—142.

Dowd, J.E. and Riggs, D.S., 1965. A comparison of estimates of Michaelis—Menten kinetic constants from various linear tranformations. Journal of Biological Chemistry, 240: 863—869.

Drăgan-Bularda, M. and Kiss, S., 1972. Dextranase activity in soil. Soil Biology and Biochemistry, 4: 413—416.

Drobnik, J., 1955. Spaltung der Stärke durch den enzymatischen Komplex der Böden. Folia Biologica (Praha), 1: 29—40.

Dubach, P. and Mehta, N.C., 1963. The chemistry of soil humic substances. Soils and Fertilisers, 26: 293—300.

Duchaufour, P., 1976. Dynamics of organic matter in soils of temperate regions: its action on pedogenesis. Geoderma, 15: 31—40.

Durand, G., 1964a. Modifications de l'activité de l'uréase en présence de bentonite. Comptes Rendus Hebdomadairès des Séances de l'Académie des Sciences, 259: 3397—3400.

Durand, G., 1964b. Influence des minéraux argileux sur la décomposition de quelques dérivés nucléiques. Annales de l'Institut Pasteur, 107: 136—147.

Durand, G., 1965. Les enzymes dans le sol. Revue d'Écologie et de Biologie du Sol, 2: 141—205.

Edsall, J.T., 1953. The size, shape, and hydration of protein molecules. In: H. Neurath and K. Bailey (Editors), The Proteins, Vol. 1 (Part B). Academic Press, New York, N.Y., pp. 549—726.

Eeckman, J.P. and Laudelout, H., 1961. Chemical stability of hydrogen montmorillonite suspensions. Kolloid Zeitschrift, 178: 99—107.

Ensminger, L.E. and Gieseking, J.E., 1939. The adsorption of proteins by montmorillonitic clays. Soil Science, 48: 467—471.

Ensminger, L.E. and Gieseking, J.E., 1941. The adsorption of proteins by montmorillonitic clays and its effect on base-exchange capacity. Soil Science, 51: 125—132.

Ensminger, L.E. and Gieseking, J.E., 1942. Resistance of clay-adsorbed proteins to proteolytic hydrolysis. Soil Science, 53: 205—209.

Erickson, A.E., 1948. The Stability of Organic—Clay Complexes to Enzymatic and Chemical Reactions. Ph.D. Thesis, University of Illinois, Urbana, Illinois.

Estermann, E.F. and McLaren, A.D., 1959. Stimulation of bacterial proteolysis by adsorbents. The Journal of Soil Science, 10: 64—78.

Estermann, E.F., Peterson, G.H. and McLaren, A.D., 1959. Digestion of clay—protein, lignin—protein, and silica—protein complexes by enzymes and bacteria. Soil Science Society of America Proceedings, 23: 31—36.

Felbeck, G.T., Jr, 1965. Structural chemistry of soil humic substances. Advances in Agronomy, 17: 327—368.

Felbeck, G.T., Jr, 1971. Chemical and biological characterization of humic matter. In: A.D. McLaren and J. Skujinš (Editors), Soil Biochemistry, Vol. 2. Marcel Dekker, New York, N.Y., pp. 36—59.

Fenstermaker, C.A., Grant, W.H., Morrissey, B.W., Smith, L.E. and Stromberg, R.R., 1974. Interaction of plasma proteins with surfaces. National Bureau of Standards, Washington, D.C. Report No. 74-470, 72 pp.

Fevold, H.L., 1951. Egg proteins. Advances in Protein Chemistry, 6: 188—252.

Fieldes, M. and Claridge, G.G.C., 1975. Allophane. In: J.E. Gieseking (Editor), Soil Components, Vol. 2. Inorganic Components. Springer Verlag, Berlin, pp. 351—393.

Finer, E.G., Franks, F., Phillips, M.C. and Suggett, A., 1975. Gel formation from solutions of single-chain gelatin. Biopolymers, 14: 1995—2005.

Flaig, W., Beutelspracher, H. and Rietz, E., 1975. Chemical composition and physical properties of humic substances. In: J.E. Gieseking (Editor), Soil Components, Vol. 1. Organic Components. Springer Verlag, Berlin, pp. 1—211.

Fontana, B.J. and Thomas, J.R., 1961. The configuration of adsorbed alkyl methacrylate polymers by infrared and sedimentation studies. The Journal of Physical Chemistry, 65: 480—487.

Fraser, M.J., 1957. Behaviour of proteins at interfaces. Journal of Pharmacy and Pharmacology, 9: 497—521.

Fripiat, J.J. and Cruz-Cumplido, M.I., 1974. Clays as catalysts for natural processes. Annual Review of Earth and Planetary Science, 2: 239—256.

Fripiat, J.J. and Poncelet, G., 1974. Clays and prebiotic synthesis. Advances in Organic Geochemistry Proceedings 6th International Meeting, pp. 875—881.

Fruton, J.S. and Simmonds, S., 1961. General Biochemistry, 2nd Edition. John Wiley and Sons, Inc., New York, N.Y., 1071 pp.

Galstyan, A.S. and Havoundjian, Z.S., 1970. Dehydrogenases in the clay fraction of soils. Doklady Akademii Nauk S.S.S.R., 195: 707—709.

Galstyan, A.S. and Tatevosyan, G.S., 1964. Ferment activity as characteristic of soil types. Transactions of the 8th International Congress of Soil Science, Bucharest, 3: 711—718.

Galstyan, A.S., Tatevosyan, G.S. and Havoundjian, Z.S., 1968. Fixation of enzymes by soil fractions of different particle sizes. Transactions of the 9th International Congress of Soil Science, Adelaide, 3: 281—288.

Garvey, M.J., Tadros, Th.F. and Vincent, B., 1976. A comparison of the adsorbed layer thickness obtained by several techniques of various molecular weight fractions of poly(vinyl alcohol) on aqueous polystyrene latex particles. Journal of Colloid and Interface Science, 55: 440—453.

Gieseking, J.E. (Editor), 1975. Soil Components, Vol. 1. Organic components. Springer Verlag, Berlin, 534 pp.

Giles, C.H., 1970. Interpretation and use of sorption isotherms. In: J.G. Gregory (Editor), Sorption and Transport Processes in Soils. Monograph No. 37, Society of Chemical Industry, London, pp. 14—31.

Goh, K.M., 1972. Amino acid levels as indicators of paleosols in New Zealand profiles. Geoderma, 7: 33—47.

Greenland, D.J., 1972. Interactions between organic polymers and inorganic soil particles. Mededelingen Fakulteit Landbouwwetenschappen, Rijksuniversiteit Gent, 37: 897—914.

Greenland, D.J., Laby, R.H. and Quirk, J.P., 1962. Adsorption of glycine and its di-, tri-, and tetra-peptides by montmorillonite. Transactions of the Faraday Society, 58: 829—841.

Greenland, D.J., Laby, R.H. and Quirk, J.P., 1965. Adsorption of amino acids and peptides by montmorillonite and illite. Part 1. Cation exchange and proton transfer. Transactions of the Faraday Society, 61: 2013—2023.

Gustavson, K.H., 1966. The function of the basic groups of collagen in its reaction with vegetable tannins. Journal of the Society of Leather Trades' Chemists, 50: 144—160.

Haider, K., Martin, J.P. and Filip, Z., 1975. Humus biochemistry. In: E.A. Paul and A.D. McLaren (Editors), Soil Biochemistry, Vol. 3. Marcel Dekker, New York, N.Y., pp. 195—244.

Haig, A.D., 1955. Some Characteristics of Esterase and Urease-like Activity in Soil. Ph.D. Thesis, University of California, Davis, California.

Handley, W.R.C., 1961. Further evidence for the importance of residual leaf protein complexes in litter decomposition and the supply of nitrogen for plant growth. Plant and Soil, 15: 37—73.

Harada, T., 1959. The mineralization of native organic nitrogen in paddy soils and the mechanism of its mineralization. Bulletin of the National Institute of Agricultural Sciences (Tokyo), 9B: 123—199.

Harter, R.D., 1975. Effect of exchange cations and solution ionic strength on formation and stability of smectite—protein complexes. Soil Science, 120: 174—181.

Harter, R.D. and Ahlrichs, J.L., 1967. Determination of clay surface acidity by infrared spectroscopy. Soil Science Society of America Proceedings, 31: 30—33.

Harter, R.D. and Stotzky, G., 1971. Formation of clay—protein complexes. Soil Science Society of America Proceedings, 35: 383—389.

Harter, R.D. and Stotzky, G., 1973. X-ray diffraction, electron microscopy, electrophoretic mobility, and pH of some stable smectite—protein complexes. Soil Science Society of America Proceedings, 37: 116—123.

Hartley, B.S., 1964. The structure and activity of chymotrypsin. In: T.W. Goodwin, J.I. Harris, and B.S. Hartley (Editors), Structure and Activity of Enzymes. Academic Press, London, pp. 47—60.

Hartley, G.S. and Roe, J.W., 1940. Ionic concentrations at interfaces. Transactions of the Faraday Society, 36: 101—109.

Hendricks, S.B., 1941. Base exchange of the clay mineral montmorillonite for organic cations and its dependence upon adsorption due to van der Waals forces. The Journal of Physical Chemistry, 45: 65—81.

Hétier, J.M., Jerez, F.G. and Bruckert, S., 1974. Morphoscopie et composition des complexes organo-minéraux des andosols. Comptes Rendus Hebdomadaires des Séances de l'Académie des Sciences, 278: 2735—2737.

Hobson, R.P. and Page, H.J., 1932. Studies on the carbon and nitrogen cycles in the soil. VII. The nature of the organic nitrogen compounds of the soil: "humic" nitrogen. Journal of Agricultural Science (Cambridge), 22: 497—515.

Hoffmann, G., 1959. Verteilung und Herkunft einiger Enzyme im Boden. Zeitschrift für Pflanzenernährung, Düngung, Bodenkunde, 85: 97—104.

Hofmann, E. and Hoffmann, G., 1966. Die Bestimmung der biologischen Tätigkeit in Böden mit Enzymmethoden. Advances in Enzymology, 28: 365—390.

Hornby, W.E., Lilly, M.D. and Crook, E.M., 1966. The preparation and properties of ficin chemically attached to carboxymethyl cellulose. Biochemical Journal, 98: 420—425.

Hornby, W.E., Lilly, M.D. and Crook, E.M., 1968. Some changes in the reactivity of enzymes resulting from their chemical attachment to water-insoluble derivatives of cellulose. Biochemical Journal, 107: 669—674.

Hurst, H.M. and Burges, N.A., 1967. Lignin and humic acids. In: A.D. McLaren and G.H. Peterson (Editors), Soil Biochemistry, Vol. 1. Marcel Dekker, New York, N.Y., pp. 260—286.

Irving, G.C.J. and Cosgrove, D.J., 1976. The kinetics of soil acid phosphatase. Soil Biology and Biochemistry, 8: 335—340.

Jackman, R.H., 1955. Organic phosphorus in New Zealand soils under pasture. II. Relations between organic phosphorus content and some soil characteristics. Soil Science, 79: 293—299.

Katchalski, E., Silman, I. and Goldman, R., 1971. Effect of the microenvironment on the mode of action of immobilized enzymes. Advances in Enzymology, 34: 445—536.

Khaziev, F.Kh., 1975. Thermodynamic characteristics of enzymic reactions in soil. Biologicheskie Nauki (Moscow), 18: 121—127. See: Chemical Abstracts, 84: 42394 (1976).

Kiss, S., 1958a. Invertase activity in clay mineral—soil mixtures. Nature (London), 182: 203—204.

Kiss, S., 1958b. Untersuchungen über die Produktion von Saccharase im Boden. Zeitschrift für Pflanzenernährung, Düngung, Bodenkunde, 81: 117—125.

Kiss, S., 1961. Über die Anwesenheit von Lävansucrase im Boden. Naturwissenschaften, 48: 700.

Kiss, S., Drăgan-Bularda, M. and Rădulescu, D., 1975. Biological significance of enzymes accumulated in soil. Advances in Agronomy, 27: 25—87.

Kobayashi, Y. and Aomine, S., 1967. Mechanism of inhibitory effect of allophane and montmorillonite on some enzymes. Soil Science and Plant Nutrition (Tokyo), 13: 189—194.

Kononova, M.M., 1966. Soil Organic Matter, 2nd Edition. Pergamon Press, Oxford, 544 pp.

Kragh, A.M., 1964. The adsorption of gelatin by suspended solids and the mechanism of flocculation. Journal of Photographic Science, 12: 191—201.

Kroll, L. and Kramer, M., 1955. Der Einfluss der Tonmineralien auf die Enzym-Aktivität der Bodenphosphatase. Naturwissenschaften, 42: 157—158.

Kuprevich, V.F. and Shcherbakova, T.A., 1971. Comparative enzymatic activity in diverse types of soil. In: A.D. McLaren and J. Skujinš (Editors), Soil Biochemistry, Vol. 2. Marcel Dekker, New York, N.Y., pp. 167—201.

Ladd, J.N., 1972. Properties of proteolytic enzymes extracted from soil. Soil Biology and Biochemistry, 4: 227—237.

Ladd, J.N. and Butler, J.H.A., 1975. Humus—enzyme systems and synthetic organic polymer—enzyme analogs. In: E.A. Paul and A.D. McLaren (Editors), Soil Biochemistry, Vol. 3. Marcel Dekker, New York, N.Y., pp. 143—194.

Lagaly, G. and Weiss, A., 1975. The layer charge of smectite layer silicates. Proceedings of the International Clay Conference, Mexico City, pp. 157—172.

Laura, R.D., 1974. Effects of neutral salts on carbon and nitrogen mineralization of organic matter in soil. Plant and Soil, 41: 113—127.

Laura, R.D., 1975. On the protective effect of clay minerals. Soil Science, 120: 241—243.

Laura, R.D., 1976. On the interlamellar spacings of protein—montmorillonite complexes. Soil Science, 121: 189—190.

Lespinat, P.A., Hétier, J.M., Thomann, C., Chone, T. and Dimon, A., 1976. Utilisation de mais mûr uniformément marqué an ^{14}C pour l'étude de la matière organique de trois sols (andosol, sol brun, sol ferrallitique). Bulletin de l'Association française pour l'Étude du Sol — Science du Sol, pp. 53—66.

Lewis, J.A. and Starkey, R.L., 1968. Vegetable tannins, their decomposition and effects on decomposition of some organic compounds. Soil Science, 106: 241—247.

Lloyd, A.B., 1975. Extraction of urease from soil. Soil Biology and Biochemistry, 7: 357—358.

Loeb, G.I., 1965. Surface chemistry of proteins and polypeptides. Naval Research Laboratory, Washington, D.C. — Report No. 6318, 81 pp.

Lynch, D.L. and Cotnoir, L.J., 1956. The influence of clay minerals on the breakdown of certain organic substrates. Soil Science Society of America Proceedings, 20: 367—370.

Lynch, D.L. and Lynch, C.C., 1958. Resistance of protein—lignin complexes, lignins and humic acids to microbial attack. Nature (London), 181: 1478—1479.

MacRitchie, F., 1972. The adsorption of proteins at the solid/liquid interface. Journal of Colloid and Interface Science, 38: 484—488.

Mahler, H.R. and Cordes, E.H., 1968. Biological Chemistry. Harper International Edition. Harper and Row, New York, N.Y. and John Weatherhill, Inc., Tokyo, 872 pp.

Marshall, C.E., 1964. The Physical Chemistry and Mineralogy of Soils, Vol. 1. Soil Materials. John Wiley and Sons, Inc., New York, N.Y., 388 pp.

Marshall, K.C., 1975. Clay mineralogy in relation to survival of soil bacteria. Annual Review of Phytopathology, 13: 357—373.

Martin, R.B., 1964. Introduction to Biophysical Chemistry. McGraw-Hill, New York, N.Y., 365 pp.

Martin-Smith, M., 1963. Uricolytic enzymes in soil. Nature (London), 197: 361—362.

Mattson, S., 1932. The laws of soil colloidal behavior: VII. Proteins and proteinated complexes. Soil Science, 23: 41—72.

Mattson, S. and Koutler-Andersson, E., 1942. The acid—base condition in vegatation, litter, and humus: V. Products of partial oxidation and ammonia fixation. Lantbruks-Högskolans Annaler, 10: 284—332.

Mattson, S. and Koutler-Andersson, E., 1943. The acid—base condition in vegetation, litter, and humus: VI. Ammonia fixation and humus nitrogen. Lantbruks-Högskolans Annaler, 11: 107—134.

Mayaudon, J. and Sarkar, J.M., 1974. Chromatographie et purification des diphenol oxydases du sol. Soil Biology and Biochemistry, 6: 275—285.

Mayaudon, J., El Halfawy, M. and Bellinck, C., 1973a. Decarboxylation des acides aminés aromatiques-1-^{14}C par les extraits de sol. Soil Biology and Biochemistry, 5: 355—367.

Mayaudon, J., El Halfawy, M. and Chalvignac, M.A., 1973b. Propriétés des diphenol oxydases extraites des sols. Soil Biology and Biochemistry, 5: 369—383.

McBride, M.B. and Mortland, M.M., 1973. Segregation and exchange properties of alkylammonium ions in a smectite and vermiculite. Clays and Clay Minerals, 21: 323—329.

McLaren, A.D., 1954a. The adsorption and reactions of enzymes and proteins on kaolinite. I. The Journal of Physical Chemistry, 58: 129—137.

McLaren, A.D., 1954b. The adsorption and reactions of enzymes and proteins on kaolinite. II. The action of chymotrypsin on lysozyme. Soil Science Society of America Proceedings, 18: 170—174.

McLaren, A.D., 1960. Enzyme action in structurally restricted systems. Enzymologia, 21: 356—364.

McLaren, A.D., 1963. Enzyme activity in soils sterilized by ionizing radiation and some comments on micro-environments in nature. In: N.E. Gibbons (Editor), Recent Progress in Microbiology. University of Toronto Press, pp. 221—229.

McLaren, A.D., 1969. Radiation as a technique in soil biology and biochemistry. Soil Biology and Biochemistry, 1: 63—73.

McLaren, A.D., 1970. Temporal and vectorial reactions of nitrogen in soil: A review. Canadian Journal of Soil Science, 50: 97—109.

McLaren, A.D., 1972. Consecutive biochemical reactions in soil with particularl reference to the nitrogen cycle. C.N.R. Laboratorio per la Chimica del Terreno, Pisa, Conferenze, 1: 5—18.

McLaren, A.D., 1975. Soil as a system of humus and clay immobilized enzymes. Chemica Scripta, 8: 97—99.

McLaren, A.D. and Ardakani, M.S., 1972. Competition between species during nitrification in soil. Soil Science Society of America Proceedings, 36: 602—606.

McLaren, A.D. and Babcock, K.L., 1959. Some characteristics of enzyme reactions at surfaces. In: T. Hayashi (Editor), Subcellular Particles. The Ronald Press, New York, N.Y., pp. 23—36.

McLaren, A.D. and Barshad, I., 1976. Clays, proteins, and nonsense. Soil Science, 121: 188.

McLaren, A.D. and Estermann, E.F., 1956. The adsorption and reactions of enzymes and proteins on kaolinite. III. The isolation of enzyme—substrate complexes. Archives of Biochemistry and Biophysics, 61: 158—173.

McLaren, A.D. and Estermann, E.F., 1957. Influence of pH on the activity of chymotrypsin at a solid—liquid interface. Archives of Biochemistry and Biophysics, 68: 157—160.

McLaren, A.D. and Lewis, C., 1950. Concerning the inactivation—denaturation of pepsin. Journal of Polymer Science, 5: 379—383.

McLaren, A.D. and Packer, L., 1970. Some aspects of enzyme reactions in heterogeneous systems. Advances in Enzymology, 33: 245—308.

McLaren, A.D. and Peterson, G.H., 1961. Montmorillonite as a caliper for the size of protein molecules. Nature (London), 192: 960—961.

McLaren, A.D. and Peterson, G.H., 1965. Physical chemistry and biological chemistry of clay mineral—organic nitrogen complexes. In: W.V. Bartholomew and F.E. Clark (Editors), Soil Nitrogen. American Society of Agronomy, Monograph 10, Madison, Wisconsin, pp. 259—284.

McLaren, A.D. and Peterson, G.H., (Editors), 1967. Soil Biochemistry, Vol. 1. Marcel Dekker, New York, N.Y., 509 pp.

McLaren, A.D. and Pukite, A., 1975. Ubiquity of some soil enzymes and isolation of soil organic matter with urease activity. In: D. Pavoledo and H.L. Golterman (Editors), Humic Substances, Proceedings International Meeting, Nieuwersluis. Pudoc, Wageningen, pp. 187—193.

McLaren, A.D. and Skujinš, J., 1968. The physical environment of microorganisms in soil. In: T.R.G. Gray and D. Parkinson (Editors), The Ecology of Soil Bacteria. Liverpool University Press, pp. 3—24.

McLaren, A.D., Peterson, G.H. and Barshad, I., 1958. The adsorption and reactions of enzymes and proteins on clay minerals. IV. Kaolinite and montmorillonite. Soil Science Society of America Proceedings, 22: 239—244.

McLaren, A.D., Pukite, A.H. and Barshad, I., 1975. Isolation of humus with enzymatic activity from soil. Soil Science, 119: 178—180.

Messing, R.A., (Editor), 1975a. Immobilized Enzymes for Industrial Reactors. Academic Press, New York, 232 pp.

Messing, R.A., 1975b. Adsorption of proteins on glass surfaces and pertinent parameters for the immobilization of enzymes in the pores of inorganic carriers. Journal of Non-Crystalline Solids, 9: 277—283.

Milestone, N.B., 1971. Allophane, its structure and possible uses. Journal, New Zealand Institute of Chemistry, 35: 191—197.

Mills, O.E. and Creamer, L.K., 1971. The binding of casein to halloysite. New Zealand Journal of Dairy Science and Technology, 6: 61—65.

Mitz, M.A. and Summaria, L.J., 1961. Synthesis of biologically active cellulose derivatives of enzymes. Nature (London), 189: 576—577.

Morgan, H.W. and Corke, C.T., 1975. Adsorption, desorption, and activity of glucose oxidase on selected clay species. Canadian Journal of Microbiology, 22: 684—693.

Morrissey, B.W. and Stromberg, R.R., 1974. The conformation of adsorbed blood proteins by infrared bound fraction measurements. Journal of Colloid and Interface Science, 46: 152—164.

Mortland, M.M., 1968. Protonation of compounds at clay mineral surfaces. Transactions of the 9th International Congress of Soil Science, Adelaide, 1: 691—699.

Mortland, M.M., 1970. Clay—organic complexes and interactions. Advances in Agronomy, 22: 75—117.

Mortland, M.M. and Gieseking, J.E., 1952. The influence of clay minerals on the enzymatic hydrolysis of organic phosphorus compounds. Soil Science Society of America Proceedings, 16: 10—13.

Mosbach, K. (Editor), 1976. Immobilized Enzymes. Methods in Enzymology, Vol. 44. Academic Press, New York, N.Y., 1000 pp.

Nannipieri, P., Ceccanti, B., Cervelli, S. and Sequi, P., 1974. Use of 0.1 M pyrophosphate to extract urease from podzol. Soil Biology and Biochemistry, 6: 359—362.

Norrish, K., 1954. Swelling of montmorillonite. Discussions of the Faraday Society, 18: 120—134.

Paecht-Horowitz, M., 1973. Inorganic clays as possible prebiotic peptide templates. Israel Journal of Chemistry, 11: 369—378.

Parks, F.P., 1974. The Sorption of Escherichia Coli Alkaline Phosphatase by Selected Clay Minerals and the Phosphatase Activity of Selected Idaho Soils. Ph.D. Thesis, University of Idaho.

Parsons, J.W. and Tinsley, J., 1975. Nitrogenous substances. In: J.E. Gieseking (Editor), Soil Components, Vol. 1. Organic Components. Springer Verlag, Berlin, pp. 263—304.

Paul, E.A., 1970. Plant components and soil organic matter. Recent Advances in Phytochemistry, 3: 59—104.

Paulson, K.N. and Kurtz, L.T., 1969. Locus of urease activity in soil. Soil Science Society of America Proceedings, 33: 897—901.

Pinck, L.A., 1962. Adsorption of proteins, enzymes and antibiotics by montmorillonite. Clays and Clay Minerals, 9: 520—529.

Pinck, L.A. and Allison, F.E., 1951. Resistance of a protein—montmorillonite complex to decomposition by soil microorganisms. Science, 114: 130—131.

Pinck, L.A. and Allison, F.E., 1961. Adsorption and release of urease by and from clay minerals. Soil Science, 91: 183—188.

Pinck, L.A., Dyal, R.S. and Allison, F.E., 1954. Protein—montmorillonite complexes, their preparation and the effects of soil microorganisms on their decomposition. Soil Science, 78: 109—118.

Powlson, D.S., 1975. Effects of biocidal treatments on soil organisms. In: N. Walker (Editor), Soil Microbiology. Butterworks, London, pp. 193—224.

Ramirez-Martinez, J.R. and McLaren, A.D., 1966. Some factors influencing the determination of phosphatase activity in native soils and in soils sterilized by irradiation. Enzymologia, 31: 23—38.

Rand, B. and Melton, I.E., 1975. Isoelectric point of the edge surface of kaolinite. Nature (London), 257: 214—216.

Ross, D.J., 1975. Studies on a climosequence of soils in tussock grasslands. 7. Distribution of invertase and amylase activites in soil fractions. New Zealand Journal of Science, 18: 518—526.

Ross, D.J. and McNeilly, B.A., 1972. Some influences of different soils and clay minerals on the activity of glucose oxidase. Soil Biology and Biochemistry, 4: 9—18.

Russell, E.W., 1956. Soil Conditions and Plant Growth, 8th Edition. Longmans, Green and Co., London, 635 pp.

Samsonov, G.V., 1974. Specificity of intermolecular action in systems including polyelectrolytes and ions of organic physiologically active substances. Pure and Applied Chemistry, 38: 151—173.

Satyanarayana, T. and Getzin, L.W., 1973. Properties of a stable cell-free esterase from soil. Biochemistry, 12: 1566—1571.

Scharpenseel, H.W., 1971. Special methods of chromatographic and radiometric analysis. In: A.D. McLaren and J. Skujinš (Editors), Soil Biochemistry, Vol. 2. Marcel Dekker, New York, N.Y., pp. 96—128.

Scharpenseel, H.W., 1972. Natural radiocarbon measurement of soil and organic matter fractions and on soil profiles of different pedogenesis. Proceedings of the 8th International Conference on Radiocarbon Dating, Lower Hutt, 2: E1—E12.

Scheffer, F. and Ulrich, B., 1960. Lehrbuch der Agrikulturchemie und Bodenkunde. 3. Humus und Humusdüngung. Ferdinand Enke Verlag, Stuttgart, 266 pp.

Schnitzer, M. and Khan, S.U., 1972. Humic Substances in the Environment. Marcel Dekker, New York, N.Y., 327 pp.

Sequi, P., 1974. Gli enzimi del terreno. Italia Agricola, 111: 90—109.

Sigler, P.B., Blow, D.M., Matthews, B.W. and Henderson, R., 1968. Structure of crystalline α-chymotrypsin. II. A preliminary report including a hypothesis for the activation mechanism. Journal of Molecular Biology, 35: 143—164.

Silberberg, A., 1962a. Adsorption of flexible macromolecules. I. The isolated macromolecule at a plane surface. The Journal of Physical Chemistry, 66: 1872—1883.

Silberberg, A., 1962b. Adsorption of flexible macromolecules. II. The shape of the adsorbed molecule; the adsorption isotherm; surface tension. The Journal of Physical Chemistry, 66: 1884—1907.

Silberberg, A., 1967. Adsorption of flexible macromolecules. III. Generalised treatment of the isolated macromolecule. The Journal of Chemical Physics, 46: 1105—1114.

Skujinš, J.J., 1967. Enzymes in soil. In: A.D. McLaren and G.H. Peterson (Editors), Soil Biochemistry, Vol. 1. Marcel Dekker, New York, N.Y., pp. 371—414.

Skujinš, J.J. and McLaren, A.D., 1968. Persistence of enzymatic activities in stored and geologically preserved soils. Enzymologia, 34: 213—225.

Skujinš, J.J. and McLaren, A.D., 1969. Assay of urease activity using ^{14}C-urea in stored, geologically preserved, and in irradiated soils. Soil Biology and Biochemistry, 1: 89—99.

Skujinš, J.J., Estermann, E.F. and McLaren, A.D., 1959. Proteolytic activity of *Bacillus subtilis* in a clay—protein paste system analogous to soil. Canadian Journal of Microbiology, 5: 631—639.

Skujinš, J.J., Pukite, A. and McLaren, A.D., 1974. Adsorption and activity of chitinase on kaolinite. Soil Biology and Biochemistry, 6: 179—182.

Solomon, D.H., 1968. Clay minerals as electron acceptors and/or electron donors in organic reactions. Clays and Clay Minerals, 16: 31—39.

Sørensen, H., 1955. Xylanase in the soil and the rumen. Nature (London), 176: 74.

Sørensen, L.H., 1969. Fixation of enzyme protein in soil by the clay mineral montmorillonite. Experientia, 25: 20—21.

Sørensen, L.H., 1972. Stabilization of newly formed amino acid metabolites in soil by clay minerals. Soil Science, 114: 5—11.

Sørensen, L.H., 1974. Rate of decomposition of organic matter in soil as influenced by repeated air drying—rewetting and repeated additions of organic material. Soil Biology and Biochemistry, 6: 287—292.

Speir, T.W., 1977. Studies on a climosequence of soils in tussock grasslands. 10. Distribution of urease, phosphatase and sulphatase activities in soil fractions. New Zealand Journal of Science, 20: 151—157.

Srivastava, S.K. and Chauhan, O.S., 1975. Viscometric studies on the aqueous suspensions of clay—protein complexes. Australian Journal of Chemistry, 28: 1273—1277.

Stevenson, F.J. and Butler, J.H.A., 1969. Chemistry of humic acids and related pigments. In: G. Eglinton and M.T.J. Murphy (Editors), Organic Geochemistry. Springer Verlag, Berlin, pp. 534—557.

Stotzky, G., 1972. Activity, ecology and population dynamics of microorganisms in soil. CRC Critical Reviews in Microbiology, 2: 59—137.

Stotzky, G. and Kunc, F., 1975. Response to letter to the editor. Soil Science, 120: 244—245.

Suzuki, H., Ozawa, Y. and Maeda, H., 1966. Water-insoluble enzyme. Hydrolysis of sucrose by insoluble yeast invertase. Agricultural and Biological Chemistry (Tokyo), 30: 807—812.

Tabatabai, M.A. and Bremner, J.M., 1971. Michaelis constants of soil enzymes. Soil Biology and Biochemistry, 3: 317—323.

Talibudeen, O., 1950. Interlamellar adsorption of protein monolayers on pure montmorillonoid clays. Nature (London), 166: 236.

Talibudeen, O., 1955. Complex formation between montmorillonoid clays and amino-acids and proteins. Transactions of the Faraday Society, 51: 582—590.

Tarasevich, Yu.I., Smirnova, V.A., Manakhova, L.I., Ropot, V.M. and Sivalov, E.G., 1975. Adsorption of albumin on clay minerals. Kolloidnyi Zhurnal, 37: 912—917.

Theng, B.K.G., 1972. Adsorption of ammonium and some primary n-alkylammonium cations by soil allophane. Nature (London), 238: 150—151.

Theng, B.K.G., 1974a. Complexes of clay minerals with amino acids and peptides. Chemie der Erde, 33: 125—144.

Theng, B.K.G., 1974b. The Chemistry of Clay—Organic Reactions. Adam Hilger Ltd., London, 343 pp.

Theng, B.K.G., 1976. Interactions between montmorillonite and fulvic acid. Geoderma, 15: 243—251.

Theng, B.K.G. and Scharpenseel, H.W., 1975. The adsorption of ^{14}C-labelled humic acid by montmorillonite. Proceedings of the International Clay Conference, Mexico City, pp. 643—653.

Theng, B.K.G. and Walker, G.F., 1970. Interactions of clay minerals with organic monomers. Israel Journal of Chemistry, 8: 417—424.

Theng, B.K.G., Greenland, D.J. and Quirk, J.P., 1967. Adsorption of alkylammonium cations by montmorillonite. Clay Minerals, 7: 1—17.

Thies, C., 1965. Determination of the structure of adsorbed polymers by infrared analysis. American Chemical Society Division of Polymer Chemistry, Preprints, 6: 320—325.

Thomas, D. and Kernevez, J.P. (Editors), 1976. Analysis and Control of Immobilized Enzyme Systems, Proceedings International Symposium. North Holland, Amsterdam, 306 pp.

Thornton, J.I., Crim, D. and McLaren, A.D., 1975. Enzymatic characterization of soil evidence. Journal of Forensic Sciences, 20: 674—692.

Usame, S. and Inoue, S., 1974. Kinetics and continuous reaction of glucoamylase adsorbed on adsorbent. Asai Garasu Kogyo Gijutsu Shoreikai Kenkyu Hokuku, 25: 39—54. See: Chemical Abstracts, 83: 203228 (1975).

Van Olphen, H., 1963. Introduction to Clay Colloid Chemistry. Interscience Publishers, New York, N.Y., 301 pp.

Vansant, E.F. and Uytterhoeven, J.B., 1972. Thermodynamics of the exchange of n-alkylammonium ions on Na-montmorillonite. Clays and Clay Minerals, 20: 47—54.

Voets, J.P. and Dedeken, M., 1966. Soil enzymes. Mededelingen Fakulteit Landbouwwetenschappen, Rijksuniversiteit Gent, 31: 177—190.

Vroman, L. and Adams, A.L., 1969. Identification of rapid changes at plasma/solid interfaces. Journal of Biomedical Materials Research, 3: 43—67.

Waksman, S.A. and Iyer, R.K., 1932. Contribution to our knowledge of the chemical nature and origin of humus. Soil Science, 34: 43—69.

Weetall, H.H. and Suzuki, S. (Editors), 1975. Immobilized Enzyme Technology: Research and Applications. Plenum Press, New York, N.Y., 321 pp.

Weiss, A., 1969. Organic derivatives of clay minerals, zeolites, and related minerals. In: G. Eglinton and M.T.J. Murphy (Editors), Organic Geochemistry. Springer Verlag, Berlin, pp. 737—781.

Wingard, L.B., Jr (Editor), 1972. Enzyme Engineering, Biotechnology and Bioengineering Symposium No. 3. Interscience Publishers, New York, N.Y., 415 pp.

Yavich, P.A. Sarabunovich, A.G. and Beridze, P.Z., 1975. Consistent patterns in the formation and breakdown of tannin—protein complexes. Farmatsiya (Moscow), 24: 20—23. See: Chemical Abstracts: 83: 65370 (1975).

Young, E.G., 1963. Occurrence, classification, preparation and analysis of proteins. In: M. Florkin and E.H. Stotz (Editors), Comprehensive Biochemistry, Vol. 7. Proteins (Part 1). Elsevier, Amsterdam, pp. 1—55.

Zaborsky, O., 1973. Immobilized Enzymes. CRC Press, Cleveland, Ohio, 175 pp.

Zittle, C.A., 1953. Adsorption studies of enzymes and other proteins. Advances in Enzymology, 14: 319—374.

Zvyagintsev, D.G. and Velikanov, L.L., 1968. Influence of soils and clay minerals on the activity of glucose oxidase and invertase. Pochvovedenie, No. 6: 100—106.

Chapter 8

NUCLEIC ACIDS

8.1. INTRODUCTION

In common with soil nitrogen, a substantial proportion (30—70%) of the total phosphorus in (surface) soils is organically bound. Since this fraction of soil phosphorus ultimately derives from animals, microorganisms, and plants or their respective remains, such compounds as phosphoproteins and sugar phosphates occur in soil. Their content however, may be so low as to escape detection. On the other hand, the inositol phosphates, nucleic acids, and phospholipids have been recognised, if not positively identified, as being present in soil (Ulrich and Benzler, 1955; Anderson, 1967, 1975; Cosgrove, 1967; McKercher, 1968; Flaig, 1971). Of these three main classes of phosphate esters, the inositol phosphates ("phytin") are the most abundant, representing up to half of the total organic phosphorus. It is often difficult to obtain an accurate estimate of the nucleic acid content of soil but for most soils the amount is unlikely to exceed 10% of this total. The content in phospholipids or phosphatides is even less, rarely exceeding 1%.

The occurrence of nucleic acids as such in soil is frequently not directly demonstrable but is indicated by the isolation from soil extracts of such compounds as adenine, guanine, and cytosine together with their respective deaminated derivatives. Thus, the existence of ribonucleic acid (RNA) is inferred from the appearance in the extract of its constituent nucleotides on treating soil with alkali. Deoxyribonucleic acid (DNA) is not depolymerised by this treatment but on acid hydrolysis of soil or of its humic acid extract, purines and pyrimidines are released, the proportion of bases detectable by chromatographic and spectrophotometric means being in the order of cytosine > guanine > thymine > adenine.

Nucleic acids may be regarded as polyanions, although under moderately acid conditions the molecules can acquire positive charges by accepting protons from the solution (Jordan, 1955; Mahler and Cordes, 1968). In DNA, for example, the bases adenine and cytosine (pK_a 4—5) are the first to become protonated, followed by guanine ($pK_a \sim 3$) as the solution pH falls from neutrality. The former reaction does not appear to impair its base-pairing properties or lead to alterations in helix dimensions but as guanine is protonated, conformational changes may occur. Because protonation gives rise to the formation of cationic groups in the molecule the interactions of nucleic acids with clays are sensitive to the ambient solution pH. In this respect, the adsorption characteristics of nucleic acids resemble those of proteins. Another point of note is that the primary phosphate in polynucleotides has a pK_a near unity. For this reason, nucleic acids commonly occur as salts, that is, accompanied by the appropriate complement of cations. In these salts, monovalent cations merely serve as charge-balancing species for the phosphate residues. On the other hand, polyvalent cations, particularly those of the transition metal series such as Co^{2+}, Mn^{2+}, Ni^{2+}, and Zn^{2+}, tend to be bound stoichiometrically, and hence, their effect on the solution conformation of nucleic acid is relatively greater. These properties of nucleic acids influence their behaviour at clay mineral surfaces.

When added to soil, nucleic acids are quite rapidly decomposed by microbial enzymes to yield purines, pyrimidines, and inorganic phosphate (e.g. Durand, 1966; Cosgrove, 1967; Greaves and Wilson, 1970). Since the rate of breakdown is normally estimated from the amount of phosphate released, the term "dephosphorylation" has been used to denote the mineralisation of nucleic acids in soil systems. The available evidence indicates that the dephosphorylation process, which is influenced by soil pH conditions, seldom, if ever, goes to completion (Dyer and Wrenshall, 1941; Pearson et al., 1941; Bower, 1949; Cosgrove, 1967). The fraction which persists in soil is probably of microbial origin but as Anderson (1958) has pointed out, its amount (in DNA) is greater than can be accounted for by the number of living microorganisms. This and other observations on the fate of organic phosphorus compounds in soil (Pierre, 1948; Bower, 1949, Goring and Bartholomew, 1950) suggest that nucleic acids in soil may be stabilised by attachment to inorganic soil colloids of which the clay fraction plays an important part. Although clay—nucleic acid complexes have not been isolated from soils, Stevenson et al. (1952) were able to detect nucleic acid constituents in the hydrolysate of clay-adsorbed organic compounds extracted from a soil. As with proteins, the concept that clays are actively involved in protecting nucleic acids in soil against microbial degradation has stimulated research into the mechanisms underlying the clay—nucleic acid interaction.

8.2. FORMATION AND PROPERTIES OF COMPLEXES

One of the earliest systematic studies on the reactions of clay minerals with nucleic acids and their constituents is due to Bower (1949) who found that these compounds (from yeast) were rapidly taken up by kaolinite and montmorillonite. The extent of adsorption, expressed as a percentage of the amount added, was sensitive to variations in suspension pH, uptake declining sharply as the pH was raised from 2.5 (for kaolinite) and 4.5 (for montmorillonite) to 6.5, beyond which very little, if any, adsorption occurred (cf. Fig. 8.1). The amount adsorbed also depended on the clay mineral species, montmorillonite taking up about three times as much nucleic acid as did kaolinite at pH 4.5.

These observations were explained in terms of electrostatic interactions between the basic (cationic) groups of the nucleic acid and the negatively charged sites at the mineral surface. The higher (total) exchange capacity and surface area of montmorillonite as compared with kaolinite would account for the former's ability to remove a relatively greater proportion of the nucleic acid from solution (below pH 7). The effect of pH on nucleic acid adsorption by clays has been fully substantiated by later workers (Goring and Bartholomew, 1952; Flaig et al., 1955; Greaves and Wilson, 1969) and will be further discussed below.

Although X-ray diffraction analyses were not carried out on the montmorillonite complexes, Bower's adsorption data implied that some intercalation had taken place. The finding that the rate of dephosphorylation of adsorbed nucleic acid was closely related to the amount adsorbed was consistent with this view. In this connection, Bower noted that whereas enzyme activity was reduced by 20% in the presence of montmorillonite, the decrease in dephosphorylation was appreciably greater than this amount. This led him to infer that adsorption of the enzyme by the clay contributed but little to the observed reduction in nucleic acid mineralisation. At first sight this view might appear to argue against Mortland and Gieseking's (1952) observations on the inactivation of phosphatase due to adsorption onto various clay minerals, described in the preceding chapter. It should be recalled, however, that the relative activity of adsorbed enzymes is influenced by the order in which the enzyme and its respective substrate are added to the mineral (cf. Table 7.4). Bower's figure referred to the situation where the enzyme (nuclease) was added to the clay—nucleic acid complex. A far greater reduction in the relative activity of nuclease would have been obtained had he added the nucleic acid to the clay—enzyme system.

Goring and Bartholomew (1952) subsequently investigated in more detail the factors affecting the interactions of DNA and RNA with acid-washed samples of an illite, a kaolinite, and a montmorillonite. They confirmed Bower's (1949) finding that adsorption progressively declined with an increase in suspension pH from 4 to 6. For a given pH value, uptake fell in the order montmorillonite > illite > kaolinite, which was also the sequence of their exchange capacity.

Although a cation exchange mechanism was clearly implicated in the binding of nucleic acids to clays as Bower had suggested, the acid-washed minerals used by Goring and Bartholomew (1952) would invariably have contained Al^{3+} ions in their exchange complex. In addition to electrostatic interactions, the nucleic acids would presumably be attached by a cation-bridge type of bonding by which the phosphate group of the molecules is linked to the clay surface through an exchangeable Al^{3+} ion. The involvement of phosphate groups in this way was, in fact, indicated by their supplementary measurements on acid-washed clays whose exchange complex had partially been filled by divalent cations. Such a link is facilitated by the tendency of polyvalent cations to form stoichiometric compounds with nucleic acid, referred to previously. That monovalent cations fail to act likewise is also apparent from the results of Flaig et al. (1955). These workers found that the cation exchange capacity of the complex, CEC_{comp}, was greater than that of the parent montmorillonite, CEC_m. Only when the phosphate groups in the nucleic acid have previously been neutralised was $CEC_{comp} < CEC_m$ and fell with the amount of nucleic acid adsorbed. Incidentally, a cation-bridge mechanism has also been proposed by Gordon and Kleinschmidt (1970) for the adsorption of double-stranded DNA onto the cleavage surface of Al^{3+}-muscovite.

Further support for this type of linkage in Goring and Bartholomew's (1952) system was provided by the observation that only 20—30% of the adsorbed nucleic acid could be desorbed by washing the montmorillonite complex with 1 M NaCl at pH 6. Almost complete desorption, however, could be achieved by treating the complex with 1 M NH_4OH or less concentrated electrolyte solutions at pH > 8.4. It should perhaps be pointed out that such treatments are likely to cause extensive denaturation, if not hydrolysis, of the adsorbed nucleic acids.

Another feature of the clay—nucleic acid interaction is that adsorption, when it can occur, is generally enhanced by the addition of neutral electrolytes to the system. In this respect, the salts of polyvalent cations are more effective than those of monovalent species (Goring and Bartholomew, 1952; Flaig et al., 1955; Greaves and Wilson, 1969; Gordon and Kleinschmidt, 1970). The effect of ionic strength on nucleic acid adsorption by montmorillonite will be taken up again later.

In common with proteins, Goring and Bartholomew (1952) noted that a rapid initial rise in uptake was followed by a slow (up to 24 h) approach to equilibrium. The rate of adsorption was concentration- and temperature-dependent but the equilibrium amount adsorbed was little affected by temperature. These observations would indicate that the adsorption of nucleic acids by clays is a diffusion-controlled process and that the adsorbed molecules undergo a slow conformational rearrangement at the clay surface.

Depending on the amount adsorbed, basal spacings of 1.4—1.8 nm were recorded for montmorillonite complexes with DNA and RNA which had previously been washed with acetone and dried at 363 K. Such low values would indicate partial interlayer penetration by denatured, fully collapsed, randomly coiled chains rather than the presence of native nucleic acid molecules in the interlayer space.

The quantitative aspects of the interaction process have been re-examined by Greaves and Wilson (1969) using calf thymus DNA, yeast RNA, and montmorillonite saturated with Na^+, K^+, Ca^{2+}, and Mg^{2+} ions. In agreement with Goring and Bartholomew (1952), the initial rate of adsorption was found to be extremely rapid with about 85—90% of the equilibrium amount taken up during the first 15 min. After 60 min, nearly all of what could be adsorbed was removed from solution beyond which time uptake proceeded very slowly until an apparent equilibrium was attained (16 h).

The effect of suspension pH and the valency of the exchangeable cation on DNA adsorption by montmorillonite is illustrated in Fig. 8.1. From pH 9 to 5.5 (region 1), adsorption remained at a constant low level which, in the unbuffered system, depended on the valency of the saturating cation at the clay surface. When the pH was lowered from 5.5 to 3.5 (region 2), there was a steep, almost linear increase in adsorption, as Bower (1949) and Goring and Bartholomew (1952) had reported earlier.

The behaviour in region 1 largely reflects the fact that nucleic acids carry very few positive charges over this pH range (Jordan, 1955) and hence, electrostatic interaction between adsorbate and clay surface is minimal. Both solute and surface are, in fact, negatively charged in this region. That a small but measurable positive adsorption was nonetheless recorded would indicate the operation of secondary attractive forces, such as Van der Waals and cation-bridge interactions. In an aqueous environment, bonding by means

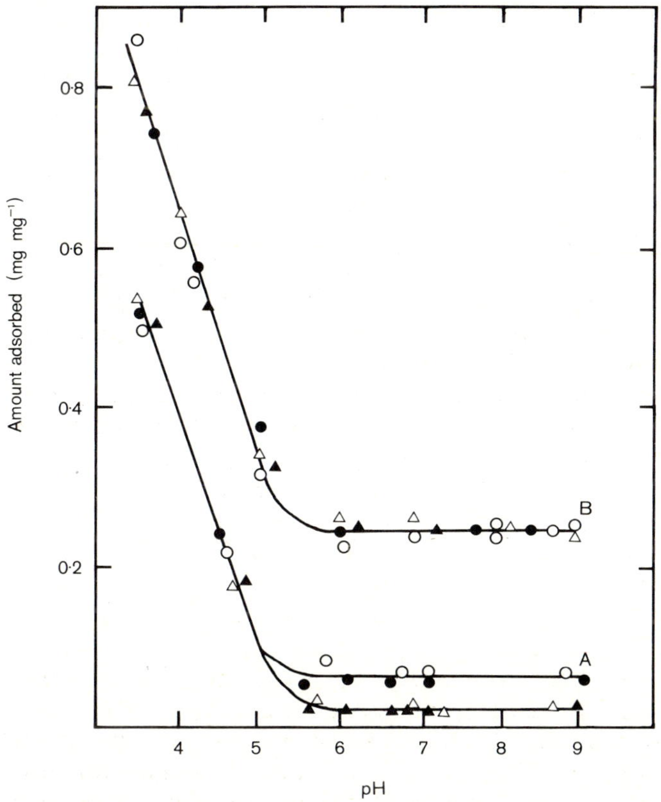

Fig. 8.1. The effect of pH on the adsorption of DNA by montmorillonite saturated with different cations. (○), Mg^{2+}-montmorillonite; (●), Ca^{2+}-montmorillonite; (△), K^{+}-montmorillonite; (▲), Na^{+}-montmorillonite. A, aqueous suspension; B, buffered suspension (after Greaves and Wilson, 1969).

of a "water-bridge" is also likely to occur. This is a special type of cation-bridging in which the anionic groups of the polymer are hydrogen bonded to water molecules in the primary hydration shell around the exchangeable cation. Such a mechanism has been postulated for the interactions of humic and fulvic acids with montmorillonite (Theng and Scharpenseel, 1975; Theng, 1976). Either way, divalent ion clays would adsorb more than their monovalent counterparts and this accords with observation (Fig. 8.1). In addition, Greaves and Wilson (1969) suggested that some of the added DNA might be precipitated in the form of its calcium or magnesium nucleate when these divalent cations were being displaced from their exchange sites by Na^+ ions which accompanied the DNA polyanion or were added to raise the pH.

On the other hand, electrostatic interactions would be primarily responsible for adsorption in region 2 since, with falling pH between pH 5.5 and 3.5, the purine and pyrimidine groups of DNA become progressively protonated. The steep rise in uptake on lowering the suspension pH also reflects the concomitant decrease in nucleic acid solubility associated with a reduction in molecular size. This effect of acidity has been reported as early as 1953 by Reichmann et al. who found that the mean molecular radius of DNA at pH 2.6 was approximately half that at pH 6.5.

Fig. 8.1 also shows that irrespective of pH, adsorption was enhanced in the presence of (universal) buffer. An analogous situation obtained when a neutral electrolyte solution (0.02 M KCl) was added to the system, confirming the finding of earlier workers (Goring and Bartholomew, 1952; Flaig et al., 1955) and recalling the results of Armstrong and Chesters (1964) for lysozyme. Greaves and Wilson (1969) suggested the possibility of ion-pairing between the phosphate residues of the nucleic acid and K^+ (or Na^+) ions to account for this behaviour. However, as we have indicated, such a mechanism would be relatively unimportant with monovalent cations. Rather, by screening the negative charges on the nucleic acid molecules, added electrolytes would reduce intra- and inter-molecular repulsion, leading to a more compact molecular conformation. An increase in ionic strength would also compress the diffuse electrical double layers around the clay particles, making for a closer approach of the nucleic acid to the mineral surface. Both these effects would favour adsorption.

As might be expected, the intercalation of nucleic acids by montmorillonite is pH-dependent, no penetration of the interlayer space being observed at pH > 6 (Greaves and Wilson, 1969). The extent of interlayer expansion or Δ-value, as a function of the amount of DNA and RNA adsorbed, is shown in Fig. 8.2. The linear increase in Δ-value as adsorption progressed and the presence of two expansion stages or steps, in the case of DNA, again paralleled the behaviour of lysozyme towards montmorillonite (cf. Fig. 7.3).

The first stage terminating at a Δ-value of 1.8—1.9 nm may therefore be ascribed to segregation of DNA-rich and inorganic ion-rich layers within a single montmorillonite crystal. The absence of a rational series of basal

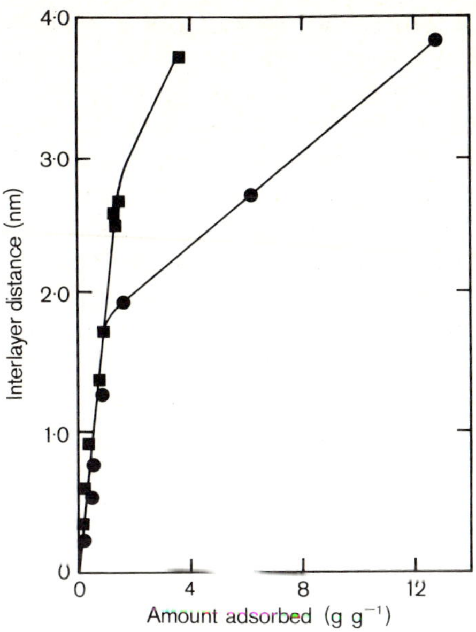

Fig. 8.2. Variation of interlayer distance or Δ-value with the amount of nucleic acid adsorbed. (○) DNA; (■) RNA (after Greaves and Wilson, 1969).

reflections in the X-ray diffraction pattern lends further support to this suggestion. The amount adsorbed corresponding to this level of interlayer expansion (~1.2 g per g montmorillonite) presumably represents the "monolayer" capacity. At this point a single layer of DNA molecules with their shorter axis lying perpendicular to the silicate layer, was present over the entire surface of the clay. The observed expansion of 1.8—1.9 nm due to the intercalation of a monolayer of DNA also accords with the thickness of the Watson—Crick double helix (~2 nm). In the case of RNA the point where the first expansion stage ended and the second began was not so clearly defined. There was an inflexion at a Δ-value of ~2.7 nm, at which point the amount adsorbed was ~1.8 g g^{-1}. By analogy with DNA and on the basis of the direct proportionality between expansion and adsorption, this amount may represent a monolayer coverage by RNA. If so, the corresponding extent of expansion may be identified with the average thickness of the intercalated (RNA) random coil.

The explanation for the occurrence of a second expansion stage is not obvious since the corresponding adsorption isotherms were linear and the points for both DNA and RNA fall on the same line. For DNA the second expansion stage terminated at a Δ-value of ~3.8 nm. At first sight, it might appear that layer segregation has occurred by which single and double layers of molecules were randomly interstratified within a single clay crystal. The amount adsorbed at this expansion level (~12 mg mg^{-1}), however, was about 10 times as large as the monolayer capacity. A greater part of the adsorbed DNA over and above this capacity must, therefore, have been present on external crystal surfaces, possibly in the form of molecular aggregates or as multimolecular layers. On the other hand, the amount of RNA adsorbed to give a comparable interlayer expansion (Δ-value ~ 3.6 nm) was only in the order of 3.8 g g^{-1}. Thus, a relatively large proportion of the material taken up beyond the "first" expansion stage could be accommodated in the interlayer space.

It is interesting to note that montmorillonite —RNA complexes which have a strong basal reflection near 3.4 nm (Δ-value ~ 2.4 nm) also show a subsidiary line at 1.8 nm (Greaves and Wilson, 1970). When such complexes were incubated with soil for different lengths of time the former progressively weakened in intensity and sometimes disappeared altogether from the X-ray diffraction trace whilst the subsidiary reflection became stronger. (Table 8.1). Accordingly, the 1.8 nm line could not reasonably be assigned to the second-order basal reflection. Rather, these observations are consistent with the formation of two intercalation zones within a single crystal: a more greatly expanded zone at the periphery and a less expanded one towards the centre.

Like their protein counterparts, nucleic acids complexed with clays are by no means immune to decomposition by microorganisms. In the case of montmorillonite, the ease with which the adsorbed nucleic acid is degraded by this means depends on the type of polymer and its location with respect to the silicate layers. Not unexpectedly perhaps, Greaves and Wilson (1970) found that the material present in the interlayer space was more resistant to microbial attack than that which was attached to external crystal surfaces. As indicated in the preceding paragraph, the degree of this resistance for RNA complexes was also influenced by the extent to which the nucleic acid

TABLE 8.1

Changes in basal spacing (nm) of montmorillonite—nucleic acid complexes on incubation with different soils for the stated periods (from Greaves and Wilson, 1970)

Incubation period (days)	Soil sample					
	I (Ardfork)		II (Hardslacks)		III (Sandend)	
	DNA	RNA	DNA	RNA	DNA	RNA
0	2.57	3.20 (1.76)	2.56	3.45 (1.88)	2.56	3.45 (1.88)
4	1.86	2.57	1.98	3.23 (1.81)	1.68	2.45
7	1.79	1.67 (2.74)	1.95	1.72 (3.20)	1.70	2.05
11	—	—	1.92	1.86 (3.40)	1.70	1.63

Numbers in brackets denote reflections of lesser intensity.

penetrated the interlayer space. On the other hand, the DNA analogues showed a uniform interlayer collapse when similarly incubated with soil, the greatest extent of contraction occurring during the first few days after incubation (Table 8.1).

Greaves and Wilson also noted that the addition to soil of montmorillonite—nucleic acid complexes might actually lead to a large increase in microbial numbers. Interestingly, externally adsorbed RNA was capable of inducing a larger microbial response than a greater amount of the nucleic acid added as such to the soil. Incubation studies using γ-ray sterilised soils further indicated that intra- rather than extra-cellular nucleases were responsible for degrading clay-adsorbed nucleic acids in soil. These enzymes could presumably enter the interlayer space of the complex and act on the intercalated substrate material, bringing about a contraction in basal spacing, as shown in Table 8.1.

Despite the complexity of the system, it seems clear that complex formation with clay minerals confers a measure of stability to nucleic acids. How long this protective effect of clays would last under field conditions is not known since, by their nature, laboratory experiments are of relatively short duration. However, the fact that nucleic acids or, at least, their residues and derivatives can be extracted from soil, suggests that these materials do possess long-term stability.

REFERENCES

Anderson, G., 1958. Identification of derivatives of deoxyribonucleic acid in humic acid. Soil Science, 86: 169—174.

Anderson, G., 1967. Nucleic acids, derivatives and organic phosphates. In A.D. McLaren and G.H. Peterson (Editors), Soil Biochemistry, Vol. 1. Marcel Dekker, New York, N.Y., pp. 67—90.

Anderson, G., 1975. Other organic phosphorus compounds. In: J.E. Gieseking (Editor), Soil Components, Vol. 1. Organic Components. Springer Verlag, Berlin, pp. 305—331.

Armstrong, D.E. and Chesters, G., 1964. Properties of protein-bentonite complexes as influenced by equilibration conditions. Soil Science, 98: 39—52.

Bower, C.A., 1949. Studies on the forms and availability of soil organic phosphorus. Iowa Agricultural Experiment Station, Research Bulletin 362, pp. 963—996.

Cosgrove, D.J., 1967. Metabolism of organic phosphates in soil. In: A.D. McLaren and G.H. Peterson (Editors), Soil Biochemistry, Vol. 1. Marcel Dekker, New York, N.Y., pp. 216—228.

Durand, G., 1966. Contribution a l'étude de la biologie du sol: sur la catabolisme des acides nucléiques et de leurs dérivés. Thesis, University of Toulouse No. 260, 233 pp.

Dyer, W.J. and Wrenshall, G.L., 1941. Organic phosphorus in soils: III. The decomposition of some organic phosphorus compounds in soil cultures. Soil Science, 51: 323—329.

Flaig, W., 1971. Organic compounds in soil. Soil Science, 111: 19—33.

Flaig, W., Kuron, H. and Kaul, R., 1955. Über die Sorption von Nucleinstoffen an Tonkolloiden. Zeitschrift für Pflanzenernährung, Düngung, Bodenkunde, 71: 141—154.

Gordon, C.N. and Kleinschmidt, A.K., 1970. Electron microscopic observation of DNA adsorbed on aluminum-mica. Journal of Colloid and Interface Science, 34: 131—141.

Goring, C.A.I. and Bartholomew, W.V., 1950. Microbial products and soil organic matter: 2. The effect of clay on the decomposition and separation of the phosphorus compounds in microorganisms. Soil Science Society of America Proceedings, 14: 152—156.

Goring, C.A.I. and Bartholomew, W.V., 1952. Adsorption of mononucleotides, nucleic acids, and nucleoproteins by clays. Soil Science, 74: 149—164.

Greaves, M.P. and Wilson, M.J., 1969. The adsorption of nucleic acids by montmorillonite. Soil Biology and Biochemistry, 1: 317—323.

Greaves, M.P. and Wilson, M.J., 1970. The degradation of nucleic acids and montmorillonite-nucleic acid complexes by soil microorganisms. Soil Biology and Biochemistry, 2: 257—268.

Jordan, D.O., 1955. The physical properties of nucleic acids. In: E. Chargaff and J.N. Davidson (Editors), The Nucleic Acids, Vol. 1. Academic Press, New York, N.Y., pp. 447—492.

Mahler, H.R. and Cordes, E.H., 1968. Biological Chemistry. Harper International Edition. Harper and Row, New York, N.Y. and John Weatherhill, Inc., Tokyo, 872 pp.

McKercher, R.B., 1968. Studies on soil organic phosphorus. Transactions of the 9th International Congress of Soil Science, Adelaide, 3: 547—553.

Mortland, M.M. and Gieseking, J.E., 1952. The influence of clay minerals on the enzymatic hydrolysis of organic phosphorus compounds. Soil Science Society of America Proceedings, 16: 10—13.

Pearson, R.W., Norman, A.G. and Ho, C., 1941. The mineralization of the organic phosphorus of various compounds in soil. Soil Science Society of America Proceedings, 6: 168—175.

Pierre, W.H., 1948. The phosphorus cycle and soil fertility. Journal of the American Society of Agronomy, 40: 1—14.

Reichmann, M.E., Bunce, B.H. and Doty, P., 1953. The changes induced in sodium deoxyribonucleate by dilute acid. Journal of Polymer Science, 10: 109—119.

Stevenson, F.J., Marks, J.D., Varner, J.E. and Martin, W.P., 1952. Electrophoretic and chromatographic investigations of clay-adsorbed organic colloids: 1. Preliminary investigations. Soil Science Society of America Proceedings, 16: 69—73.

Theng, B.K.G., 1976. Interactions between montmorillonite and fulvic acid. Geoderma, 15: 243—251.

Theng, B.K.G. and Scharpenseel, H.W., 1975. The adsorption of ^{14}C-labelled humic acid by montmorillonite. Proceedings of the International Clay Conference, Mexico City, pp. 643—653.

Ulrich, B. and Benzler, J.H., 1955. Der organisch gebundene Phosphor im Boden. Eine Literaturübersicht. Zeitschrift für Pflanzenernährung, Düngung, Bodenkunde, 70: 220—249.

Chapter 9

NUCLEOPROTEINS AND VIRUSES

9.1. INTRODUCTION

As with the clay—nucleic acid interaction but perhaps more so, there is a paucity of data concerning the behaviour of nucleoproteins at clay mineral surfaces. The reason for this lack of information may be traced to the fact that the amount of these substances in soil is extremely low, if they are at all detectable. Accordingly, nucleoproteins are not considered to play an important part in humus formation and plant nutrition.

On the other hand, the interactions of clay with bacteriophages and, more particularly, with viruses, both of which are essentially nucleoproteins capable of self-replication, have received an increasing amount of attention. This is so because many phytopathogenic viruses are found in soil (e.g. Slykhuis, 1976) the clay fraction of which is the most active adsorbent, acting as both carrier and transmitter of viral infectivity (Myamoto, 1959a,b). That viruses can persist and survive in soil for prolonged periods has been known for a long time (e.g. Johnson and Ogden, 1929). Here again, complex formation with clay minerals has been invoked to account for viral stability (Van der Want, 1952; Thung, 1955; Murphy et al., 1958; David and Gardiner, 1967; Gerba and Schaiberger, 1975). This does not mean, however, that the silt and sand fractions of soil are inert in this respect. Robeck et al. (1962) and Lefler and Kott (1974), for example, have shown that sand could take up some viruses but because of its low surface area and surface reactivity, the capacity of sand to hold viruses is rather limited (Dieterich, 1953). Another point of interest is that adsorption onto clays does not invariably lead to virus inactivation in that the clay—virus complex may be as infectious as the unadsorbed species (Schaub et al., 1974).

The ability of clays to remove bacteriophages and viruses from waste water provides a useful means of concentrating these substances in aquatic systems (Moore et al., 1974). Equally relevant in this context is that viruses can be effectively immobilised when they pass through soil. This has important implications for such practices as the infiltration of sewage through soil, the landfill disposal of human and animal waste, and the irrigation of agricultural land with municipal and industrial effluents (Duboise et al., 1974; Wellings et al., 1974, 1975). A symposium on virus survival in water and waste water systems has been convened (Malina and Sagik, 1974) whilst the fate of viruses in the soil environment has been discussed in two recent reviews by Bitton (1975) and Gerba et al. (1975).

9.2 FORMATION AND PROPERTIES OF COMPLEXES

As is to be expected, the reactions of nucleoproteins and viruses with clay minerals are, in many ways, similar to those given by nucleic acids and proteins, discussed in the preceding chapters. For this reason, we need only summarise the main points of the adsorption process, drawing attention to those features which may be peculiar to the clay—nucleoprotein (virus) interaction.

Like nucleic acids and proteins before them, nucleoproteins and viruses are rapidly adsorbed by clays, most (~90%) of the added amount being removed from solution within the first 10 min or so after bringing the components together (Carlson et al., 1968; Schaub et al., 1974). What meagre data are available indicate that adsorption isotherms tend to be of the H-type although conformity to either the Freundlich or the Langmuir equation has been found empirically (see Bitton, 1975; Gerba et al., 1975).

Being polyelectrolytes, nucleoproteins (and viruses) are sensitive to variations in the pH of the system when they interact with clays. Fig. 9.1 shows the extent of adsorption of southern bean mosaic virus (SBMV) by kaolinite as a function of suspension pH (McLaren and Peterson, 1965). The close resemblance between this curve and that given by DNA (cf. Fig. 8.1) suggests that the underlying mechanism is similar to both systems. Thus, the steep rise in up-

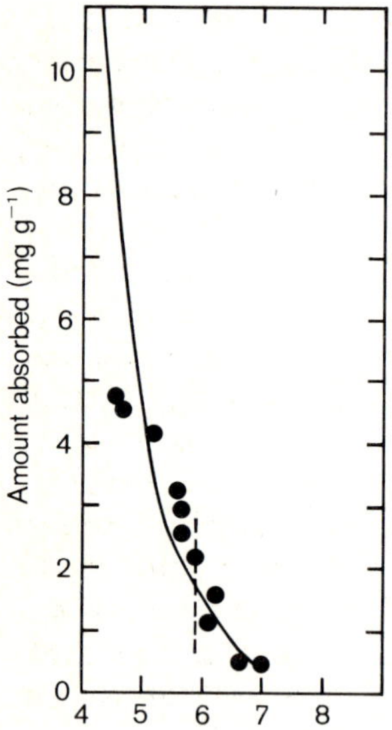

Fig. 9.1. Adsorption of southern bean mosaic virus by kaolinite in function of pH. Vertical dashed line indicates the isoelectric point of the virus (after McLaren and Peterson, 1965).

take as the pH falls below the isoelectric point of the virus would indicate that electrostatic interactions are largely involved in the process, since in this pH range SBMV particles would carry a net positive charge. That appreciable adsorption occurs above the isoelectric pH may be ascribed to the operation of secondary attractive interactions such as Van der Waals attraction and possibly entropy effects as well. The contribution from this source to the overall adsorption energy would, if anything, be greater in this instance because of the large molecular dimensions of viruses as compared with proteins. This and their relative rigidity may also account for the absence in Fig. 9.1 of a maximum at or near the isoelectric point which characterises the pH—uptake curves of proteins (cf. Figs. 7.2 and 7.5).

The high affinity of nucleoproteins and viruses for clays was indicated by the early work of Bower (1949) who found that (at a given clay/adsorbate ratio) there was nearly complete adsorption over the pH range of 2 to 7. Carlson et al. (1968) and Schaub et al. (1974) have subsequently reported similarly. That this property is related to molecular size (weight) is illustrated by the finding that, all things being equal, adsorption decreases in the order nucleoprotein > nucleic acid > nucleotide > nucleoside (Goring and Bartholomew, 1952; Flaig et al., 1955).

Bower (1949) also noted that under comparable conditions montmorillonite was a better adsorbent for nucleoproteins than kaolinite, presumably because the former mineral had a larger accessible surface area. Similarly, Van der Want (1952) reported that montmorillonite and illite could take up larger amounts of tobacco viruses as compared with kaolinite. Since virus molecules are too bulky to penetrate the montmorillonite interlayers, adsorption is confined to the external crystal surfaces. This being so, the extent of adsorption for illite may actually exceed that given by montmorillonite because illite has a higher surface density of charge (Carlson et al., 1968).

The question as to whether both the basal and edge surfaces of the clay crystals are involved in virus attachment could, in principle, be decided by examining the complex under an electron microscope. In practice, however, the location of the virus at the clay surface is often difficult to demonstrate unambiguously using this technique (Jakubowski, 1969). In the case of tobacco mosaic virus (TMV) adsorbed onto kaolinite at low pH, McLaren and Peterson (1965) have been able to show that the virus particles adhere — in a somewhat irregular array — more to the basal surfaces than to the crystal edges. This observation may be ascribed to the fact that below the isoelectric point (pH ~3.5) of TMV, both the kaolinite edge surface and the virus carry a net positive charge and so would tend to repel each other. For Na^+-montmorillonite at pH 3.2 and in the presence of citrate buffer (ionic strength 0.0375), McLaren and Peterson recorded an adsorption of 1.83 mg TMV per mg clay. Taking a value of 40×10^6 for the particle weight and 15×300 nm^2 for the dimension of the rod-shaped virus (Mahler and Cordes,

1968) and assuming that the rods lie with their long axis parallel to the basal surface ("face-to-face"), the area occupied amounts to 124 m^2 per g clay. On this basis and in the absence of interlayer adsorption, the surface requirement of TMV exceeds the available (external) crystal area of montmorillonite by some 30—40%. This would suggest that either there was some considerable overlap between virus particles or a portion of the rods is oriented end-on ("edge-to-face") at the surface. That viruses may adopt both a face-to-face and an edge-to-face orientation has been shown by Roper and Marshall (1974) in electron micrographs of *Escherichia coli* bacteriophage adsorbed onto montmorillonite surfaces.

The presence of neutral electrolytes or buffers in the system markedly enhances the adsorption of bacteriophages and viruses by clays (Fildes and Kay, 1963; Carlson et al., 1968; Jakubowski, 1969; Schaub et al., 1974), an observation which is again reminiscent of the behaviour of nucleic acids (cf. Fig. 8.1). In this respect the salts of polyvalent cations are much more effective than those of monovalent species. This effect of ionic strength on adsorption may be ascribed to the compression of the diffuse double layers around clay particles (reduction of the zeta potential) and to the screening of the negative charges on the virus, allowing the adsorbate to approach the mineral surface more closely. In addition, polyvalent cations may act as a bridge between the anionic groups on the virus and the negatively charged sites at the clay surface, a linkage which may be either direct or indirect through hydrogen bonding to the water molecules in the primary hydration shell around the cation ("water-bridging"). Further evidence for cation- and water-bridge formation is provided by the finding that the addition of compounds capable of sequestering metal ions, such as EDTA, pyrophosphate, and oxalate, leads to a reduction in uptake (Hukuhara and Wada, 1972). Related to this is the ability of small organic substances and proteinaceous materials to compete with viruses for adsorption sites at the clay surface (Thung, 1955; Carlson et al., 1968; Jakubowski, 1969).

Finally, we wish to mention the potential pharmaceutical applications of the clay—virus interaction. Because of the ability of clays to take up viruses, they have been considered as active ingredients in intestinal-adsorbent preparations. The work of Bartell et al. (1960) has shown that a variety of human enteroviruses, such as poliovirus, ECHO-9, and coxsackie B-3, can be adsorbed by kaolinite and palygorskite ("attapulgite"), with the latter having a higher binding capacity. The superiority of palygorskite in this respect is undoubtedly related to its open structure (cf. Fig. 1.9) enclosing channels into which the virus can presumably be accommodated.

REFERENCES

Bartell, P., Pierzchala, W. and Tint, H., 1960. The adsorption of enteroviruses by activated attapulgite. Journal of the American Pharmaceutical Association, Scientific Edition, 49: 1—4.

Bitton, G., 1975. Adsorption of viruses onto surfaces in soil and water. Water research, 9: 473—484.

Bower, C.A., 1949. Studies on the forms and availability of soil organic phosphorus. Iowa Agricultural Experiment Station, Research Bulletin 362, pp. 963—996.

Carlson, G.F., Jr, Woodard, F.E., Wentworth, D.F. and Sproul, O.J., 1968. Virus inactivation on clay particles in natural waters. Journal of the Water Pollution Control Federation, 40: R89—R106.

David, W.A.L. and Gardiner, B.O.C., 1967. The persistence of a granulosis virus of *Pieris brassicae* in soil and sand. Journal of Invertebrate Pathology, 9: 342—347.

Dieterich, B.H., 1953. The behaviour of bacterial viruses in contact with ordinary and uniform filter sand. Masters Thesis, Harvard University, Cambridge, Massachusetts.

Duboise, S.M., Sagik, B.P. and Moore, B.E.D., 1974. Virus migration through soils. In: J.F. Malina, Jr and B.P. Sagik (Editors), Virus Survival in Water and Wastewater Systems. Center for Research in Water Resources, The University of Texas at Austin, pp. 233—240.

Fildes, P. and Kay, D., 1963. The conditions which govern the adsorption of a tryptophan-dependent bacteriophage to kaolin and bacteria. Journal of General Microbiology, 30: 183—191.

Flaig, W., Kuron, H. and Kaul, R., 1955. Über die Sorption von Nucleinstoffen an Tonkolloiden. Zeitschrift für Pflanzenernährung, Düngung, Bodenkunde, 71: 141—154.

Gerba, C.P. and Schaiberger, G.E., 1975. The effects of particulates on virus survival in seawater. Journal of the Water Pollution Control Federation, 41: 93—103.

Gerba, C.P., Wallis, C. and Melnick, J.L., 1975. Fate of wastewater bacteria and viruses in soil. Journal of the Irrigation and Drainage Division, pp. 157—174.

Goring, C.A.I. and Bartholomew, W.V., 1952. Adsorption of mononucleotides, nucleic acids, and nucleoproteins by clays. Soil Science, 74: 149—164.

Hukuhara, T. and Wada, H., 1972. Adsorption of polyhedra of a cytoplasmic-polyhedrosis virus by soil particles. Journal of Invertebrate Pathology, 20: 309—316.

Jakubowski, W., 1969. Adsorption of poliovirus in seawater by clay minerals and marine sediment. Bacteriological Proceedings, V198, pp. 179—180.

Johnson, J. and Ogden, W.B., 1929. Overwintering of tobacco mosaic virus. Bulletin Wisconsin Agricultural Experiment Station 95: 1—25.

Lefler, E. and Kott, Y., 1974. Virus retention and survival in sand. In: J.F. Malina, Jr and B.P. Sagik (Editors), Virus Survival in Water and Wastewater Systems. Center for Research in Water Resources, The University of Texas at Austin, pp. 84—91.

Mahler, H.R. and Cordes, E.H., 1968. Biological Chemistry. Harper International Edition. Harper and Row, New York, N.Y. and John Weatherhill, Inc., Tokyo, 872 pp.

Malina, J.F., Jr, and Sagik, B.P., (Editors), 1974. Virus Survival in Water and Wastewater Systems. Center for Research in Water Resources, The University of Texas at Austin, 264 pp.

McLaren, A.D. and Peterson, G.H., 1965. Physical chemistry and biological chemistry of clay mineral—organic nitrogen complexes. In: W.V. Bartholomew and F.E. Clark (Editors), Soil Nitrogen. American Society of Agronomy, Monograph 10, Madison, Wisconsin, pp. 259—284.

Moore, B.E.D., Funderburg, L. and Sagik, B.P., 1974. Application of viral concentration techniques to field sampling. In: J.F. Malina, Jr and B.P. Sagik (Editors), Virus Survival in Water and Wasterwater Systems. Center for Research in Water Resources, The University of Texas at Austin, pp. 3—15.

Murphy, W.H., Jr, Eylar, O.R., Schmidt, E.L. and Syverton, J.T., 1958. Adsorption and translocation of mammalian viruses by plants. I. Survival of mouse encephalomyelitis and poliomyelitis viruses in soil and plant root environment. Virology, 6: 612—622.

Myamoto, Y., 1959a. The nature of soil transmission in soil-borne plant viruses. Virology, 7: 250—251.

Myamoto, Y., 1959b. Further evidence for the longevity of soil-borne plant viruses adsorbed by soil particles. Virology, 9: 290—291.

Robeck, G.G., Clarke, N.A. and Dostal, K.A., 1962. Effectiveness of water treatment processes in virus removal. Journal of the American Water Works Association, 54: 1275—1290.

Roper, M.M. and Marshall, K.C., 1974. Modification of the interaction between *Escherichia coli* and bacteriophage in saline sediment. Microbial Ecology, 1: 1—13.

Schaub, S.A., Sorber, C.A. and Taylor, G.W., 1974. The association of enteric viruses with natural turbidity in aquatic environments. In: J.F. Malina, Jr and B.P. Sagik (Editors), Virus Survival in Water and Wastewater Systems. Center for Research in Water Resources, The University of Texas at Austin, pp. 71—83.

Slykhuis, J.T., 1976. Virus and virus-like diseases of cereal crops. Annual Review of Phytopathology, 14: 189—210.

Thung, T.H., 1955. Soil-borne tobacco viruses. Congrès Scientifique International du Tabac, 1er, Paris-Bergerac, 2: 684—685.

Van der Want, J.P.H., 1952. Some remarks on soil-borne virus diseases. Proceedings Conference on Potato Virus Diseases, Wageningen-Lisse, pp. 71—75.

Wellings, F.M., Lewis, A.L. and Mountain, C.W., 1974. Virus survival following wastewater spray irrigation of sandy soils. In: J.F. Malina, Jr and B.P. Sagik (Editors), Virus Survival in Water and Wastewater Systems. Center for Research in Water Resources, The University of Texas at Austin, pp. 253—260.

Wellings, F.M., Lewis, A.L., Mountain, C.W. and Stark, L.M., 1975. Virus consideration in land disposal of sewage effluents and sludge. Florida Scientist, 38: 202—207.

Chapter 10

POLYSACCHARIDES

10.1. INTRODUCTION

By way of introduction, we wish to make a few remarks on the occurrence in soil and some of the properties of polysaccharides which may be relevant to an understanding of their reactions with clays. More comprehensive accounts of the amount and nature of carbohydrates in soil, together with their isolation and identification, have been given by Mehta et al. (1961), Gupta (1967), Decau (1968), Swincer et al. (1969), Finch et al. (1971), and Greenland and Oades (1975).

Of the large number and variety of naturally occurring polysaccharides which are known to exist, a majority enters the soil system being introduced by plants, animals, and as a result of microbial synthesis. In the soils literature, the term "polysaccharides" is often used to refer to that part of the total soil carbohydrates which is extractable by relatively mild reagents, such as water and dilute solutions of acids, alkalis, and buffers. In order to avoid confusion, we shall use this term in the conventional (chemical) sense, that is, to denote molecules consisting of ten or more monosaccharide or sugar units. The terms "carbohydrates" and "saccharides" will be reserved for the whole group of monomeric, oligomeric, and polymeric species.

Like proteins and nucleic acids, carbohydrates, being an excellent source of energy for soil organisms, are fairly rapidly decomposed when added to soil unless they enter into complex formation with soil colloids. The essential (biological) instability of added saccharides is also reflected in the finding that only a very small proportion (<1%) of the total carbohydrate content of soils occurs in the free or uncomplexed form at any one time. The bulk of soil carbohydrates, either as such or in association with humic substances, is presumably attached to inorganic soil constituents. The unavailability to microorganisms and their extracellular enzymes of soil polysaccharide, rather than its structural composition, is at the basis of its stability in soil (Cheshire, 1977). Complex formation with clays, metals, and humic substances would also account for the difficulty in isolating saccharides from soil, little more than half of the carbohydrates occurring in soil having so far been separated from inorganic materials.

The carbohydrate content of soils may be estimated by treating the soil with a variety of reagents (HCl, HCOOH, H_2SO_4, and NaOH) under different conditions, and analysing the amount of monosaccharides in the hydrolysate by colorimetric and chromatographic methods. For most soils this amount lies between 5 and 25% of the organic matter content, to which it is generally related. The occurrence of a great variety of polysaccharides in soil may be inferred from the large number of hexoses, pentoses, uronic acids, and hexosamines found in the hydrolysate. This approach, however, gives little indication as to the way in which the monomeric sugar units are linked in the parent polymers, even if the chemical nature of such species could be identified. Many attempts have therefore been made to extract polysaccharides as such from soil. Unlike those involving soil proteins, these efforts have largely been successful. However, the polysaccharides so obtained are, not unexpectedly, rather "crude", containing an appreciable proportion of non-saccharide materials. Nevertheless, the fact that they can be isolated in polymeric form leaves little room for doubt as to their presence and persistence in soil.

This observation, together with the finding that in forming a complex with clays and other soil minerals, polysaccharides are involved in stabilising soil aggregates, have provided much impetus to studying the nature of the clay—polysaccharide interaction. Research into the fate of carbohydrates in soil is further stimulated by the finding that these materials participate in humus formation. The role they play in the humification process may be both direct in that sugars form a component, if not an integral part of humic substances, and indirect in that they serve as substrates for soil microorganisms.

Because of this intimate association with the mineral and organic fractions of soil, polysaccharides extracted from soil are, as already mentioned, usually contaminated with other soil constituents. This means that the extract has to be taken through a series of rather laborious purification steps. As a consequence the final yield of material is small relative to the original weight of the soil sample. In addition, the purified material is frequently very heterogeneous in terms of its structural, charge, and molecular weight characteristics.

For all these reasons an increasing number of more recent adsorption studies has been carried out using commercially available polysaccharides which can be produced on a relatively large scale from simple sugar precursors by selected bacteria. Well-known examples are the synthesis of uncharged dextrans, the α-(1 → 6)-linked D-glucans from sucrose by *Leuconostoc mesenteroides* NRRL B-512F, and of the polyanionic xanthan from dextrose by *Xanthomonas campestris* NRRL B-1459 (see Appendix). The initials NRRL stand for Northern Regional Research Laboratory, USDA, Preoria, Illinois. From such dextrans or their lower molecular weight fractions, a variety of cationic (containing amino sugar residues) and anionic (containing carboxyl or uronic acid residues) derivatives may be prepared by chemical means. The structural properties of these industrially produced polysaccharides together with their practical applications in various chemical and industrial processes, have been summarised by Jeanes (1974). These polymers have the advantage over their soil counterparts not only in being obtainable in quantity but, more importantly, in having relatively well-defined structural and solution properties to which adsorption behaviour may be related.

10.2. GENERAL ASPECTS OF THE CLAY—POLYSACCHARIDE INTERACTION

Complexes of clay minerals with polysaccharides may conveniently be classified and discussed in terms of the charge characteristics of the polymers concerned.

By analogy with synthetic, uncharged, linear macromolecules, described in Chapter 3, polysaccharides possessing no charged groups in their structure are adsorbed by ion-dipole and Van der Waals interactions (cf. Table 12.3). Again, the driving force of the adsorption process would be provided by entropy effects arising from a net desorption of water molecules and the resulting gain in translational entropy by the system (Greenland, 1972). The presence of hydroxyl and other polar groups in the molecule also offers scope for hydrogen bonding of polysaccharides to surface oxygens of the silicate layer. These factors and the large number of surface—segments contacts that may be established generally lead to strong interactions between unchanged polysaccharides and clays.

On the other hand, the adsorption of charged polysaccharides is primarily influenced by charge charge (coulombic) interactions although secondary adsorption forces of the type mentioned above may also contribute to the overall process. Positively charged polysaccharides are adsorbed primarily by a cation exchange mechanism. For this reason, their overall affinity for the clay is, if anything, greater than that shown by the uncharged species, leading to an essential collapse of the entire polymer chain on the surface. As might be expected, little, if any, adsorption occurs with anionic polysaccharides. In some instances these materials are negatively adsorbed, that is, they are essentially repelled from the negatively charged clay surface. However, appreciable (positive) uptake may take place at low solution pH or when polyvalent, exchangeable cations are present at the mineral surface. As we have indicated in several places, the pH effect is associated with the appearance of Al^{3+} ions or of positively charged hydroxy-aluminium compounds at the exchange complex of the clay and to a lesser extent with the development of positive charges at the crystal edges. In addition, a low pH would depress the degree of dissociation of the anionic (carboxyl) groups on the polymer. Polyvalent, exchangeable cations favour adsorption in that they can act as bridges between such anionic groups and the surface because a single polyvalent ion is unlikely to neutralise, to the same extent, a number of discretely spaced, negatively charged sites on the silicate layer. Cation-bridging may be either direct or indirect through water molecules in the primary hydration shell of the cation ("water-bridge" formation). The nature of the exchangeable cation also affects the degree of dispersion of the clay system and hence, the relative accessibility of the interlayer space in expanding layer silicates.

The rate of uptake of uncharged and anionic polysaccharides may be quite low. Like most polymeric compounds, however, polysaccharides once adsorbed are difficult to desorb by a mere lowering of their concentration in solution.

10.3. FORMATION AND PROPERTIES OF COMPLEXES

10.3.1. Uncharged polysaccharides

An early attempt at measuring the ability of different clay minerals to take up various uncharged polysaccharides, such as cellulose derivatives and dextrans, is that of Lynch and co-workers (Lynch and Cotnoir, 1956; Lynch et al., 1956, 1957). They found that the isotherms for the adsorption of these polymers by montmorillonite generally conformed to the L-type although some compounds gave curves approaching the H-type. The maximum extent of adsorption, as measured by the plateau value of the respective isotherms, depended on the type of polysaccharide used. In many cases this value was smaller for the simpler than for the structurally more complex molecules. For example, the disaccharide sucrose, representing the simplest compound in the series, was only minimally adsorbed by montmorillonite and not at all by kaolinite.

As with the plateau adsorption, the relative affinity of the cellulose compounds for montmorillonite, deduced from the slope of the initial linear portion of the isotherm, decreased in the order cellulose dextrin > methylcellulose > hydroxyethylcellulose > carboxymethylcellulose. Being a polyanion, the last-mentioned polymer should strictly not be included here; its position in the series is undoubtedly related to its anionic character. They also stated that the nature of the exchangeable cation (H^+ and Ca^{2+}) had but little influence on the adsorption of these and related polymers, such as glycogen, corn polysaccharide, and starch dextrin.

Lynch et al (1956) did not elaborate on the probable mechanisms underlying the adsorption process, apart from suggesting that hydrogen bonding interactions might be involved. This suggestion was based on infrared spectroscopic examination of the complexes but their evidence was not entirely convincing. They noted, however, that the adsorbed material could not be readily released from its complex with clay by means of weak bases, salts, and acids although the simpler compounds were less strongly held than molecules with longer chains.

The adsorption of a given polysaccharide (methylcellulose, dextran, hydroxyethylcellulose) was subsequently shown to depend on its intrinsic viscosity, or viscosity number (η). Assuming that the shape of the polymer in solution is independent of concentration, the magnitude of (η) is a direct measure of molecular weight (Pasika and Cragg, 1962). On this basis, one might expect an increase in (η) to lead to greater adsorption, as Clapp and Emerson (1972) have found. Lynch et al. (1957), however, observed quite the opposite relationship between (η) and uptake. They suggested that this might be due to a decrease in solubility and mobility as chain length increased. However, a fall in polymer solubility normally promotes rather than depresses adsorption, as exemplified by the behaviour of proteins when they adsorb onto clays near the isoelectric pH (cf. Chapter 7). It seems likely, therefore, that here the increase in (η) is associated with a bulkier conformation, that is, an increase in coil dimension due to strong water—polymer interactions. The observed decrease in uptake as (η) rises may thus be ascribed to the greater difficulty that the bulkier polymers experience in gaining access to the interlayer space of montmorillonite as compared with their lower molecular weight, more compact counterparts. In addition, the species with a high (η) would, by virtue of their low mobility, conceivably take relatively longer to adjust to a stable conformation at the clay surface. In this connection, Dubach et al. (1955) reported earlier that the adsorption of polysaccharides onto clays was influenced by the duration of polymer—clay contact. Nevertheless, a decrease in polysaccharide uptake with an increase in molecular weight or chain length is more the exception than the rule, as we shall presently see.

On the other hand, the lower adsorption by kaolinite as compared with montmorillonite can be ascribed to the surface requirement of the polymer. Not only is the available crystal area in montmorillonite more extensive than that in kaolinite, but the interlayer surface of the former mineral is, in addition, partially accessible to the polymer.

The ability of uncharged polysaccharides to penetrate the montmorillonite interlayers has been demonstrated by several workers using X-ray diffractometry. Lynch et al. (1956), for example, recorded a difference in basal spacing between the complex and the parent (hydrated) montmorillonite of less than 0.28 nm. If it were assumed, however, that most of the interlayer water was displaced by the intercalated polymer, the resultant interlayer separation or Δ-value varied between 0.32 and 0.53 nm. According to Greenland (1956a) the minimum molecular thickness for a number of intercalated monosaccharides, derived from the dimension of the pyranose and furanose rings in some sugar crystals, lies in the range of 0.47 to 0.53 nm. Recorded Δ-values of less than 0.5 nm would thus indicate either that intercalation was incomplete or, more probably, that layer segregation had taken place. The occurrence of the latter process is further supported by the data of Clapp et al. (1968) and Olness and Clapp (1973) who showed that the basal spacing of Na^+-montmorillonite complexes with dextran B-512F and polytran rose as the amount adsorbed increased, recalling the behaviour of proteins and nucleic acids (cf. Figs. 7.3. and 8.2). Greenland (1956b) has reported similarly for a montmorillonite—soil polysaccharide complex.

The formation and properties of montmorillonite complexes with polyglucoses (dextrans and amylose) have been investigated in some detail by a number of workers (Clapp et al., 1968; Clapp and Emerson, 1972; Parfitt and Greenland, 1970a; Olness and Clapp, 1973, 1975). Of the commercial polymers, dextran B-512F, referred to earlier, has featured prominently. This and its relative, polytran, have a molecular weight of $\geqslant 2 \times 10^6$ as determined by Sephadex gel filtration. B-512F contains 95% α-(1 → 6) and 5% α-(1 → 3) linkages whereas polytran has 75% β-(1 → 3) and 25% β-(1 → 6)-linked glucose units.

Although the maximum amount adsorbed depended on the montmorillonite sample, its pretreatment, and other experimental conditions, the isotherms for the adsorption of B-512F were invariably of the H-type. This was also true for polytran but here a second rise occurred beyond a solute concentration of 0.3 mg ml^{-1}, as shown in Fig. 10.1 for Na^+-montmorillonite (Olness and Clapp, 1975). As usual, the amount adsorbed was estimated from the difference in polymer concentration between the initial and final solutions. Likewise, Al^{3+}- and Ca^{2+}-montmorillonites were shown by Parfitt and Greenland (1970a) to yield H-type isotherms with B-512F, the initial slope and plateau adsorption of which decreased in the order $Na^+ > Al^{3+} > Ca^{2+}$. The amount adsorbed obtained by the difference method, was found

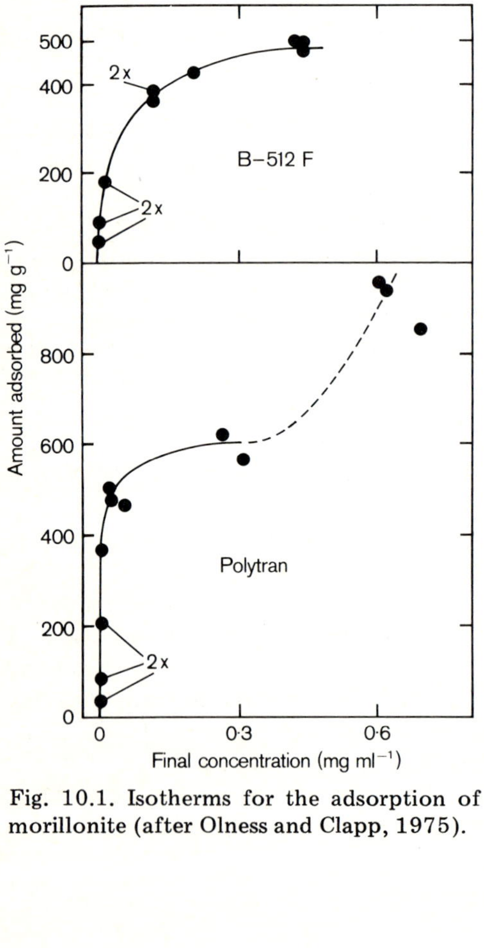

Fig. 10.1. Isotherms for the adsorption of dextran B-512F and polytran by Na^+-montmorillonite (after Olness and Clapp, 1975).

to be less than that estimated from the glucose released on acid hydrolysis (M/2 H_2SO_4 at 378 K for 16 h) or from the carbon content of the corresponding clay—dextran complexes, but agreed quite well with that calculated from the weight loss on ignition.

With reference to Fig. 10.1 the plateau adsorption for B-512F and polytran (before the second rise) was ~445 and ~660 mg g^{-1}, respectively. However, for both dextrans, the extent of interlayer expansion was limited to ~0.53 nm. Bearing in mind that their molecular weight was comparable, this observation would immediately suggest that a greater proportion of polytran was adsorbed on external crystal surfaces as compared with B-512F. This suggestion was further supported by comparing their respective volume requirements. Assuming a total and interlayer surface area of 800 and 720 $m^2\ g^{-1}$, respectively, for montmorillonite, Olness and Clapp (1973) calculated a value of 42×10^{-3} and 191×10^{-3} $cm^3\ g^{-1}$ for the external crystal and interlayer volume, respectively, or a total of 233×10^{-3} $cm^3\ g^{-1}$. Taking a volume of 0.16 nm^3 for an anhydroglucose unit, the volume requirement for B-512F and polytran at plateau adsorption was estimated as 265×10^{-3} and 369×10^{-3} $cm^3\ g^{-1}$, respectively. Thus, these estimates implied that whereas B-512F could be accommodated as a monolayer in an extended (flat) conformation, a substantial proportion (~40%) of the externally adsorbed polytran was apparently not in physical contact with the clay surface.

On this basis, the fraction of bound polymer segments (p) for polytran was relatively small. This would mean that the adsorbed layer consists of short trains alternating with long loops, an arrangement which was consistent with the finding that only the material adsorbed on external crystal surfaces was susceptible to enzymic degradation, leading to a complex containing 440 mg g^{-1}. Such a surface conformation may also account for the second rise in the isotherm for polytran when the solution concentration was raised beyond ~0.3 mg ml^{-1} (Fig. 10.1). The additional material taken up might simply be enmeshed within the "loopy" structure of the existing surface species, as the gently rising slope of the curve would indicate. However, inferences regarding the probable conformation of adsorbed polymers, drawn from adsorption—desorption data above, must remain somewhat speculative. Independent measurements of adsorbed layer thickness by such means as ellipsometry are required if our interpretation of surface conformation is to be less equivocal.

The question arises as to why B-512F behaved differently from polytran in the way it was attached to the montmorillonite crystal. As Olness and Clapp (1975) have suggested, the answer must be sought in the primary structure of these molecules in that B-512F has a relatively high ratio of primary to secondary hydroxyl groups. This presumably enabled B-512F to adsorb more readily and strongly onto the clay by means of ion—dipole interactions so that nearly all of its segments became attached to the surface. The more or less two-dimensional collapse of the polymer chain on interlayer surfaces must also be the reason behind the biological and chemical stability of uncharged polysaccharides adsorbed onto montmorillonite. We have already cited the behaviour of polytran complexes towards enzymic hydrolysis as well as the fact that the adsorption capacity measured by acid treatment of the complex was lower than that obtained by the difference method.

Some time ago, Lynch and Cotnoir (1956) noted that the enzymic breakdown of some cellulosic compounds by soil microorganisms could be attenuated, or at least delayed, by adding montmorillonite to the system. On the other hand, a comparable quantity of illite or kaolinite had but a negligible effect on the rate of decomposition of the same materials, as measured by the amount of carbon dioxide evolved during a given incubation period. Although no adsorption measurements were carried out in conjunction with degradation studies, it seemed reasonable on their part to propose that interlayer uptake, in the case of montmorillonite, partly accounted for the above observations.

By analogy with the action of enzymes on clay-adsorbed proteins, we may suppose that formation of an enzyme—polysaccharide complex at the mineral surface is required before the enzyme-catalysed hydrolysis can proceed. Because of the limited interlayer expansion of the montmorillonite—dextran complexes (Δ-value < 0.6 nm), the interlayer polymer was presumably inaccessible to microbial exoenzymes. Steric protection of this type, however, could not be invoked to account for the apparent resistance of interlayer dextran molecules to acid hydrolysis (Olness and Clapp, 1975). As illustrated in Fig. 10.2 the recovery of adsorbed B-512F and polytran (as glucose) by

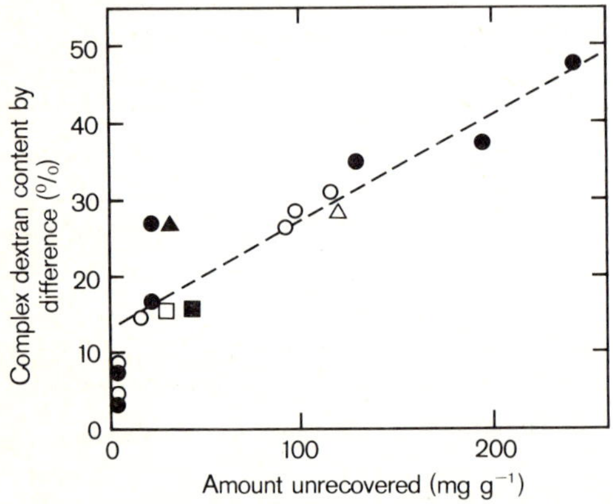

Fig. 10.2. The influence of dextran content in the complex on the amount retained against hydrolysis by M/2 H_2SO_4 at 398 K for 6 h. (●) polytran complex; (■) polytran complex coagulated with Na_2SO_4; (▲) polytran complex hydrolysed for 16 h under reflux: (○) B-512F complex; (□) B-512F complex coagulated with Na_2SO_4; (△) complex hydrolysed for 16 h under reflux (after Olness and Clapp, 1975).

this treatment was less than quantitative for complexes containing more than 13% (130 mg g^{-1}) polysaccharide. This observation may plausibly be explained in terms of any one or a combination of the following processes: acid-induced degradation and structural modification of the monomer; re-adsorption of the hydrolysis products by the mineral residue, or inability of the proton to approach the glycosidic link sufficiently closely because of the particular conformation adopted by the polymer between two opposing silicate layers. Whatever the underlying mechanism, the stability of adsorbed polysaccharides has important implications for the presumed function they fulfil in stabilising clay and soil systems, an aspect which we shall describe more fully later.

Reference has already been made to the finding by Lynch et al. (1957) that the adsorption by clays of some cellulosic compounds tended to decrease with an increase in intrinsic viscosity of the polymer. In view of the fact that, for uncharged linear macromolecules, this solution property is directly related to molecular weight, their results would suggest that the low molecular weight species were relatively better adsorbed. This behaviour is, as we have said, seemingly at variance with the general observation that the adsorption of uncharged polymers increases with molecular weight, at least up to a certain optimum degree of polymerisation unless factors such as polymer conformation, flexibility, and solubility intervene and overrule this relationship. Indeed, using uncharged bacterial polysaccharides of similar structure but different intrinsic viscosity (cf. Table 10.2), Clapp and Emerson (1972) noted that their adsorption by Ca^{2+}-montmorillonite rose with an increase in (η) and hence, in molecular weight.

The applicability of this "rule" for the polysaccharide—clay interaction has been demonstrated on other occasions. Clapp et al. (1968), for example, observed that, at comparable clay/polymer ratios, the uptake of two B-512F fractions of molecular weight 71 000 and 127 000 by Na^{+}-montmorillonite was much (1/7—1/8) less than that of the "parent" polymer (cf. Fig. 10.1). Similarly, Parfitt and Greenland (1970a), using two dextrans with a weight-average molecular weight of 11 200 and 100 000, failed to measure any adsorption by montmorillonite in the sodium form. It is interesting to note, however, that an amylose preparation of comparable molecular weight (100 000) was adsorbed in substantial amounts by the same montmorillonite sample, saturated with Na^{+}, Ca^{2+}, and Al^{3+} ions. The difference in adsorption behaviour between dextran and amylose probably reflects their respective solution conformation which is known to exert a profound effect on the hydration properties of polysaccharides.

However likely, this interpretation must remain somewhat equivocal, for despite the considerable amount of work which has been carried out on the conformation of amylose in solution (Stone, 1969; Suggett, 1975), no self-consistent picture has emerged from the results. The view that amylose in neutral aqueous systems is composed of short helical segments forming a more or less stiff, worm-like coil, has been questioned by Banks and Greenwood (1967) on hydrodynamic grounds. They considered amylose to behave rather like a (normal) random coil with no pronounced helical character under this condition. If so, a basal spacing of ~1.55 nm obtained for the complex with Ca^{2+}-montmorillonite, would imply extensive uncoiling and flattening of the polymer chain in the interlayer space. By the same token, the failure of the low molecular weight dextrans to become adsorbed may be ascribed to strong solute—solvent interactions, possibly by virtue of the particular conformation that they adopt in solution.

As indicated earlier, the relative affinity of B-512F dextran for montmorillonite saturated with different cations, decreased in the order $Na^+ > Al^{3+} > Ca^{2+}$. Since this is analogous to the behaviour of polyethylene glycols at the montmorillonite surface (Parfitt and Greenland, 1970b; cf. Chapter 3), we may infer that the amount of water desorbed from the respective clay system during adsorption decreases in the same order. If so, the relative affinity or adsorption energy arises, to a large extent, from the attendant gain in translational entropy. The influence of the exchangeable cation on adsorption must therefore be related to the amount of water which the corresponding (parent) clays contain in their interlayer space. However, it is difficult on this basis alone, to account for the observation that Na^+-montmorillonite adsorbs more than twice the amount of B-512F dextran at the plateau as compared with the calcium- and aluminium-saturated samples. It seems probable that, in addition, the sodium clay offers a greater amount of accessible interlayer surface to the polymer than do the polyvalent clay systems because the former is initially more dispersed (Clapp et al., 1968).

Finally we wish to make the remark that ion—dipole interactions, so crucially important in determining the adsorption of small hydroxylated organics, seemingly play but a subsidiary part in the uptake from aqueous solutions of uncharged polysaccharides containing polar hydroxy groups by montmorillonite. If such interactions were predominant, the relative affinity of the solute for the clay would be directly related to the polarising power of the exchangeable cation. In other words, the order $Na^+ < Ca^{2+} < Al^{3+}$ would obtain, as exemplified by the behaviour of fulvic and humic acids under similar conditions (cf. Fig. 12.4). However, the feasibility of hydrogen bonding between the hydroxyl groups of the polymer and the surface oxygens of the silicate layer should not be dismissed. Such bonding is clearly possible and, indeed, probable but its occurrence has not yet been unambiguously demonstrated.

10.3.2. Negatively charged polysaccharides

In examining the interactions of some pectinates and alginates of plant origin with kaolinite and montmorillonite, Lynch et al. (1957) observed that the calcium clays took up more of these compounds than did the hydrogen forms. Again montmorillonite generally showed a greater sorption capacity as compared with kaolinite, confirming Beutelspracher's (1955) earlier finding. They further noted that the amount of pectin adsorbed decreased as the degree of methylation increased.

These observations led them to suggest that the carboxyl groups of the polymers were involved in the adsorption process, the exchangeable polyvalent cations acting as a "bridge" between such groups and the negatively charged clay surface. Since monovalent cations would only show a very limited, if any, tendency to function in this way, a cation-bridge mechanism could not adequately account for the appreciable uptake by the hydrogen clay systems. We must therefore presume that some Al^{3+} ions (or their polyhydroxy derivatives) were present at the exchange sites of the H^+-montmorillonite, used by Lynch et al., such ions being released from their positions in the octahedral sheet of the silicate layer by proton attack. This acid-induced transformation of hydrogen clays into the corresponding H^+/Al^{3+} forms may even lead to a reversal in sorption capacity, that is, the hydrogen clay becoming a better adsorbent for anionic polysaccharides as compared with its calcium saturated counterpart.

Besides being dependent on the nature of the exchangeable cation, the interactions of clays with negatively charged polysaccharides are also influenced by the pH of the system. This is so because pH controls the degree of dissociation of the carboxyl groups which, in turn, affects the magnitude of the negative charge on the polymer. Assuming that the mean pK_a of poly(galacturonic acid) is similar to that of uronic acids which have a pK_a of 3.2 (Matthews, 1959), an appreciable proportion of the carboxylate groups on the polymer would be protonated at pH 3. This means that as the solution pH is lowered from neutrality, the polysaccharide would become progressively less negative and hence adsorption would increase. The latter tendency would also be reinforced by the presence at low pH of positively charged aluminium species at the exchange complex of the clay. These factors can account, at least qualitatively, for the adsorption behaviour of anionic polysaccharides. Parfitt and Greenland (1970a), and Parfitt (1972), for example, observed that poly(galacturonic acid) (see Appendix) was effectively repelled by Na^+- and Ca^{2+}-montmorillonite and none was adsorbed by Al^{3+}-montmorillonite at pH 6. Lowering of the solution pH led to an appreciable increase in uptake by all clay samples. As might be expected, the greatest relative increase occurred with the aluminium clay.

Fig. 10.3 shows the isotherm for the adsorption by Al^{3+}-montmorillonite at pH 3.6. The H-type curve reached a maximum or plateau value of 280 mg g^{-1}. Although the affinity for the clay was high as was the amount adsorbed, no interlayer expansion of the montmorillonite crystal was recorded when the complexes were examined by X-ray diffraction. This failure of poly(galacturonic acid) to penetrate the interlayer space is a feature of polyanion

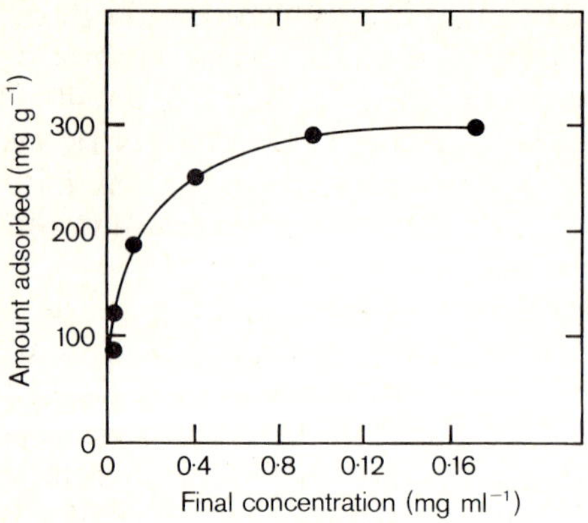

Fig. 10.3. Isotherm at 275 K for the adsorption of poly(galacturonic acid) by Al^{3+}-montmorillonite at pH 3.6 (after Parfitt and Greenland, 1970a).

adsorption by expanding layer silicates. Relevant in this context are the recent measurements by Cortez (1977) on the adsorption of lipopolysaccharides by kaolinite, illite, and montmorillonite. As the name suggests, these polymers contain an appreciable (~15%) proportion of lipid materials in their molecule, about half of which appears to be phosphorylated. Under neutral or slightly acid conditions, the polymers would therefore be negatively charged. He was able to show that the extent of adsorption was approximately proportional to the external crystal area of the clay samples, indicating the absence of interlayer penetration. Only in those situations where the carboxyl groups of the polymer are largely undissociated, as in the case of fulvic acid at pH 2.5, can intercalation occur (Schnitzer and Khan, 1972).

Using a similar approach to that outlined for B-512F dextran and polytran, we may estimate the spatial requirement of poly(galacturonic acid). On the basis that adsorption is confined to the external crystal surface of montmorillonite, it can be shown that only a quarter of the plateau amount (at pH 3.6) would be accommodated if the polymer were to adopt a fully extended conformation. It may therefore be inferred that the molecule is attached to the clay surface by relatively few of its segments, the major fraction of which being present in the form of loops extending away from the surface. This contrasts with the behaviour of dextran B-512F whose chains tend to collapse and spread at the surface. The difference in preferred surface conformation between poly(galacturonic acid) and dextran extends to polyanions and uncharged polymers in general. This fact has important implications for the use and effectiveness of such materials as soil aggregate stabilisers ("soil conditioners"), a topic which will be discussed more fully in section 10.4.

At this point we should perhaps interpolate what little information there is on the flocculating ability of negatively charged polysaccharides, including those extracted from soil. To be an effective flocculant a polymer must be capable of adsorbing onto and acting as a bridge between several clay particles. The extent of interparticle bridging is, for this reason, more a function of loop length, that is, of the unbound fraction, than of the total amount of polymer adsorbed.

Using two negatively charged polysaccharides, one extracted from soil and another from *Plantago arenaria*, Dubach et al. (1955) observed that both were adsorbed in appreciable amounts by montmorillonite. The soil polymer, however, was ineffective as a flocculant of clay dispersions, presumably because the loops it formed at the surface were insufficiently long to effect interparticle bridging. On the other hand, the soil polysaccharide used by Saini and MacLean (1966) was capable of flocculating dispersed kaolinite particles. It is in this situation where flocculation occurs that the flocculating efficiency of a given polymer becomes dependent on the amount present. This was illustrated by measuring the relative filtration rate which increased to a maximum at some optimum value of the initial polymer concentration, and then declined as this concentration was further increased (cf. Fig. 2.5). This observation is of general applicability to dispersed solid—polymer systems (cf. section 2.4) and may be explained in terms of a concurrent rise in the extent of flocculation and interparticle bridging below the maximum filtration rate, which may or may not coincide with maximum flocculation. However, beyond this point very little free adsorbing surface is available to the solute and the polymer loops surrounding each particle interfere with those on neighbouring particles, preventing them from coming together (cf. Fig. 2.4).

Further support for the involvement of the anionic (carboxyl) groups and of the exchangeable cations in the interaction between montmorillonite and poly(galacturonic acid) is provided by the infrared spectra (Fig. 10.4) of the clay—polysaccharide complexes formed at different solution pH values (Parfitt, 1972). The spectrum of the polymer in the acid form (A) was characterised by the presence of a strong band near 1723 cm^{-1} and weaker bands at 1630, 1410, and 1328 cm^{-1}. Adsorption at 1723 and 1328 cm^{-1} could be assigned to unionised carboxyl groups and that at 1630 and 1410 cm^{-1} to the ionised (carboxylate) forms. As might be expected, conversion of the acid into its corresponding aluminium salt gave rise to a spectrum in which the 1723 cm^{-1} band was virtually absent, being replaced by strong absorption at 1630 cm^{-1} (B). The spectrum of the Al^{3+}-montmorillonite complex formed at pH 2.9 (C), showed strong bands at 1723, 1630, 1410 and 1328 cm^{-1}. This would indicate that both dissociated and unionised carboxyl groups were present on the polymer. It should be pointed out that the clay—polysaccharide complexes were examined in the air-dry state. The intensity of the 1630 cm^{-1} band in spectra D to F would therefore be partly due to the bending vibration of adsorbed water. For the same reason, it is difficult

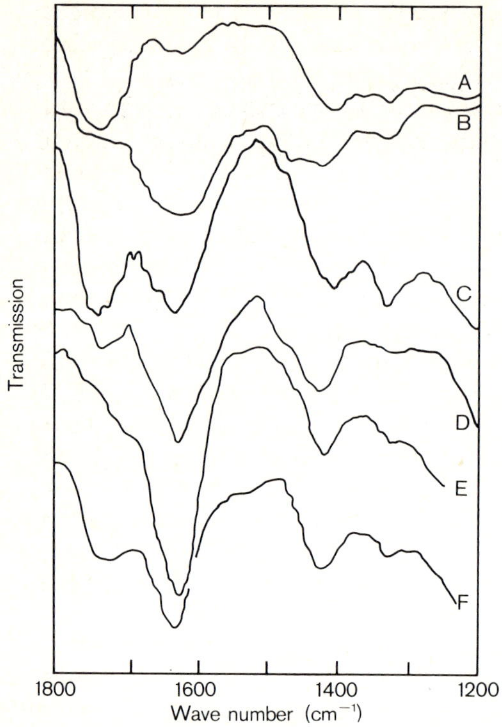

Fig. 10.4. Infrared spectra of A, poly(galacturonic acid); B, its aluminium salt; C, poly-(galacturonic acid) adsorbed onto Al^{3+}-montmorillonite at pH 2.9; D, as for C but at pH 5.0; E, adsorbed onto Ca^{2+}-montmorillonite at pH 2.9; F, adsorbed onto Na^{+}-montmorillonite at pH 2.9 (after Parfitt, 1972).

to determine whether there is a direct association between anionic group and exchangeable cation or whether this interaction occurs through a water bridge. At pH 5, the carboxyl groups of poly(galacturonic acid) would be largely dissociated. The spectrum of the aluminium clay complex at this pH (D) accordingly showed a considerable weakening of the 1723 and 1328 cm^{-1} bands, whilst those at 1630 and 1410 cm^{-1} intensified. One might expect that the spectra of the complexes with Ca^{2+}- and Na^{+}-montmorillonite at pH 2.9 (E and F) respectively, would show features similar to those of spectrum C and not, as they did, to spectrum D. No explanation was given to account for this behaviour, although a reaction of the type

$$Ca^{2+}\text{-clay} + 2\ R{-}COOH \rightarrow 2\ H^{+}\text{-clay} + Ca(OOC{-}R)_2$$

could have occurred at the montmorillonite surface, as Yariv et al. (1966) had suggested for benzoic acid.

The infrared data are fully consistent with the proposal that the anionic groups of the polysaccharide are associated more with the exchangeable cations than with the silicate surface. As we have said, it is not possible to deduce the precise mode of this association from Fig. 10.4 due to the interference from adsorbed water. However, by analogy with the behaviour of fulvic and humic acids (Theng and Scharpenseel, 1975; Theng, 1976), both of which are polyanionic, the adsorption of poly(galacturonic acid) from aqueous solutions probably occurs by hydrogen bonding of its carboxylate groups to water molecules directly coordinated to the exchangeable cation.

Earlier, Kohl and Taylor (1961) had also used infrared spectroscopy to obtain information on the mode of bonding between polyanions and montmorillonite surfaces. Their spectrum of poly(galacturonic acid) on H^+-montmorillonite in the 1800—1500 cm^{-1} range closely resembles that of Fig. 10.4D, indicating that a portion of the carboxyls was dissociated. This resemblance was probably due to the presence of positively charged aluminium species at the surface. In addition to the shoulder near 1720 cm^{-1}, at which frequency the acid form strongly absorbed, there was an inflection at ~1700 cm^{-1}, as was also indicated in spectrum D. This inflection was ascribed to hydrogen bonding of C = O of carboxyls to O—H groups at the crystal edge. However, on the basis of what has been said of such systems, this observation rather reflects the presence of some intramolecular hydrogen bonding between carboxyl and hydroxyl groups on the polymer loops.

This suggestion does not imply that the crystal edges are not involved in adsorption. Indeed, with 1 : 1 type minerals for which the edge area represents an appreciable proportion (~20%) of the total, the crystal edges play an important part in the uptake of polyanions at low pH. In some instances adsorption even appears to be predominantly, if not exclusively, confined to such areas, as the electron micrographs of Thiessen (1942) have so elegantly demonstrated for the kaolinite—gold sol system. In this regard, Bitton et al. (1976) have reported that the ability of K^+-montmorillonite to flocculate the non-capsulated strain of *Klebsiella aerogenes* could be substantially reduced or inhibited by adding the capsular polysaccharide to the system. From this they inferred that the polymer which contained carboxylate groups attached to, and so neutralised, the positive charges at the montmorillonite crystal edges.

What seems clear is that in situations where positive uptake is recorded, anionic polysaccharides can and do interact through the exchangeable cations which are located at the basal surfaces. This view is also supported by the observation that, like that of humic substances (cf. Fig. 12.4), the affinity of these compounds for the clay surface increases with the polarising power of the exchangeable cation. This point is illustrated by the results of Moavad et al. (1974) who have determined the isotherms for the adsorption of extracellular, negatively charged polysaccharides by kaolinite saturated by different cations. They found that the relative affinity or adsorption energy, derived from the initial slope of the respective isotherms, increased in the order $Na^+ < Ba^{2+} < H^+ < Al^{3+} < Fe^{3+}$. Earlier, Saini and MacLean (1966) had reported similarity for a soil polysaccharide although their plateau values in absolute terms were much smaller than those obtained by Moavad et al. (1974). The possibility that exchangeable polyvalent cations may form an insoluble precipitate with anionic polysaccharides at the clay surface should not be discounted. However, Parfitt's (1972) data and the observed variations in adsorption capacity among polyvalent cation clay systems (Table 10.1) indicate that surface precipitation reactions are not important.

Bearing in mind that the external crystal area of kaolinite is, if anything, less extensive than that of montmorillonite, the high plateau values listed in Table 10.1 would mean that only a minor fraction of the polymer segments would have been attached to the surface. In other words, the polymer was likely to be adsorbed with short trains and long loops. This concept is also in keeping with the observation that the temperature coefficient of the adsorption process is generally positive. This is so because loop size is expected to increase (with rising temperature) if there is no concomitant increase in the area occupied by the trains as adsorption increases with temperature (Silberberg, 1962; Rowland and Eirich, 1966).

Reference has already been made to the behaviour of soil polysaccharides at clay surfaces. These substances consist of a mixture of uncharged, anionic, and cationic species although the positively charged component normally makes up but a very minor proportion of the whole. Accordingly, it seems best to include what little information there is on the clay—soil polysaccharide interaction in this section.

Being so heterogeneous as to charge and molecular weight, it is perhaps not surprising to find, as Finch et al. (1966) did, that under a given set of

TABLE 10.1

Adsorption of an extracellular polysaccharide from *Lipomyces lipofer* by kaolinite saturated with different cations at two temperatures (from Moavad et al., 1974)

Saturating cation	Extent of adsorption ($\mu g/mg^{-1}$)	
	298 K	328 K
H^+/Al^{3+}	108	95
Na^+	40	43
K^+	40	nd
Ba^{2+}	80	nd
Ca^{2+}	70	75
Mn^{2+}	75	nd
Al^{3+}	300	>640
Fe^{3+}	>1200	>1200

nd = not determined.

conditions, some fractions of soil polysaccharides are preferentially adsorbed by clays. For the same reason, the overall mechanism underlying their adsorption would be expected to reflect the presence of both uncharged and anionic components.

In an attempt at assessing the relative contribution of each component to the adsorption process, Parfitt and Greenland (1970a) have used the carbazole and the anthrone methods to analyse saccharide concentrations in solution. Their results, using montmorillonite saturated with Na^+, Ca^{2+}, and Al^{3+} ions, indicated that although both components were taken up by all samples at pH 2.4, the uronic acids were more strongly adsorbed. Interestingly, adsorption appeared to be little affected by the nature of the saturating cation under this condition. At pH 5.8, on the other hand, uptake of soil polysaccharides was strongly influenced by the polarising power of the exchangeable cation. Thus, little adsorption of the uronic acid fraction occurred with Ca^{2+}-montmorillonite, and none with the sodium clay whereas the uncharged component was adsorbed in amounts comparable to those obtained at the lower pH value. For both components, uptake by Al^{3+}-montmorillonite was greater than by either the sodium or the calcium clay systems. Similarly, using a sample of ^{14}C-labelled soil polysaccharide and montmorillonite saturated with Fe^{3+}, H^+/Al^{3+}, and Ca^{2+} ions, Guckert et al. (1975) found that adsorption decreased in the order $Fe^{3+} > H^+/Al^{3+} > Ca^{2+}$. These observations are in keeping with the concept that the anionic groups of the polysaccharide and the exchangeable cations are involved in the adsorption process.

The influence of the exchangeable cation and of pH on the reaction of two soil polysaccharide fractions with kaolinite and montmorillonite, was earlier demonstrated by Finch et al. (1966). The higher uptake by montmorillonite as compared with kaolinite at comparable solute concentrations in solution may again be ascribed to the larger extent of available surface in the former system, particularly so when intercalation can occur. They also found that H^+-montmorillonite was more reactive than the untreated clay (Fig. 10.5) since, judging from its origin, the latter sample would have been predominantly sodium saturated whereas some of the exchange positions in the hydrogen clay were likely to have been occupied by Al^{3+} ions and by positively charged hydroxy-aluminium species. The low pH of the hydrogen system would also depress the dissociation of the anionic groups as their electrophoretic measurements indeed indicated. Since this would have the effect of reducing the negative charge on the polysaccharides, electrostatic repulsion between solute and surface would be lessened. In addition, the untreated clay, as they pointed out, contained some free electrolytes which would tend to lower adsorption in this system.

On the presupposition that adsorption should decline with a decrease in pH, Finch et al. (1966) have raised the feasibility of hydrogen bonding between polymer and clay surface. However, on the basis of the above anal-

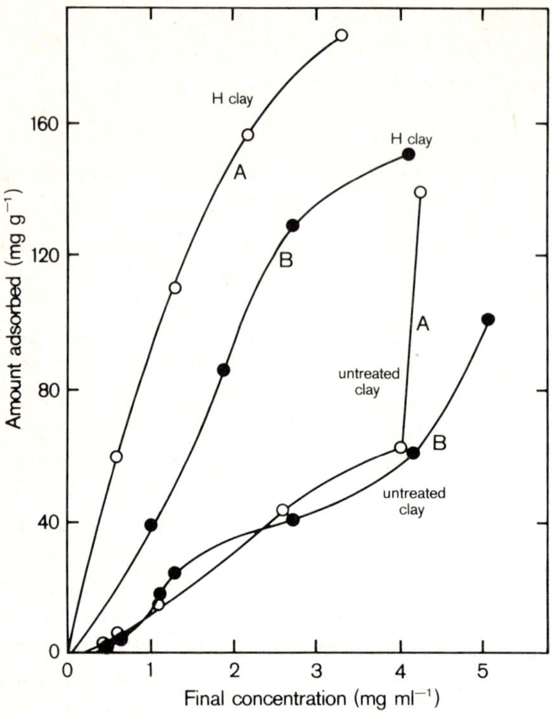

Fig. 10.5. Isotherms for the adsorption of two soil polysaccharide fractions (A and B) by 2% (w/v) of untreated and H^+-montmorillonite (after Finch et al., 1966).

ysis and of previous considerations, it seems unnecessary to invoke this type of bonding here to account for the observations. Hydrogen bonding interactions of the type suggested may, of course, contribute to the overall process.

The isotherms for the untreated montmorillonite showed an inflection when adsorption reached ~60 mg g^{-1} at a concentration of 4 mg ml^{-1}, beyond which there was a second rise reminiscent of the polytran—montmorillonite system (cf. Fig. 10.1). Assuming a fully extended conformation the amount adsorbed at the inflection point corresponds to a complete coverage of the external crystal surface. The isotherms for the H^+-montmorillonite, on the other hand, are of the L-type. The "monolayer" capacity estimated from the Langmuir plot corresponded to a coverage of both the external and the interlayer surfaces by the polymer. However, this is only circumstantial evidence for interlayer adsorption by the hydrogen system although the X-ray diffraction data of Parfitt and Greenland (1970a) do indicate that soil polysaccharides, or at least some of their (presumably uncharged) components, can be intercalated by montmorillonite. For one fraction, the plateau value obtained with Na^{+}-kaolinite likewise corresponded to a monolayer coverage of extended chains. This would suggest that the basal surfaces of kaolinite crystals are involved in the adsorption process.

10.3.3. Positively charged polysaccharides

Relatively little information is available on complex formation between clays and polysaccharides containing positively charged amino sugar units in their structure. By analogy with the behaviour of glucosamine (Parfitt, 1972), however, the adsorption of cationic polysaccharides by layer silicates would be primarily controlled by an exchange reaction between the cationic groups of the polymer and the cations occupying exchange sites at the mineral surface. Further support for such a mechanism is provided by the relative strength of their adsorption.

In comparing the behaviour of a number of polysaccharides towards Ca^{2+}-montmorillonite (cf. Table 10.2) Clapp and Emerson (1972), for example, observed that the cationic polymers were taken up to the same extent as the most strongly adsorbed neutral species and this despite the relatively low intrinsic viscosity of the former materials. As discussed in section 10.2, this finding can be explained in terms of the relative magnitude of the surface—segment interaction energy. When a positively charged segment is adsorbed onto a negatively charged site at the clay surface, this energy would far exceed that due to Van der Waals or hydrogen bond interactions between surface and an uncharged polymer segment. Cationic polysaccharides would therefore be expected to collapse rapidly on the silicate surface with nearly all of their segments in contact with it.

For the chitosan and deacetylated chitin, Clapp and Emerson recorded a maximum adsorption of 35 mg g^{-1}. This value corrresponded to about 0.18 mmol g^{-1} of amino groups, all of which would have been protonated under the prevailing pH regime. Since the cation exchange capacity of the calcium clay was 0.45 mmol g^{-1}, all cationic groups of the polymer would have been capable of exchanging with the Ca^{2+} ions at the surface unless steric effects intervened. It is, of course, implicit in this proposal that the polysaccharide can at least partially penetrate the montmorillonite interlayers. Such an assumption seems reasonable in view of the fact that polycations, in general, are readily intercalated (e.g. Ruehrwein and Ward, 1952). The strong adsorption of chitosan and the chitin derivatives was also reflected in the ability of their respective montmorillonite complexes to withstand periodate oxidation and dispersion in sodium salt solutions. The relative stability of clay—polysaccharide complexes when subjected to such treatments is further elaborated below.

10.4. POLYSACCHARIDES AND SOIL/CLAY AGGREGATE STABILITY

It is not intended here to give a comprehensive summary of the literature dealing with the effect of polysaccharides on soil/clay aggregation. Rather, by way of rounding off the discussion on clay—polysaccharide reactions we wish to briefly mention the stabilising ability of various polysaccharides in relation to their adsorption characteristics.

The beneficial influence of organic matter on soil physical conditions has been known for a very long time. Of the many and varied polymeric organic substances present in soil, the polysaccharides, particularly those of microbial origin, have been considered as being the most important aggregate-stabilising agents. Much of the evidence for this hypothesis is of an indirect nature, being gathered from experiments in which different polysaccharides are added to soil/clay systems and the resulting increase in the size and water-stability of the aggregates measured (cf. Fig. 10.6). It is only relatively recently that attempts have been made to assess the effect on aggregate stability of the selective removal of the polysaccharide fraction from soil.

For reasons mentioned in the preceding sections, most of the data on the aggregating effect of polysaccharides refer to uncharged and negatively charged polymers. It should also be pointed out that we are here concerned with a situation where the polysaccharide is added to an already aggregated soil/clay system, in which the particles are in close proximity to one another. This is to be distinguished from a situation where the individual particles are separated by relatively large distances, as in a soil/clay dispersion ("suspension"). When this distinction is recognised and the difference in adsorption behaviour between uncharged and negatively charged polysaccharides is borne in mind, most of the published data do submit to a rational and integrated interpretation. The use of synthetic polymers, in general, as flocculants of clay dispersions and as soil conditioners, is described more fully in Chapter 6.

As with flocculation, aggregate stabilisation by polymers essentially involves the formation of bonds (bridges) between particles or groups of particles making up an aggregate. Having a linear, flexible chain structure, polysaccharides are clearly well suited to act in this way. Chain length, as measured by the degree of polymerisation or the related parameters of molecular weight and intrinsic viscosity, is therefore a prime factor influencing aggregating ability. For a given chain length, an uncharged polysaccharide which tends to adsorb by many surface—segments contacts would understandably be more effective than an anionic polymer which apparently is tied to the solid surface by relatively few segments. For negatively charged polysaccharides, therefore, the degree of anionic substitution as well as pH and the nature of the exchangeable cation would be relatively important. The shape effect is reflected, for example, in the finding that humic and fulvic acids, being essentially spherical, are not as effective as anionic polysaccharides in stabilising soil aggregates.

As indicated above, the long-standing recognition that polysaccharides in soil are at least implicated in aggregate formation and stabilisation, has largely come from observations that addition of polysaccharides (of various types and from diverse sources) to soil, can markedly increase the amount and size of water-stable aggregates. The degree of aggregation, or simply "aggregation", can be expressed in several ways, the most common being as a percentage of (stable) aggregates greater than a certain dimension which remain after shaking in water.

Most of the principles involved in aggregate stabilisation by polysaccharides emerged from studies carried out during the two post-war decades. Not coincidentally, this period also saw the tremendous upsurge of interest in the application of synthetic, linear polymers to improve the physical condition of soils, giving rise to the term "soil conditioners" (Harris et al., 1966).

Using various levans, made up of β-(2 → 6) linked D-fructose units from *Bacillus subtilis* (see Appendix), Geoghegan and Brian (1948) for example, found that for a given polymer, aggregation increased with the amount applied up to a point beyond which it tended to level off. For a given rate of application, aggregation generally rose with increasing intrinsic viscosity and hence, molecular weight or chain length of the polymer. These observations, depicted in Fig. 10.6, accord with the known mode of bonding of uncharged polysaccharides which, being strongly adsorbed, tend to spread over the surfaces of adjacent particles like a "coat of paint" (Greenland, 1963). Beyond a certain polymer/soil ratio, a further increase in polymer application would presumably increase the frequency of interparticle bonding and/or lead to a more thickly adsorbed coat but have a relatively small effect on overall stability. Similarly, Gaur and Rao (1975) have observed a positive correlation between viscosity and aggregate stabilising ability of polysaccharides isolated from various soil bacteria.

Geoghegan and Brian also noted that dextrans synthesised by *Leuconostoc* spp. were slightly superior to the best levans examined. This may reflect a difference in the conformation adopted at the soil surface between polyglucoses and polyfructoses and hence, possibly, in the extent of interparticle bonding.

Compared with the levans a polysaccharide preparation from *Rhizobium trifolii*, containing 67% glucose and 20% uronic acid residues, was ineffective as an aggregate stabiliser (Geoghegan, 1954). The reason for this should be sought in the relatively few surface—segment links formed due to electrostatic repulsion between polymer and negatively charged soil particles. An anionic polysaccharide would therefore require a longer chain and/or a higher rate of application to attain a comparable effect. A similar explanation would apply to the inefficiency of alginic acid and pectin for stabilising calcium and sodium-rich soils. The marked aggregating effect of these materials in hydrogen soils can be ascribed to the presence of surface Al^{3+} ions or

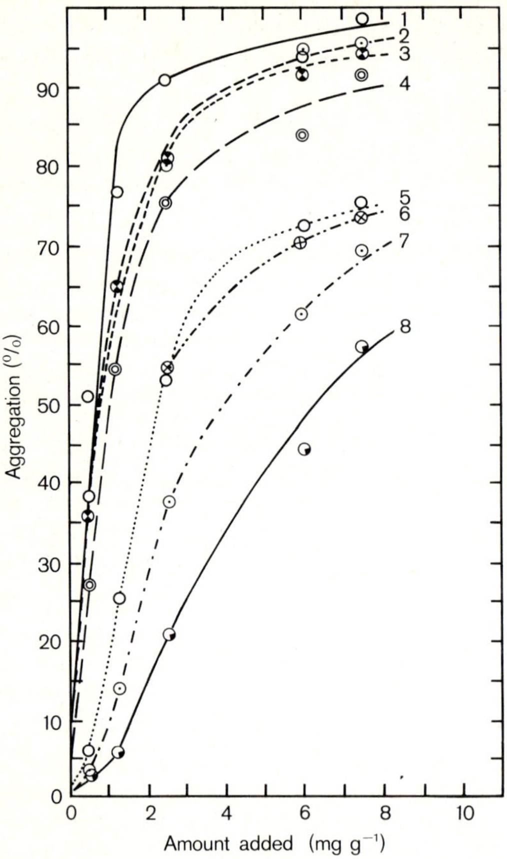

Fig. 10.6. Effect of increasing quantities of various levans on aggregation of soil particles. The molecular weight of the levans, estimated from viscosity data, was: (1) 26 800; (2) 18 800; (3) 14 000; (4) 16 000; (5) 7 200; (6) 8 000; (7) 6 200; (8) 6 400 (after Geoghegan and Brian, 1948).

positively charged hydroxy-aluminium species, and to a decrease in the number of ionised carboxyl groups on the polymer. In contrast to the anionic polysaccharides, a sample of positively charged gastric mucin mucopolysaccharide was found to be as effective as the levans and the dextrans. Being adsorbed by a cation exchange mechanism, the polymer chain presumably adopts a fully extended (flat) conformation with all, or nearly all, of its segments attached to the particle surface.

With only minor variations, the results of previous and subsequent studies by numerous workers (see Harris et al., 1966; Martin, 1971; Greenland and Oades, 1975; Hepper, 1975) may be interpreted along similar lines, that is, in terms of the charge, size, and conformation of the adsorbed polysaccharide on the one hand, and the nature of the exchangeable cation and pH conditions on the other. Of the many data in the literature which illustrate these points, we shall only mention a few.

Martin and Aldrich (1955), for example, observed that for acid soils, aggregation might actually rise as the proportion of uronic acid units in the polysaccharide increased. Quite the opposite effect was found for soils whose exchange sites become progressively occupied by Na^+ and K^+ ions. We may again suppose that in the former system adsorption and interparticle bonding are favoured by the presence of polyvalent aluminium species at the particle surfaces which can act as a bridge between polymer and soil. In addition, a low pH would reduce the effective negative charge on the polysaccharide. Monovalent cations, on the other hand, cannot fulfil either function.

Similar considerations can also account for the seemingly conflicting finding by Clapp et al. (1962) that the uronic acid content of high molecular weight rhizobial polysaccharides is inversely rather than directly related to their aggregating ability. However, the soils they used ranged in pH from 4.7 to 8.0 and might be likened to the monovalent cation systems of Martin and Aldrich. Apart from the increase in repulsion between polymer and soil, a rise in the degree of carboxyl substitution on the molecule would also reduce the ease by which the polysaccharide could be intercalated by any expanding layer silicates which might occur in the soil. At a degree of substitution sufficient to produce water-soluble derivatives, the aggregating effectiveness of cellulosic compounds, like that of dextrans and levans, would generally increase with chain length.

Another, more direct, approach to assessing the stabilising effect of polymers and of polysaccharides, in particular, is to measure the stability of natural or artificially prepared soil/clay aggregates which have previously been subjected to a variety of treatments designed to either remove or degrade the adsorbed polymer.

Emerson (1956, 1963) and Clapp and Emerson (1965a, b) have developed a so-called sodium saturation—dispersion technique by which a clay—polymer flake or soil crumb is first extracted with 0.05 M NaCl and then immersed in dilute solutions of NaCl or NaCl/Na-*p*-nitrophenol to assess its dispersion (swelling) characteristics. An untreated Ca^{2+}-montmorillonite flake, for example, would disperse at a concentration of 0.01 M NaCl. Depending on the type of polymer used, a flake of the clay—polymer complex may or may not disperse (slake) under identical conditions. If the flake remains intact, other solutions can be tested as extractants, such as 1 M NaCl, 0.1 M $Na_4P_2O_7$ at pH 7, and 0.05 M $NaIO_4$. It should be noted that the action of periodate is qualitatively different from that of chloride or pyrophosphate. Periodate or periodic acid is known to be capable of oxidising *cis*-glycols, breaking the bond between two (adjacent) carbon atoms carrying an hydroxyl group. In the case of most polysaccharides, this action leads to ring opening and the resulting polyaldehyde derivative is then readily degraded into low molecular weight fragments under alkaline conditions, such as in 0.1 M sodium borate at pH 10. Interparticle bonding within an aggregate is thereby destroyed.

By using 1 M NaCl and neutralised 0.15 M $Na_4P_2O_7$ solutions on Ca^{2+}-montmorillonite complexes, Emerson (1960) was able to distinguish between three groups of polymers on the basis of their principal mode of bonding to the clay: firstly, those which were readily removed by NaCl, such as alginate, humate, and pectate suggesting a polyvalent cation bridge type of linkage; secondly, those which resisted NaCl leaching but were extracted by $Na_4P_2O_7$, such as polyacrylate and polyacrylamide (both are non-polysaccharides) suggesting chelation of Ca^{2+} ions by the polymer; thirdly, those which resisted either treatment, such as dextrans and levans, suggesting that they are attached by Van der Waals interactions between surface and polymer segments, probably reinforced by entropy effects.

TABLE 10.2

Dispersion characteristics of flakes prepared from complexes of Ca^{2+}-montmorillonite with various polysaccharides, in relation to the properties and the amount adsorbed of the respective polymers. The flakes were first extracted with 0.05 M NaCl, then treated with either 1 M NaCl, 0.1 M $Na_4P_2O_7$, or 0.05 M $NaIO_4$. The dispersing solutions were either 0.01 M NaCl/Na-*p*-nitrophenol or 0.05 M NaCl (from Clapp and Emerson, 1972)

Polysaccharide notation	Source	Charge	Uronic acid (%)	Amino sugar (%)
B-512F	*Leuconostoc* sp.	Uncharged	0	0
3 DOa8	*Rhizobium* sp.	Uncharged	0	0
3 DLm32	*Rhizobium* sp.	Uncharged	0	0
3 Clk3	*Rhizobium* sp.	Negative	17	Trace
3 Dly8	*Rhizobium* sp.	Negative	16	Trace
B-1459	*Xanthomanas* sp.	Negative	20	0
Y-1401	*Cryptococcus* sp.	Negative	21	0
Brookston	Soil	Mostly uncharged and negative	9	Trace
Paulding	Soil		10	Trace
Chitosan	Commercial	Positive	0	100
Deacetylated chitin	Commercial	Positive	0	100

[a] Percentage complex calculated from the amount adsorbed per total weight of clay plus polysaccharide. I, 1 M NaCl followed by 0.01 M NaCl/Na-*p*-nitrophenol; II, 0.1 M $Na_4P_2O_7$ followed by 0.05 M NaCl; III, 0.05 M $NaIO_4$ (24 h) followed by 0.01 M NaCl/Na-*p*-nitrophenol.

In the same year, Mehta et al. (1960) reported on the behaviour of soil aggregates towards treatment with periodate—borate solutions. They showed that whereas natural aggregates remained essentially intact, those formed artificially by mixing soil with polysaccharides of microbial, plant, and soil origin dispersed in water. Subsequently, Harris and co-workers (Harris et al., 1963, 1964) observed similarly for artificial soil aggregates containing microbial gum. These observations led to the conclusion that polysaccharides were either not involved in stabilising (natural) soil aggregates, or if they were, their effect was obscured by the presence of other aggregate-binding agents. This conclusion, however, is by no means generally valid as the following examples show.

The ability of soil aggregates to withstand periodate oxidation was found by Greenland et al. (1962) to depend on the history of the site as well as the depth from which the samples were taken. Aggregates from soils which have been under pasture for many consecutive years were only slightly affected.

Intrinsic viscosity (cm^3 g^{-1})	Amount adsorbed		Dispersion characteristics		
	Nominal (%)	Final [a] (%)	I	II	III
	Clay alone		Complete	—	—
115	5	1.6	No	No	Complete
3300	5	3.4	No	No	Complete
5000	5	4.0	No	No	Complete
9000	5	0.1	No	Complete	Complete
10300	5	0.2	No	Complete	Complete
3700	5	2.7	No	No	Complete
2200	5	2.6	No	Complete	Complete
30	5	0.3	Complete	—	—
70	5	1.4	Complete	—	—
10	5	3.5	No	No	Weak
8	5	3.6	No	No	Weak

On the other hand, those from (non-calcareous) soils which have been continously cropped or under pasture for relatively short periods (<4 years), broke down under treatment, as did also subsurface samples from all sites.

Clapp and Emerson (1965a, b) have re-examined the qualitative and quantitative aspects of the periodate oxidation method in combination with the sodium-saturation—dispersion procedures as described above. They confirmed that free calcium carbonate, which incidentally was present in the (unaffected) Rendzina soil samples used by Mehta et al. (1960) and Greenland et al. (1962), interfered with and inhibited the oxidation process by periodate. The results indicated that crumbs from grassland sites owed their stability to two types of polymeric substances, one of which was removable by pyrophosphate and the other which could only be chemically degraded by the periodate—borate treatment. Periodate-oxidisable materials were largely responsible for the stability of aggregates from cultivated soils, stabilising both the clay- and silt-sized fractions. They also pointed out that the actual consumption of periodate could not be precisely related to the carbohydrate present. This is so because the (primary) glycol splitting step is followed by a progressive oxidation of the polyaldehyde in the presence of (25%) excess periodate. Furthermore, the completeness of the primary reaction is dependent on the way in which the monosaccharide units are joined in the polysaccharide. Prolonged (up to 48 h) treatment with 0.05 M $NaIO_4$ may be necessary if the polymer—clay link is to be effectively broken.

More recently, Clapp and Emerson (1972) have applied the same techniques to evaluate the relative bonding and stabilising efficiency of a variety of both charged and uncharged polysaccharides when complexed with Ca^{2+}-montmorillonite. The dispersion of flakes prepared from these complexes, in relation to polymer property and the amount adsorbed, are summarised in Table 10.2. We need only comment here that the adsorption of the *uncharged* polymers (B-512F, 3DOa8, and 3DLm32) increased in the same order as their respective intrinsic viscosity. In addition, the stability of the corresponding complexes depended on the amount of polysaccharide present, recalling Geoghegan and Brian's (1948) results (cf. Fig. 10.6). When this amount exceeded a certain level (~1.6% w/w), the duration of periodate oxidation had to be doubled (to 48 h) before the flakes would disperse. No straightforward relationship appeared to exist between adsorption capacity and viscosity for the *anionic* polymers. However, they were relatively weakly adsorbed since all of their complexes were completely dispersed after the standard (24 h) periodate treatment. The instability of the complexes with the two polysaccharides, extracted from soil with hot water, was ascribed to the polymers being incapable of forming interparticle bonds because of their low molecular weight. For complexes with 3 Clk3 and 3 Dly8, on the other hand, aggregate failure was probably attributable to lack of adsorption rather than interparticle bonding. In contrast, the *positively charged* polysaccharides were very strongly adsorbed. None of their complexes with Ca^{2+}-montmorillonite actually dispersed in the test solution after being oxidised with periodate for 24 h. Rather, the flakes swelled, becoming gel-like in consistency.

In conclusion, we wish to add the following remarks to our discussion on the role of polysaccharides in stabilising natural soil aggregates. Although the polysaccharide content of many soils is positively correlated with the degree of aggregation, such correlations are generally not very close because the fraction isolated is normally crude and does not represent all of the polysaccharides in the soil. Doubt has indeed been cast on the ability of conventional methods to obtain a quantitative estimate of microbial polysaccharides in soils, although techniques have recently been developed that apparently enable almost complete extraction of carbohydrates from soil (see Swincer et al., 1969; Greenland and Oades, 1975). However, this by itself would not necessarily improve the above correlation since only a certain fraction of the extracted polysaccharides may be actively involved in aggregate stabilisation. Furthermore, it seems probable that the manner in which this active fraction is distributed within and around soil aggregates is as important as the actual amount present.

It is also clear that other materials, both organic and inorganic, can function as aggregate stabilisers. Their presence in soil would diminish or mask any beneficial contribution of polysaccharides to aggregate stability. Polysaccharides are likely to be very important in soils whose total organic matter content is naturally low or has been depleted by cultivation.

Another relevant feature relates to the biological stability of adsorbed polysaccharides. A clue to this question is provided by the recent work of Guckert et al. (1977) who have measured the decomposition rate of ^{14}C-labelled materials adsorbed onto sand or montmorillonite. The results, summarised in Table 10.3, clearly show that even in the short term (3 weeks' incubation) the clay-adsorbed substances are markedly more resistant to

TABLE 10.3

Rate of mineralisation, as measured by the amount of CO_2 evolved, for some organic substrates adsorbed onto sand and montmorillonite (from Guckert et al., 1977)

Substrate	After 1 week (%) [a]		After 2 and 3 weeks (%) [a]		Total (%) [a]	
	Sand	Clay	Sand	Clay	Sand	Clay
Glucose	40.6	26.0	29.4	29.4	70.0	55.4
Polysaccharides	51.9	24.8	14.6	17.8	66.5	42.6
Microbial cells	18.5	4.2	12.4	3.1	30.9	7.3

[a] (^{14}C-labelled compound mineralised/^{14}C-labelled compound introduced) × 100.

microbial decomposition as compared with those held by sand. This observation may be ascribed partly to the stronger binding by montmorillonite and partly to the relative inaccessibility of the adsorbed organics, some of which are presumably intercalated. The former suggestion is further supported by the fact that all of the material adsorbed onto sand is extractable by water and alkali whereas an appreciable proportion (18.1% for glucose, 27.3% for the polysaccharides, and 67.6% for the microbial cells) resides in the alkali-resistant humin fraction when montmorillonite is the adsorbent.

REFERENCES

Banks, W. and Greenwood, C.T., 1967. The hydrodynamic behaviour of native amylose. In: G.N. Ramachandran (Editor), Conformation of Biopolymers. Academic Press, New York, N.Y., pp. 739—749.

Beutelspracher, H., 1955. Wechselwirkung zwischen anorganischen und organischen Kolloiden des Bodens. Zeitschrift für Pflanzenernährung, Düngung, Bodenkunde, 69: 108—115.

Bitton, G., Henis, Y. and Lahav, N., 1976. Effect of montmorillonite and the capsular polysaccharide on the flocculation of *Klebsiella aerogenes*. Plant and Soil, 44: 37—47.

Cheshire, M.V., 1977. Origins and stability of soil polysaccharide. The Journal of Soil Science, 28: 1—10.

Clapp, C.E. and Emerson, W.W., 1965a. The effect of periodate oxidation on the strength of soil crumbs. I. Qualitative studies. Soil Science Society of America Proceedings, 29: 127—130.

Clapp, C.E. and Emerson, W.W., 1965b. The effect of periodate oxidation on the strength of soil crumbs. II. Quantitative studies. Soil Science Society of America Proceedings, 29: 130—134.

Clapp, C.E. and Emerson, W.W., 1972. Reactions between Ca-montmorillonite and polysaccharides. Soil Science, 114: 210—216.

Clapp, C.E., Davis, R.J. and Waugaman, S.H., 1962. The effect of rhizobial polysaccharides on aggregate stability. Soil Science Society of America Proceedings, 26: 466—469.

Clapp, C.E., Olness, A.E. and Hoffmann, D.J., 1968. Adsorption studies of a dextran on montmorillonite. Transactions of the 9th International Congress of Soil Science, Adelaide, 1: 627—634.

Cortez, J., 1977. Adsorption sur les argiles de deux lipopolysaccharides rhizosphériques. Soil Biology and Biochemistry, 9: 25—32.

Decau, J., 1968. Les polysaccharides du sol: origine, évolution et rôle. Annales Agronomiques, 19: 65—82.

Dubach, P., Zweifel, G., Bach, R. and Deuel, H., 1955. Untersuchungen an der Fulvosäure-Fraktion einiger schweizerischer Böden. Zeitschrift für Pflanzenernährung, Düngung, Bodenkunde, 69: 97—108.

Emerson, W.W., 1956. Synthetic soil conditioners. Journal of Agricultural Science (Cambridge), 47: 117—121.

Emerson, W.W., 1960. Complexes of calcium-montmorillonite with polymers. Nature (London), 186: 573—574.

Emerson, W.W., 1963. The effect of polymers on the swelling of montmorillonite. The Journal of Soil Science, 14: 52—63.

Finch, P., Hayes, M.H.B. and Stacey, M., 1966. Studies on soil polysaccharides and on their interaction with clay preparations. Transactions of the Meeting of Commissions II and IV of the International Society of Soil Science, Aberdeen, pp. 19—32.

Finch, P., Hayes, M.H.B. and Stacey, M., 1971. The biochemistry of soil polysaccharides. In: A.D. McLaren and J. Skujinš (Editors), Soil Biochemistry, Vol. 2. Marcel Dekker, New York, N.Y., pp. 257—319.

Gaur, A.C. and Rao, R.V.S., 1975. Note on the isolation of bacterial gums and their influence on soil aggregate stabilization. Indian Journal of Agricultural Science, 45: 86—89.

Geoghegan, M.J., 1954. Aggregate formation in soil. Influence of some microbial metabolic products and other substances on aggregation of soil particles. Transactions of the 4th International Congress of Soil Science, Amsterdam, 1: 198—201.

Geoghegan, M.J. and Brian, R.C., 1948. Aggregate formation in soil. 1. Influence of some bacterial polysaccharides on the binding of soil particles. Biochemical Journal, 43: 5—13.

Greenland, D.J., 1956a. The adsorption of sugars by montmorillonite. I. X-ray studies. The Journal of Soil Science, 1: 319—328.

Greenland, D.J., 1956b. The adsorption of sugars by montmorillonite. II. Chemical studies. The Journal of Soil Science, 1: 329—334.

Greenland, D.J., 1963. The adsorption of polyvinyl alcohols by montmorillonite. Journal of Colloid Science, 18: 647—664.

Greenland, D.J., 1972. Interactions between organic polymers and inorganic soil particles. Mededelingen Fakulteit Landbouwwetenschappen, Rijksuniversiteit Gent, 37: 897—914.

Greenland, D.J. and Oades, J.M., 1975. Saccharides. In: J.E. Gieseking (Editor), Soil Components, Vol. 1. Organic Components. Springer Verlag, Berlin, pp. 213—261.

Greenland, D.J., Lindstrom, G.R. and Quirk, J.P., 1962. Organic materials which stabilize natural soil aggregates. Soil Science Society of America Proceedings, 26: 366—371.

Guckert, A., Tok, H.H. and Jacquin, F., 1977. Biodégradation de polysaccharides bacteriens adsorbés sur une montmorillonite. Proceedings Symposium on Soil Organic Matter Studies, Braunschweig, 1: 403—411.

Guckert, A., Valla, M. and Jacquin, F., 1975. Adsorption of humic acids and soil polysaccharides on montmorillonite. Pochvovedenie, No. 2: 41—47.

Gupta, U.C., 1967. Carbohydrates. In: A.D. McLaren and G.H. Peterson (Editors), Soil Biochemistry, Vol. 1. Marcel Dekker, New York, N.Y., pp. 91—118.

Harris, R.F., Allen, O.N., Chesters, G. and Attoe, O.J., 1963. Evaluation of microbial activity in soil aggregate stabilisation and degradation by the use of artificial aggregates. Soil Science Society of America Proceedings, 27: 542—545.

Harris, R.F., Chesters, G. and Allen, O.N., 1966. Dynamics of soil aggregation. Advances in Agronomy, 18: 107—169.

Harris, R.F., Chesters, G., Allen, O.N. and Attoe, O.J., 1964. Mechanisms involved in soil aggregate stabilisation by fungi and bacteria. Soil Science Society of America Proceedings, 28: 529—532.

Hepper, C.M., 1975. Extracellular polysaccharides of soil bacteria. In: N. Walker (Editor), Soil Microbiology. Butterworths, London, pp. 93—110.

Jeanes, A., 1974. Applications of extracellular microbial polysaccharide—polyelectrolytes: Review of literature, including patents. Journal of Polymer Science, Symposium No. 45, 209—227.

Kohl, R.A. and Taylor, S.A., 1961. Hydrogen bonding between the carbonyl group and Wyoming bentonite. Soil Science, 91: 223—227.

Lynch, D.L. and Cotnoir, L.J., 1956. The influence of clay minerals on the breakdown of certain organic substances. Soil Science Society of America Proceedings, 20: 367—370.

Lynch, D.L., Wright, L.M. and Cotnoir, L.J., 1956. The adsorption of carbohydrates and related compounds on clay minerals. Soil Science Society of America Proceedings, 20: 6—9.

Lynch, D.L., Wright, L.M., Hearns, E.E. and Cotnoir, L.J., 1957. Some factors affecting the adsorption of cellulose compounds, pectins, and hemicellulose compounds on clay minerals. Soil Science, 84: 113—126.

Martin, J.P., 1971. Decomposition and binding action of polysaccharides in soil. Soil Biology and Biochemistry, 3: 33—41.

Martin, J.P. and Aldrich, D.G., 1955. Influence of soil exchangeable cation ratios on the aggregating effects of natural and synthetic soil conditions. Soil Science Society of America Proceedings, 19: 50—54.

Matthews, M.B., 1959. Macromolecular properties of isomeric chondroitin sulfates. Biochimica et Biophysica Acta, 35: 9—17.

Mehta, N.C., Dubach, P. and Deuel, H., 1961. Carbohydrates in the soil. Advances in Carbohydrate Chemistry, 16: 335—355.

Mehta, N.C., Streuli, H., Müller, M. and Deuel, H., 1960. Role of polysaccharides in soil aggregation. Journal of the Science of Food and Agriculture, 11: 40—47.

Moavad, H., Guzev, V.S., Bab'yeva, I.P. and Zvyagintsev, D.G., 1974. Adsorption of extracellular polysaccharide of the yeast *Lipomyces lipofer* on kaolinite. Pochvovedenie, No. 11: 79—84.

Olness, A. and Clapp, C.E., 1973. Occurrence of collapsed and expanded crystals in montmorillonite—dextran complexes. Clays and Clay Minerals, 21: 289—293.

Olness, A. and Clapp, C.E., 1975. Influence of polysaccharide structure on dextran adsorption by montmorillonite. Soil Biology and Biochemistry, 7: 113—118.

Parfitt, R.L., 1972. Adsorption of charged sugars by montmorillonite. Soil Science, 113: 417—421.

Parfitt, R.L. and Greenland, D.J., 1970a. Adsorption of polysaccharides by montmorillonite. Soil Science Society of America Proceedings, 34: 862—866.

Parfitt, R.L. and Greenland, D.J., 1970b. The adsorption of poly(ethylene glycols) on clay minerals. Clay Minerals, 8: 305—315.

Pasika, W.M. and Cragg, L.H., 1962. The viscosity behaviour of linear and branched dextran sulfates. Journal of Polymer Science, 57: 301—310.

Rowland, F.W. and Eirich, F.R., 1966. Flow rates of polymer solutions through porous disks as a function of solute. II. Thickness and structure of adsorbed polymer films. Journal of Polymer Science, A-1, 4: 2401—2421.

Ruehrwein, R.A. and Ward, D.W., 1952. Mechanism of clay aggregation by polyelectrolytes. Soil Science, 73: 485—492.

Saini, G.R. and MacLean, A.A., 1966. Adsorption—flocculation reactions of soil polysaccharides with kaolinite. Soil Science Society of America Proceedings, 30: 697—699.

Schnitzer, M. and Khan, S.U., 1972. Humic Substances in the Environment. Marcel Dekker, New York, N.Y., 327 pp.

Silberberg, A., 1962. The adsorption of flexible macromolecules. Part II. The shape of the adsorbed molecule; the adsorption isotherm, surface tension, and pressure. The Journal of Physical Chemistry, 66: 1884—1907.

Stone, A.L., 1969. Conformations of hexose polysaccharides in solution. In: S.N. Timasheff and G.D. Fasman (Editors), Structure and Stability of Biological Macromolecules. Marcel Dekker, New York, N.Y., pp. 353—415.

Suggett, A., 1975. Polysaccharides. In: F. Franks (Editor), Water: A Comprehensive Treatise, Vol. 4. Plenum Press, New York, N.Y., pp. 519—567.

Swincer, G.D., Oades, J.M. and Greenland, D.J., 1969. The extraction, characterization, and significance of soil polysaccharides. Advances in Agronomy, 21: 195—235.

Theng, B.K.G., 1976. Interactions between montmorillonite and fulvic acid. Geoderma, 15: 243—251.

Theng, B.K.G. and Scharpenseel, H.W., 1975. The adsorption of ^{14}C-labelled humic acid by montmorillonite. Proceedings of the International Clay Conference, Mexico City, pp. 643—653.

Thiessen, P.A., 1942. Wechselseitige Adsorption von Kolloiden. Zeitschrift für Elektrochemie, 48: 675—681.

Yariv, S., Russell, J.D. and Farmer, V.C., 1966. Infrared study of the adsorption of benzoic acid and nitrobenzene in montmorillonite. Israel Journal of Chemistry, 4: 201—213.

Chapter 11

LIGNOSULPHONATES

11.1. INTRODUCTION

Interest in the clay—lignosulphonate interaction arose from the long-standing use of these polyanionic substances as additives of drilling muds where they control the rheological properties of the system, such as viscosity, gel strength, and fluid loss. The patent literature on this and other industrial applications of raw and chemically modified lignosulphonates, has recently been summarised by Browning (1975).

Attempts at investigating the behaviour of lignosulphonates towards clay and inorganic solids are of relatively recent origin. Moreover, what meagre data are available on this topic, are not always concordant. As with humic substances, this situation may largely be ascribed to the inherent heterogeneity of lignosulphonates with respect to molecular weight and structure.

As the name suggests, lignosulphonates are essentially lignin in sulphonated form, being obtained from spent sulphite liquors (which are approximately 10% solutions of lignosulphonates and carbohydrates) after removal of fermentable sugars and non-lignin constituents. Unlike lignin itself, however, lignosulphonates are water-soluble and highly acidic due to the presence in their molecule, of strongly dissociated sulphonic acid groups. By analogy with the nucleic acids (Chapter 8), they normally occur as salts, that is, accompanied by an equivalent amount of cations. Again, monovalent cations largely serve as charge-balancing species which therefore would be more or less completely dissociated from the polymer matrix in an aqueous medium. On the other hand, polyvalent cations, particularly those of the transition metal type, such as chromium and iron, would be closely associated with the polyanionic structure through coordination and/or chelation, involving sulphonic as well as carboxyl and hydroxyl groups. In aqueous solutions, the negative charges on the molecule are, by this means, largely "neutralised" so that the polymer will effectively be less negatively charged than it would otherwise be.

11.2. FORMATION AND PROPERTIES OF COMPLEXES

In an early paper, Browning (1955) suggested that lignosulphonates were primarily adsorbed onto clays through electrostatic interactions although he recently (Browning, 1975) listed Van der Waals and hydrogen bonding interactions as well as covalent bonding and ion exchange as probable modes of attachment to mineral surfaces. Beeckmans (1962) emphasised the importance of pH on the adsorption and desorption of lignosulphonates by kaolinite and titanium dioxide. Jessen and Johnson (1963), on the other hand, regarded the adsorption process to be largely physical although some of the exchangeable Na^+ ions at the montmorillonite surface could apparently be replaced by Cr^{3+} and Fe^{3+} ions on the lignosulphate.

By analogy with tannate anions, Van Olphen (1963) proposed that lignosulphonates were adsorbed at the edges of clay crystals by complexing with octahedrally-coordinated aluminium ions exposed at these sites. As a result, the edge charge was reversed (from being positive initially to negative), preventing edge-to-face and edge-to-edge association. This mechanism would account for the ability of lignosulphonates to act as deflocculants (dispersants) of clay systems. It is to improve this deflocculating efficiency of lignosulphonates that sulphite liquors are frequently treated with oxidising agents, such as chromic acid. This is so because trivalent chromium salts are apparently capable of forming hydrous oxides on addition of alkali. Dichromate treatment may also induce intermolecular cross-linking and so increase the effective particle weight and solution viscosity of the polyanion. In addition, it promotes the formation of catechol and pyrocatechol groups which aid adsorption through hydrogen bonding (see Browning, 1975).

The ion exchange properties of chromium-, mixed chromium/sodium-, and chromium/calcium-lignosulphonates have been examined by McAtee and Smith (1969), using cation exchange resins (in the sodium form) and aqueous dispersions of Na^+- and Ca^{2+}-montmorillonite. Whereas no exchange of Cr^{3+} for Na^+ ions took place with the resins (Rohm and Haas IR-C-200 and Dowex 50W-X8) within the pH range of 1.5 to 11, such a reaction was observed with montmorillonite. For a given ratio of lignosulphonate to clay, relatively more sodium than calcium was replaced and this despite the fact that the extent of polyanion adsorption by Ca^{2+}-montmorillonite was greater as compared with the sodium clay system. They also noted that adsorption was influenced by the nature of the charge-balancing cation on the lignosulphonate as well as by the pH of the clay—polymer system.

X-ray diffraction examination of the montmorillonite—lignosulphonate complexes gave no evidence for intercalation. This observation is, as we have already mentioned, in line with the general behaviour of polyanions towards expanding layer silicate minerals. McAtee and Smith therefore inferred that adsorption was confined to the crystal edges of montmorillonite, as suggested earlier by Van Olphen (1963) and reiterated by Fertl and Jessen (1970). However, the absence of interlayer expansion is by itself insufficient evidence against the possibility that the basal surfaces ("faces") of the crystal do not participate in the adsorption process. Indeed, as we shall presently see, both crystal edge and face are actively involved and whether the one will be preferred to the other seems to depend on the prevailing pH conditions.

The question arises as to why the Cr^{3+} ions on the polymer can replace at least part of the exchangeable Na^+ and Ca^{2+} ions located at the basal surfaces of montmorillonite, but not from their equivalent sites on the cation exchange resins. The answer seems to lie in the fact that the Cr^{3+} ions are attached to the polymer matrix by coordination to sulphonic acid and hydroxyl groups, probably in an octahedral arrangement. If, as McAtee and Smith have suggested, adsorption were largely confined to the crystal edge, Al^{3+} ions exposed at these sites would compete with Cr^{3+} ions for the same donor groups on the lignosulphonate, in keeping with the tendency of edge aluminium to satisfy its own requirement for octahedral coordination. Such a process could conceivably weaken the bond between polyanion and $Cr^{3+}(Fe^{3+})$ ions sufficiently, releasing the trivalent cations into the matrix solution from which they can exchange for the Na^{+} or Ca^{2+} ions at the basal surfaces of the mineral. Sternik et al. (1975) went so far as to state that adsorption of ferrochrome lignosulphonates onto kaolinite, montmorillonite, and palygorskite could induce the formation of $((Cr, Fe)(OH)_2)^{+}$ species leaving the polyanions bereft of Cr^{3+}/Fe^{3+} ions. Competition for donor groups of the type described was apparently not possible with the synthetic resin—lignosulphonate system and hence, no measurable exchange of Cr^{3+}/Fe^{3+} ions on the polymer for Na^{+}/Ca^{2+} ions on the resin occurred.

The involvement of crystal edges in the adsorption of a (ferrochrome) lignosulphonate by expanding 2 : 1 type layer silicates, has been elegantly demonstrated by Siffert et al. (1972). To this end they synthesised a nickel-bearing hectorite, a trioctahedral species, in which the Mg^{2+} ions in the octahedral sheet had been isomorphously replaced by Ni^{2+} ions whose structural formula may be represented by $(Ni_{2.67}Li_{0.33})(Si_4)O_{10}(OH)_2$ (cf. Table 1.2.).

The diffuse reflectance spectrum of this blue, synthetic mineral showed absorption maxima at 8850 cm^{-1} (near-infrared), 15 150 cm^{-1} (visible), and 25 640 cm^{-1} (near-ultraviolet) which were attributable to transitions of the 3*d* electrons of nickel. The spectrum of the hectorite—lignosulphonate complex had similar features except that the maximum for all three bands was shifted to a lower wavelength (higher frequency). This indicated to them that complex formation had modified the nature of the donor groups coordinated to Ni^{2+} ions. Since these cations could only be those exposed at crystal edges, the edge surfaces must have been partially, if not exclusively, involved in the interaction process. The attendant increase in stabilisation energy, estimated from the extent of the frequency shift, was approximately 0.6 kcal mol^{-1}.

Following up this work, Siffert and Ferrand (1973) have examined the effect of pH on the interaction of ferrochrome lignosulphonates with montmorillonite and hectorite. They were able to show that adsorption involving octahedrally-coordinated cations, exposed at crystal edges, was the dominant process under acidic conditions. As the solution pH was raised above 7 there was a progressive desorption of lignosulphonate (LS) molecules until at pH 10 desorption was essentially complete. The pH effect may thus be represented by the following equilibrium,

In alkaline media, adsorption of lignosulphonate would occur predominantly at the (external) basal surfaces of the crystals. From the behaviour of the hydroxyl stretching bands of surface-adsorbed water molecules in the infrared spectrum of the clay—LS complex formed at pH 9—10, they inferred that the polymer was attached by hydrogen bonding between the aquo groups coordinated to the Cr^{3+} ion and the surface oxygens of the silicate layer, or between hydroxyl groups on the polymer and water molecules in the primary hydration shell of the exchangeable cation.

More recently, Le Bell et al. (1974) determined the isotherms for the adsorption of sodium lignosulphonates by (untreated) kaolinite as a function of the molar mass (weight average molecular weight or M_w) of the polyanion, and of solution pH. The isotherms at pH 2.0 are reproduced in Fig. 11.1; those at pH 6.0 are similar in shape but the plateau values (expressed as μmol g^{-1}) are lower. Clearly, when these values (for a given pH) are expressed as mg g^{-1}, the variation in uptake with M_w tends to vanish. These observations closely parallel the earlier findings of Beeckmans (1962) using kaolinite and titanium dioxide as adsorbents.

All the isotherms were of the L-type and the data, in fact, conformed to the Langmuir equation. Although such a fit in the case of polymeric com-

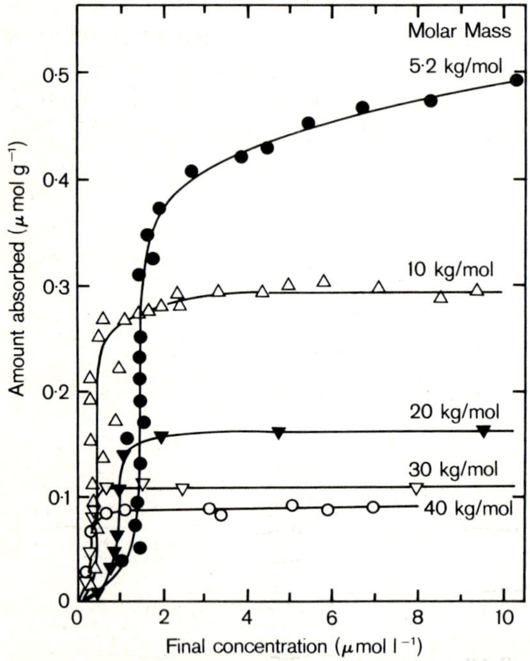

Fig. 11.1. Isotherms for the adsorption of various lignosulphonates by kaolinite at pH 2.0. The molar mass of the polyanions is indicated on the figure (after Le Bell et al., 1974).

pounds, is empirical, the application of the Langmuir approach to polymer adsorption provides a useful basis of comparison. Thus, the horizontal portion of the curves may be identified with the formation of a "monolayer" of LS molecules over the active surface. If this were so, and assuming that the molecules are rigid spheres adsorbed in hexagonal array, the fraction of surface covered at maximum uptake can be calculated. In estimating surface coverage values, Le Bell et al. used molecular diameters derived from viscosity measurements at an ionic strength of 0.01 M KCl. The results, listed in Table 11.1, showed that at pH 2.0 approximately 20% of the total surface of 13.4 $m^2\ g^{-1}$ was covered by LS, in good agreement with the proportion of surface area contained by the edges of kaolinite crystals. From this and the fact that at pH 2 the edges carried a maximum number of positive charges, they inferred that adsorption was largely confined to crystal edges. The data of Table 11.1 also illustrate the relative insensitivity of the "monolayer" capacity (column 4) to variations in molar mass when the maximum amount adsorbed is expressed in terms of the area occupied by LS per unit weight of adsorbant (column 5).

The decrease in adsorption with a rise in pH between 2 and 7 may be ascribed partly to the progressive decline in the net positive edge charge caused by adsorption of hydroxyl ions from solution, and partly to the increase in mutual repulsion between LS segments due to an increase in the number of ionised anionic groups on the polymer. It is interesting to note that positive uptake of lignosulphonate was observed above the apparent isoelectric point of the kaolinite edge which, according to Rand and Melton (1975), was close to pH 7. Only at about pH 10.5 did adsorption cease alto-

TABLE 11.1

Surface coverage of kaolinite crystals in aqueous dispersions by lignosulphonate (LS) in function of molar mass (M_w) and pH (from Le Bell et al., 1974)

M_w ($kg\ mol^{-1}$)	Molecular diameter (nm)	pH	Adsorption at monolayer capacity ($mol\ g^{-1} \times 10^8$)	Area covered by LS ($m^2\ g^{-1}$)	Surface coverage (%)
5.2	3.1	2.0	52.00	2.61	19.5
5.2	3.1	6.0	20.40	1.02	7.6
10	4.3	2.0	29.20	2.82	21.0
10	4.3	6.0	9.40	0.91	6.8
20	5.8	2.0	16.10	2.83	21.1
20	5.8	6.0	5.90	1.04	7.7
30	7.0	2.0	10.90	2.79	20.8
30	7.0	6.0	2.13	0.54	4.1
40	7.8	2.0	8.55	2.71	20.2
40	7.8	6.0	2.03	0.65	4.8

gether. These observations accord with the view that hydroxyl ions are more strongly adsorbed than lignosulphonate polyanions at crystal edge sites. By analogy with the behaviour of lignosulphonates towards 2 : 1 type layer silicates, uptake in the pH range of 7 to 10.5 must predominantly occur at the basal surfaces of kaolinite. The probable mode of attachment of (sodium) lignosulphonates to the kaolinite face is likely to occur through a "water bridge", that is, by hydrogen bonding between the functional groups (sulphonic acid, hydroxyl) of the polyanion and water molecules directly coordinated to polyvalent exchangeable ions.

Le Bell et al. (1974) also found that the addition of lignosulphonates to kaolinite under acidic conditions led to an increase in the electrophoretic mobility of the system, in keeping with the proposed mechanism of crystal edge adsorption. Of particular relevance was the situation where the mobility of kaolinite—lignosulphonate complexes at near neutral pH was comparable in value to that of the parent, untreated clay at pH 10.5, since this coincided with the pH at which kaolinite was optimally deflocculated.

In this connection, Browning (1975) has raised the feasibility of covalent bonding between the sulphonic acid group of lignosulphonates and gibbsite surfaces whereas only O—H···O bonding would occur with silica surfaces. This was thought to account for the high deflocculating efficiency of sodium lignosulphonates for kaolinite whose crystals exposed a gibbsite-like face on one side. By comparison, these substances were only mediocre dispersants for montmorillonite since, in this instance, both sides of the crystals were composed of silica sheets. Covalent bonding, however, seems unlikely and is, furthermore, difficult to reconcile with the known effect of pH on adsorption. A process which may occur with kaolinite is that of specific anion adsorption or "ligand exchange", described more fully in the following chapter. The difference in deflocculating performance of sodium lignosulphonates between kaolinite and montmorillonite may partly reflect the larger extent of external crystal area of the latter as compared with the former.

Besides being useful in drilling mud operations, lignosulphonates are also excellent stabilisers of oil-to-water emulsions, which when treated are stable in saturated salt solutions and withstand centrifugation. The ability of lignosulphonates to form a three-dimensional protective layer around some solid particles has been employed in developing microencapsulation techniques by which free-flowing powders of liquid-in-solid and solid-in-solid dispersions can be prepared with the polymers as the continuous phase (Browning, 1955, 1975). These techniques have been applied commercially to the reinforcement of synthetic elastomers and to the preparation of additives for oil well drilling. The chelating properties of lignosulphonates have found industrial and agricultural applications. The latter include the manufacture of soil conditioners and of fertiliser compositions which can slowly release microelements and nitrogen into the soil system (Horiguchi and Nakamura, 1975; Koga, 1976).

The use of lignosulphonates as stabilising agents for colloidal dispersions is another important field of application. The stability induced may be of the electrokinetic type, achieved by adding small amounts (0.2% w/w) of the polyanion to the system. For clay systems, for example, only an amount sufficient to reverse the edge charge would be required. The other type of stabilisation is normally referred to as steric. Here much larger quantities (5—6% w/w) of lignosulphonate are required since the process depends on the formation of a condensed, multimolecular layer over the entire accessible solid surface. As is to be expected, the dispersing/flocculating properties of lignosulphonates are influenced by their molecular weight, prevailing pH conditions, and the presence of extraneous electrolytes in the system. The interplay of these and other factors makes it difficult to predict the effect of lignosulphonates on dispersion stability (Chan et al., 1976; Le Bell et al., 1976).

In terms of quantity and importance, however, the application of lignosulphonates and their derivatives as additives of oil drilling fluids, still occupies a primary place. In addition to those already referred to, a variety of chemical treatments and modifications have been developed and patented in order to meet specific requirements. These include the preparation of lignosulphonates containing polyvalent cations (Al^{3+}, Cu^{2+}, Fe^{3+}), treatment with formaldehyde and ammonia, and hydrolysis with acids and alkalis, as well as graft polymerisation reactions with a number of organic monomers (Browning, 1975).

REFERENCES

Beeckmans, J., 1962. Adsorption of lignosulphonates on solids. Canadian Journal of Chemistry, 40: 265—274.

Browning, W.C., 1955. Lignosulphonic stabilized emulsions in oil well drilling fluids. Journal of Petroleum Technology, 7: 9—15.

Browning, W.C., 1975. The lignosulfonate challenge. Applied Polymer Symposium, No. 28: 109—124.

Chan, S.L.H., Baker, C.G.J. and Beeckmans, J.M., 1976. Flocculating properties of high molecular weight lignosulfonates. Powder Technology, 13: 223—230.

Fertl, W. and Jessen, F.W., 1970. Adsorption of commercial lignosulphonates on fractionated bentonitic clays. The Journal of Canadian Petroleum Technology, 9: 95—98.

Horiguchi, S. and Nakamura, M., 1975. Synthetic humus. Japanese Kokai 75, 142, 367. See Chemical Abstracts, 84: 104435 (1976).

Jessen, F.W. and Johnson, C.A., 1963. The mechanism of absorption of lignosulfonates on clay suspensions. Society of Petroleum Engineers, Journal, 3: 267—273.

Koga, M., 1976. Polycondensate compositions of lignin sulfonic acids. Japanese Kokai 76 46, 385. See Chemical Abstracts, 85: 126161 (1976).

Le Bell, J., Bergroth, B., Stenius, P. and Stenlund, B., 1974. The adsorption of sodium lignosulphonates on kaolin. Paperi ja Puu — Papper och Trä, 56: 463—466, 469—471.

Le Bell, J.C., Hurskainen, V.T. and Stenius, P.J., 1976. The influence of sodium lignosulphonates on the stability of kaolin dispersions. Journal of Colloid and Interface Science, 55: 60—68.

McAtee, J.L. and Smith, N.R., 1969. Ferrochrome lignosulfonates. I. X-ray absorption edge fine structure spectroscopy. II. Interaction with ion exchange resin and clays. Journal of Colloid and Interface Science, 29: 389—398.

Rand, B. and Melton, I.E., 1975. Isoelectric point of the edge surface of kaolinite. Nature (London), 257: 214—216.

Siffert, B. and Ferrand, C., 1973. Contribution a l'étude du mécanisme d'interaction des argiles et des lignosulfonates. Bulletin Groupe français des Argiles, 25: 135—148.

Siffert, B., Ferrand, C. and Wey, R., 1972. Mise en evidence par spectrophotometrie d'une liaison semi-polaire dans un complexe argile—lignosulfonate. Bulletin Groupe français des Argiles, 24: 41—47.

Sternik, B.A., Ostrovskaya, A.B., Medvedeva, T.V., Godovanaya, O.N. and Esel'son, M.P., 1975. Adsorption of ferrochromolignosulfonates on argillaceous minerals. Kolloidnyi Zhurnal, 37: 806—808.

Van Olphen, H., 1963. An Introduction to Clay Colloid Chemistry. Interscience Publishers, New York, N.Y., 301 pp.

Chapter 12

HUMIC SUBSTANCES

12.1. INTRODUCTION

So far we have considered the interactions of clays with organic macromolecules whose chemical structure and constitution are relatively well characterised. It is on the basis of this attribute that such compounds as enzymes, nucleic acids, polysaccharides, and proteins as well as lignin, lipids, pigments, resins, and tannins together with a variety of organic micromolecules which are to be found in soil, are conventionally regarded as non-humic substances. On entry or incorporation into the soil, these substances are fairly rapidly (<1 year) decomposed and metabolised by the soil fauna and flora, that is, they have a rather high turn-over rate. A small proportion of non-humic substances or their derivatives, however, can and do escape microbial attack, particularly in those regions of the soil environment where anaerobic conditions prevail and/or when this fraction becomes stabilised ("protected") by adsorption onto soil colloids. Because of their complex structure or toxicity to microorganisms, some non-humic compounds such as lignin and plant phenolics are also resistant to decomposition and tend to accumulate in the soil as partially degraded or microbially altered forms. These together with the relatively stable products of microbial synthesis are collectively referred to as humic substances. The biochemistry of humus formation in soil has recently been reviewed by Haider et al. (1975). It ought to be added that the distinction between non-humic and humic substances is not as clear-cut as it might, at first sight, appear. Thus, many non-humic constituents in soil either form an integral part of or are closely associated with humic substances which, in turn, enter into complex formation with clays and other inorganic soil colloids.

It is generally accepted that humification is essentially a biological and aerobic process involving both degradation and synthesis of many and varied organic compounds following some of the most complex and still incompletely understood pathways in nature. A simplified diagram showing the formation in soil of humic substances from their likely precursors is given in Fig. 12.1. The joining together of the various constituents ("building blocks" or "units") to yield humic substances, on the other hand, is probably a chemical process involving a free radical-type condensation rather than one which is enzymically controlled. As mentioned in section 7.1, the term "heteropolycondensation", coined by Dragunov in 1948 (Kononova, 1966), has been used to denote the process of random polymerisation of various phenolic acids and (usually) lignin degradation products, and in which nitrogenous compounds and probably carbohydrate moieties participate. This concept

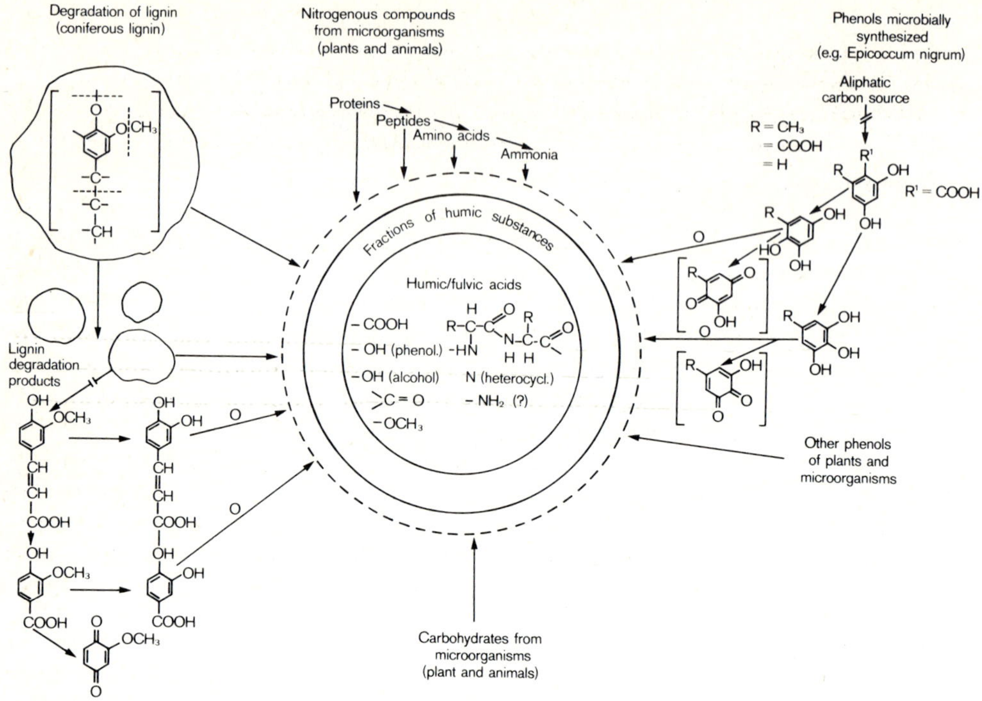

Fig. 12.1. Diagram showing the mode of formation of soil humic substances from their probable precursors (after Flaig et al., 1975).

would account, at least partly, for the known biological stability of humic substances as well as their overall similarity in physico-chemical properties, such as their yellow-to-brown and brown-to-black colour, acidity, polydispersity, high molecular weight, and lack of long-range structural order. Incidentally, humic substances are the most important class of organic materials in the environment in terms of both quantity and ubiquity. Although quantitative estimates differ depending on the underlying assumptions (Bohn, 1976), there is little doubt that the organic carbon content of humic substances far exceeds that of living organisms (e.g. Szalay, 1964).

Recorded observations on the nature of humic substances date back to the late 18th century (see Kononova, 1966; Schnitzer and Khan, 1972). Despite this long history of interest and research, these materials still defy a precise chemical description because of their inherent complexity and heterogeneity. The application, in recent years, of sophisticated instrumental methods of analysis to the elucidation of their chemical structure only serves to illustrate the immense complexity and diversity of their constituent units.

Conventionally, humic substances are divided into three principal categories or fractions according to their aqueous solubility at different pH values: the alkali-soluble humic acid, the alkali- and acid-soluble fulvic acid, and the alkali- and acid-insoluble humin. Sometimes one finds the term hymatomelanic acid denoting the ethanol-soluble fraction of humic acid which, however, may be an artefact of the fractionation process (Stevenson and Butler, 1969). In the German literature, the terms grey and brown humic acids are also encountered; these refer to the fractions of humic acid, which after redissolution in alkali, can and cannot be precipitated on adding electrolyte to the system, respectively. The scheme of fractionation defining each category, is given in Fig. 12.2. The extraction and fractionation of humic and fulvic acids from soil have been extended, modified, and refined over the years with the aim of increasing the yield, reducing the ash content, and minimising the creation of artefacts (Schnitzer and Khan, 1972).

As Felbeck (1965) has pointed out, the definition of humic acid, fulvic acid, and humin is more operational than chemical since differences in properties among them are more of degree than of kind. This view is also supported by the results of extraction and fractionation studies. Thus, there is a gradual change in spectral and other properties of the humic materials

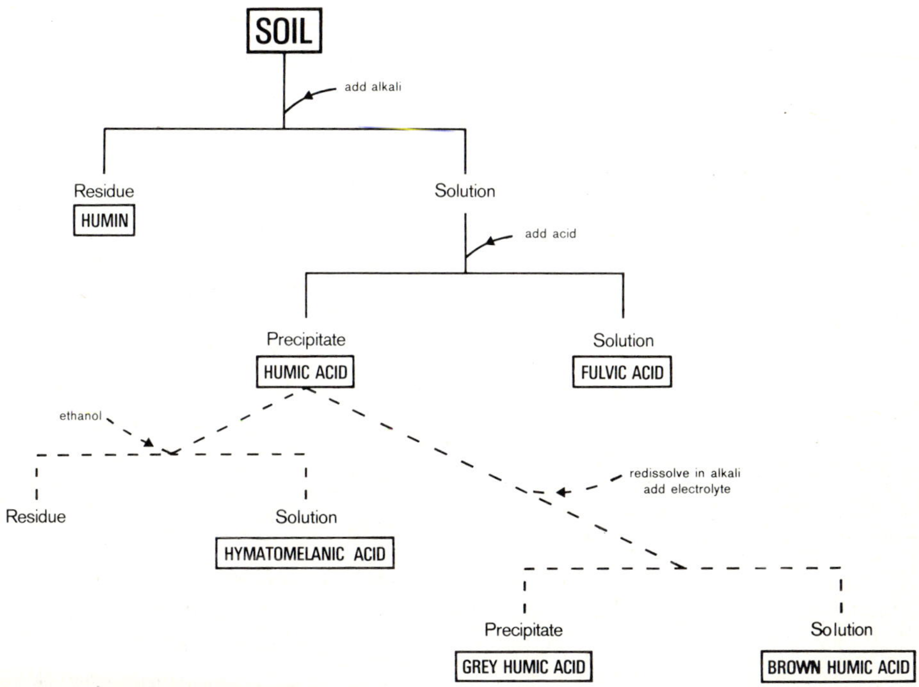

Fig. 12.2. Flow chart for the extraction and fractionation of soil organic matter.

extracted from soil using conventional and other extractants (Posner, 1966; Theng et al., 1967; Schnitzer and Khan, 1972). The gradation in properties is perhaps more clearly seen when a given extract is further fractionated according to, say, solubility in electrolyte solutions or molecular weight (Theng et al., 1968; Swift et al., 1970).

It is nevertheless useful to think in terms of the "classical" categories of humic acid, fulvic acid, and humin and that these fractions are in mutual equilibrium under natural soil conditions. Such a concept implies that any factor or process which affects or modifies the biogeochemical environment of one fraction would be reflected in and have repercussion on the others. Humin, for example, may be regarded as humic acid which has been effectively immobilised due to its intimate and strong association with the inorganic soil colloids. On the other hand, fulvic acid, being relatively uncontaminated by mineral matter, is the more mobile (soluble) form of humic acid from which it is possibly derived by depolymerisation involving oxidation and demethylation (e.g. Alexandrova, 1966). Thus, fulvic acid is generally lower in molecular weight, containing less carbon and hydroxyls but more oxygen and carboxyl groups than humic acid. These points are summarised in Table 12.1. For more details on the chemistry of humic substances than can be included here, the reader is referred to the reviews inter

TABLE 12.1

Some physico-chemical properties of soil humic substances illustrating the diffuse nature of the lines separating fulvic acid, humic acid, and humin

Property	Fulvic acid	Humic acid	Humin
Colour	Yellow-to-brown	Brown-to-black	Black
Molecular weight	Low (<2 000)	Medium (5 000—100 000)	High (>300 000?)
Carbon content (%)	40—50	55—60	55—>60
Nitrogen content (%)	<4	4	>4
Oxygen content (%)	45—48	33—36	32—34
Total acidity (meq g^{-1})	12—14	6—10	5—6
Carbonyl (meq g^{-1})	1—3	1—5	5—6
Carboxyl (meq g^{-1})	8—9	2—5	3—4
Methoxyl (meq g^{-1})	<0.5	<0.5	<0.5
Alcoholic OH (meq g^{-1})	3—5	<1—4	—
Phenolic OH (meq g^{-1})	3—6	2—6	2
Binding to clays and soil minerals	Increasing ——————→		
Free radical content	Increasing ——————→		
Mobility in soil	Decreasing ——————→		
Solubility in water	Decreasing ——————→		
Typical for soils	Podzols	Brown Earths	Black Earths

alia by Dubach and Mehta (1963), Felbeck (1965, 1971), Kononova (1966), Stevenson and Butler (1969), Schnitzer and Khan (1972), and Flaig et al. (1975).

Of the large number and variety of organic compounds, both simple and polymeric, which are known to form complexes with clay minerals (Greenland, 1965a, b; Theng, 1972a), humic substances were among the first to have been described as capable of acting in this capacity. Thus, as early as 1874 Schloesing observed that clays possessed a certain tendency to unite with soil humates (Russell, 1973). It was not until the nineteen-thirties, however, that the existence of clay—humic complexes in soil became widely recognised, their occurrence being responsible for the maintenance of soil structure and fertility (see Allison, 1973). It seems therefore ironic that the most important and ubiquitous clay—organic complexes in nature are also the least well characterised, a condition which is ascribable to our imperfect knowledge of the chemical constitution of humic substances. Related to this problem is the difficulty or well-nigh practical impossibility of obtaining discretely homogeneous fractions, that is, materials consisting of homopolymeric species different only in terms of molecular weight. Dubach and Mehta (1963) went so far as to state that "... perhaps no two molecules of humic substances are exactly alike".

The study of complex formation between clays and humic substances has been approached from essentially two directions. The first involves the isolation and separation from soil of such complexes based on size and density, followed by their characterisation. Fractionation is normally achieved by sedimentation, flotation, and centrifugation with or without the aid of ultrasonic dispersion (Arshad and Lowe, 1966; Chichester, 1969; Parasher and Lowe, 1970; Satoh and Yamane, 1972a, b; Turchenek and Oades, 1974; Watson and Parsons, 1974; Satoh, 1976; Spycher and Young, 1977). The importance of clay—organic complex formation in immobilising soil carbon is illustrated in Table 12.2, compiled by Greenland (1965a) from earlier data of various workers using different soils and separation techniques.

Alternatively, the interaction process may be studied by allowing humic substances or their fractions to react with known clay mineral species and examining the resulting complexes. Because the reactants here are relatively well characterised and the experimental conditions can be readily controlled, this approach has yielded more meaningful information on probable bonding mechanisms as compared with the first. Much of the available data on the clay—humic interaction has therefore been obtained using the second approach and it is with this that we are primarily concerned in the following discussion.

TABLE 12.2

Separation of clay—organic complexes from different soils together with their organic carbon content (from Greenland, 1965a)

Soils	Method of separation [a]	Total carbon in soil (%)	Carbon in clay—organic complex (% of total)	References
Renzina	Sedimentation in benzene-bromoform, s.g. 1.75	—	66.5	Hénin and Turc (1950)
Podzol	Sedimentation in Toulet solution, s.g. 1.8	1.6	89.6	Khan (1959)
Chernozem		4.4	85.2	
Silt under old pasture	Sedimentation in ethanol-bromoform, s.g. 2.0	2.34	77.5	Monnier et al. (1962)
Renzina	Flotation sieving	5.8	54.3	Roulet et al. (1963)
Brown earth		3.2	68.1	
Red-brown earth	Ultrasonic dispersion and sedimentation in bromoform-petroleum spirit, s.g. 2.0	2.23	71.5	Greenland and Ford (1964)
Renzina		5.8	68.4	
Lateritic red earth		1.7	97.8	
Solodized solenetz		1.04	76.4	
Solodized brown soil		0.58	51.6	

[a] s.g. = specific gravity.

12.2. BONDING MECHANISMS

It might be expected that being polyanionic (at pH >6), humic and fulvic acids would at best be weakly adsorbed by negatively charged clay mineral surfaces. This accords with experimental data described in the following section. However, appreciable uptake can take place when the clay is saturated with polyvalent cations which can act as a bridge between the anionic groups of the polymer and the mineral surface. Cation bridge formation may be regarded as a special case of ion—dipole interactions which determine the adsorption of uncharged polar organics by clays (Mortland, 1970; Theng, 1974). This is so because such interactions essentially involve the replacement of water molecules from the primary or inner coordination (hydration) shell of the exchangeable cation. For this reason, complexes formed in this way are sometimes referred to as inner coordination complexes. In hydrated clay systems, the anionic and polar groups of the polymer are normally associated with the exchangeable cation by means of hydrogen bonding to water molecules in this primary hydration shell. The mode of bonding and the resulting complexes are then referred to as water bridge formation and outer coordination complexes, respectively. The gain in entropy associated with the replacement of many water molecules (from the primary or secondary shell) by a single polymer, would also favour adsorption.

It is relevant here to recall Edwards and Bremner's (1967) concept of micro-aggregate formation in soils. They suggested that the process is one of linking organic matter (OM) to clay particles (C) through polyvalent cations (P). A micro-aggregate consisting of an assemblage of C—P—OM complexes may thus be represented as $[(\text{C—P—OM})_x]_y$, where x and y are whole integers depending on the size of the primary clay particles. Their evidence for this view came from the behaviour of natural soil micro-aggregates when treated with either exchange resins containing monovalent cations or sonic vibration, both of which led to aggregate dispersion. The former treatment presumably resulted in the displacement of polyvalent cations from the C—P—OM complexes by monovalent species from the resin, thereby weakening the C—OM links. Similarly, sonification was thought to sever the bonds linking the C—P—OM complexes into larger $(\text{C—P—OM})_x$ and $[(\text{C—P—OM})_x]_y$ assemblages. Mortland (1970) has pointed out that if the P—OM link were of the direct coordination type, sonic vibration may not be sufficiently energetic to cause bond rupture. He suggested that the complex should rather be formulated as C—P—H_2O—OM, the functional groups of the organic matter being linked to polyvalent cations through a water bridge. Such a mechanism would accord not only with the dispersion behaviour described above but also with the general mode of bonding between clay mineral surfaces and polar organic compounds in hydrated systems.

In soil, however, hydrous oxides of iron and aluminium are probably as

TABLE 12.3

Possible modes of bonding between humic substances and clay mineral surfaces

	Groups on polymer	Groups and sites on mineral	Remarks
A. Primary interactions Ion—dipole, including			
(a) Cation bridging	Anionic (carboxylate) and uncharged polar groups, such as amino, carbonyl, carboxyl, hydroxyl	Exchangeable cations at external, basal surfaces. At low pH (<4.5) also those on interlayer surfaces	Weakly hydrated and non-hydrated monovalent cations. Polyvalent cations in dehydrated systems. Desorbable by sodium salts
(b) Water bridging	Anionic (carboxylate) and uncharged polar groups, such as amino, carbonyl, carboxyl, hydroxyl	Exchangeable cations at external, basal surfaces. At low pH (<4.5) also those on interlayer surfaces	Polyvalent cations in hydrated systems. Desorbable by sodium salts, water, and ultrasonic treatment.
Anion exchange	Anionic groups	Crystal edge surfaces. Polyhydroxy compounds of iron and aluminium at external, basal surfaces	Acidic pH conditions and when exchange sites are occupied by iron and aluminium ions. Desorbable by other anions in solution, such as chloride, nitrate

Ligand exchange	Anionic groups	Crystal edge surfaces. Polyhydroxy compounds of iron and aluminium at external, basal surfaces	When exchange sites are occupied by iron and aluminium ions. In soils, on hydrous oxides of iron and aluminium. Desorbable by other specifically but more strongly adsorbed anions, such as hydroxyl
B. Secondary interactions			
Hydrogen bonding	Amino, carboxyl, hydroxyl	Oxygens of external, basal surfaces (siloxane sheet)	Weak
Van der Waals	Uncharged portions of polymer chain	Mainly at external, basal surfaces.	Not very important in hydrated systems. They become significant at high ionic strength and when the system is allowed to dehydrate. Short-range, weak but additive.
C. Indirect interactions			
Entropy effects			Arising from the gain in entropy by the system due to displacement of many water molecules from the clay surface by a single polymer chain

important as, if not more so than, layer silicates in acting as adsorbing surfaces. Since these compounds acquire positive charges at acid pH values, they can bind humic and fulvic acids by anion exchange or electrostatic attraction under these conditions. Anions held in this way can be replaced by other species in solution, such as chloride or nitrate. If, in addition, the anionic group can penetrate the coordination shell of the iron and aluminium ion and so become incorporated in the surface hydroxyl sheet, the mode of bonding is referred to as specific or ligand exchange (Hingston et al., 1967, 1968). Such anions can only be desorbed by species which themselves are specifically but more strongly adsorbed, such as the hydroxyl ion. Anion- and ligand-exchange reactions may occur at crystal edge surfaces of layer silicates but otherwise would not be important for "pure" clay mineral systems unless the exchange sites are occupied by iron or aluminium ions which tend to form polyhydroxy compounds at the (basal) surface.

In addition to the above modes of bonding, hydrogen bond formation and Van der Waals interactions between adsorbed molecules or between these and the clay surface, may be operative. Being short-range and additive, the latter would only assume significance at high ionic strength or when the system is allowed to dry out, enabling intimate contact to be established between solute and surface.

The possible mechanisms of interaction between humic substances and clays, summarised in Table 12.3, have been discussed more fully in the reviews by Greenland (1971) and Orlov et al. (1973).

In view of the structural complexity of humic substances and the many variables which determine their association with layer silicates, the interpretation of experimental data on the clay—humic interaction must remain somewhat tentative. Although a self-consistent picture of these systems is slowly emerging, there is still a great deal of scope for further refinement and clarification.

12.3. FORMATION AND PROPERTIES OF COMPLEXES

Demolon and Barbier (1929) were among the first to have systematically investigated the factors affecting the clay—humic interaction. They noted that uptake increased with a fall in suspension pH, the steepest rise occurring at low (<0.5 mg ml^{-1}) humate concentration and that the presence of Ca^{2+} ions in the system caused a marked increase in adsorption. These and the results of other early studies (e.g. Mattson, 1935; Myers, 1937; Meyer, 1941) indicated some type of chemical bonding ("polar interaction") in which the exchangeable cations were involved. On the other hand, Tyulin (1938) emphasised the part played by physical (Van der Waals) interactions in the adsorption process. Likewise, Sideri (1938) inferred from optical observations that bonding occurred because the humic molecules adopted a preferred orientation at or near clay mineral surfaces.

Jung (1943) subsequently reported that potassium humate (from peat) was strongly adsorbed by K^+- and Ca^{2+}-montmorillonite in that for a given clay/humate ratio, most of the added polymer was removed from solution. Since uptake was sensitive to variations in solution temperature, Van der Waals type interactions were also invoked as the major mode of bonding. However, as we have already indicated, this mechanism may only be of secondary importance particularly for aqueous systems. Rather, the explanation should be sought in the direction of ion—dipole interactions and water bridge formation between the functional groups of the polyanion and the exchangeable cations.

That the nature of the exchangeable cations has a determining influence on the interaction process has been confirmed and substantiated by many workers using diverse techniques, as the following examples will illustrate.

Thus, Flaig and Beutelspracher (1951) and Beutelspracher (1955), using electron microscopy, were able to demonstrate that the presence of polyvalent cations in the system was necessary for association between humic acid and clays to occur. Among these cations, those of iron and aluminium are particularly effective in promoting adsorption (Martin and Reeve, 1960; Sen, 1960; Kobo and Fujisawa, 1963; Scharpenseel, 1966; Pohl and Schwiete, 1969). It should be recalled, however, that Fe^{3+} and Al^{3+} ions are rather special in that they tend to form polyhydroxy complexes at the mineral surface, offering scope for bonding mechanisms other than cation- and water-bridge formation to occur. The possibility of anion- and ligand-exchange reactions to proceed in these systems may explain the observation that despite the large amounts adsorbed, the affinity of humic and fulvic acids for the clay surface is less than would be expected on the basis of the polarising power of these cations (Theng and Scharpenseel, 1975; Theng, 1976). Another point of note is Beutelspracher's (1955) finding that in the presence of Fe^{3+}-montmorillonite, the estimation of humate concentrations in solution by spectrophotometric means, may yield variable results depending on the wavelength selected.

A more detailed investigation into the reactions of soil humic and fulvic acids with montmorillonite saturated with H^+, K^+, and Ca^{2+} ions, was made by Evans and Russell (1959). To this end, the organic matter extracts were passed through a column of H^+-exchange resin and it was in the acid form rather than as the corresponding humate and fulvate forms, that they were added to the clay.

Uptake was rapid, at least 90% of the added amount being adsorbed within the first hour of contacting the humic substances with the clay, and was insensitive to temperature. Linear (C-type) isotherms were obtained over the concentration range used (0—6 mg ml^{-1}) except for the potassium clay system which gave L-type curves. The extent of adsorption for humic acid decreased in the order $Ca^{2+} > H^+ > K^+$; that for fulvic acid was $H^+ > Ca^{2+} > K^+$. The position of H^+- with respect to K^+-montmorillonite was undoubtedly due to the presence of some Al^{3+} ions at the exchange sites of the hydrogen saturated clay sample. Indeed, little or no uptake was observed for the native (sodium) montmorillonite sample unless the clay had previously been treated with acid. These results suggest that ion—dipole interactions, either directly or through a water bridge, involving carboxyls and other polar functional groups of the polyanion were the principal mechanism of adsorption.

In contrast to earlier findings (Demolon and Barbier, 1929; Jung, 1943), only a relatively small proportion (<25% as carbon) of the added solute was adsorbed. This together with the desorption data suggested to Evans and Russell that certain components of the humic substances were preferentially adsorbed. Although the relationship between the adsorbable portion and the whole was not clearly defined, the more recent data of Tan (1976) indicated that these components consisted of low molecular weight species. If so, this would imply that the high molecular weight constituents were, by virtue of their relatively larger size, prevented from penetrating the pore space of the clay system.

Theng (1976) and Theng and Scharpenseel (1975) have suggested that these pores may be identified with the spaces separating individual montmorillonite crystals within a domain (Aylmore and Quirk, 1960). The concept of intradomain penetration by solute molecules, discussed more fully below, is also consistent with the linearity of the isotherms. An analogous situation exists with soil allophane which consists of aggregates of spherical silico-alumina particles so that the pores or channels, in this instance, refer to the interparticle or interstitial spaces in the matrix (Kitagawa, 1971; Theng, 1972b). Here pore entry seems relatively unrestricted and the adsorption of low molecular weight humified clover extracts by allophane shows a "normal" behaviour in that the larger size components are apparently preferred to their smaller counterparts (Inoue and Wada, 1968).

Relevant in this context is the finding by Sen (1964) that the extent of adsorption of (the parent) humic acid (by illite) at a given clay/organic ratio was less than that shown by either the hymatomelanic acid fraction or the ethanol-insoluble residue of the sample (Fig. 12.2). The inference here was that even a "crude" fractionation of humic acid (into two components) apparently led to an increase in the number of (exposed) functional groups capable of binding to the metal ions at the mineral surface.

The influence of different inorganic cations in solution on the formation and stability of clay—humic complexes, has been examined by Scharpenseel (1966) using radiometric precipitation. Briefly, the method involved shaking clay suspensions with neutral solutions of ^{14}C-labelled humic (or fulvic) acid containing various metal ions, taking aliquots at intervals, centrifuging, and measuring the radioactivity of the supernatant as well as of the precipitate (complex). The latter was then subjected to gradient elution using dilute (<0.2 M) solutions of NaOH after which the activity of both the eluates and the residue was determined.

The results for montmorillonite were in accord with previous findings in that the extent to which the humate (or fulvate) sample was adsorbed generally rose with an increase in cation valency. This would again indicate that the (added) cations acted as a bridge between the functional groups on the humate (or fulvate) molecule and the negatively charged clay surface. Similar results were obtained for kaolinite but, due to its low exchange capacity, the dependence of adsorption on the nature of the cation was much less in evidence as compared with montmorillonite. Of interest is the observation that despite the large variation in adsorption (by montmorillonite), the amount of humate retained against elution with NaOH was remarkably uniform. This would mean that the extent of desorption followed essentially the same sequence as that of uptake, that is, the higher the cation valency the less stable was the complex. Valla et al. (1972b) have reported similarly for H^+-, Ca^{2+}-, and Fe^{3+}-montmorillonite using NaOH and $Na_4P_2O_7$ solutions as eluants.

The desorption (stability) behaviour of clay—humic complexes towards sodium hydroxide (and pyrophosphate) may be explained in terms of the ability of Na^+ ions to replace the polyvalent cations initially introduced into or occupying exchange sites at the clay surface. This process leads to a disruption of the clay—polyvalent ion—(water)—humate assemblage, thus releasing the (adsorbed) humate into solution. This is analogous to what Edwards and Bremner (1967) have suggested for the breakdown of their "micro-aggregate" on treatment with a Na^+-exchange resin, referred to earlier. At the same time, the added hydroxide (or pyrophosphate) ions compete with the anionic groups of the humate for adsorption sites at the mineral surface, so promoting desorption. The alkali-resistant portion of the adsorbed humate corresponds to the humin fraction (cf. Fig. 12.2). Its stability indicates a strong and intimate association with the clay, presumably because its molecular size and orientation favour the establishment of numerous points of contact with the surface. This stability would be enhanced if the humate were partly adsorbed on interlayer surfaces, as Schnitzer and Kodama (1967) have reported for a montmorillonite—fulvic acid complex. However, the possibility of interlayer uptake did not arise in the above instance since X-ray diffraction examination of the complex, before and after gradient elution, gave no indication that intercalation had taken place. Interlayer penetration, and then only to a very limited extent, might occur when the humate complex was prepared under hydrothermal conditions (Scharpenseel, 1966).

In an attempt at relating adsorption behaviour to some measurable properties of the exchangeable cation, Theng and Scharpenseel (1975) have determined the isotherms for the adsorption of ^{14}C-labelled humic acid by montmorillonite saturated with different cations. The chief advantage of using uniformly labelled materials is that uptake can be relatively accurately and conveniently measured at low solute concentrations. In addition, such measurements are not subject to the kind of error that Beutelspracher (1955) has described because adsorption is estimated from the change in overall activity (count rate) rather than on solution colour. In all instances, the humic acid was added to the clay powder as a solution of its sodium salt at pH ~7.

As Fig. 12.3 shows, all the isotherms (at 293 K) were of the C-type, adsorption increasing linearly with concentration without any indication that saturation was approached up to the limit of solute concentration used. Linear isotherms were also obtained by Levy and Francis (1976) for montmorillonite saturated with Na^+, Ca^{2+}, Al^{3+}, and Fe^{3+} ions up to a concentration of 1.5 mg humic acid per ml. It is interesting to note that Evans and Russell (1959) obtained an L-type curve for K^+-montmorillonite whilst Valla et al. (1972a), using similar techniques, observed that the curves for their Ca^{2+}- and Fe^{3+}-montmorillonite samples rose to a maximum and then declined.

Linear isotherms imply that fresh sites are being created as adsorption progresses, a process requiring the expansion of the accessible surface (Giles et al., 1960, 1974a). This requirement may be met by solute penetration of the clay interlayers leading to an expansion of the interlayer space, as Greenland

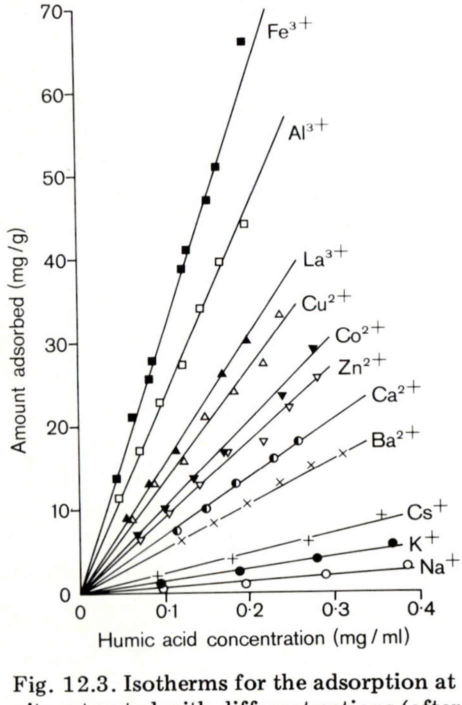

Fig. 12.3. Isotherms for the adsorption at 293 K and pH 7 of humic acid by montmorillonite saturated with different cations (after Theng and Scharpenseel, 1975).

et al. (1962) have shown for the uptake of some amino acids by Ca^{2+}-montmorillonite. Since no interlayer uptake was detectable in this instance, the linearity of the isotherms and the variation in adsorption among the samples could not be ascribed to differences in the extent of accessible surface resulting from interlayer expansion. In explanation, Theng and Scharpenseel (1975) have suggested that the humate molecule penetrated the intercrystalline or intradomain pores of the clay/water system, "domains" being regions of oriented clay crystals which together make up a clay mass (Aylmore and Quirk, 1960). In doing so, additional sites became accessible to the following solute molecules (Giles et al., 1974a). The suggested mechanism is consistent with the overall dimension and shape of humic acid molecules, derived from sedimentation and diffusion measurements (Cameron et al., 1972) as well as with the average crystal separation ("pore diameter") in domains of water-swollen montmorillonite (Aylmore and Quirk, 1962). Entry of the polyanion into intradomain pores would further be facilitated by the local accumulation of Na^+ ions (complementing the humate anions) since these may replace the cations initially present at points of solute—surface contact. Such a process would lead to an enhanced local dispersion, thus favouring pore expansion. Sommerauer et al. (1968) have proposed similarly for the interaction of synthetic polyanions with negatively charged AgBr sols.

The possibility arises that the accessible surface of the clay may be increased by a reduction in the hydrated crystal ("tactoid") size or thickness through the dissociation of some unit (silicate) layers making up an individual crystal. Such a process was suggested by Narkis et al. (1970) to account for the observed decrease in absorbance of aqueous Ca^{2+}-montmorillonite suspensions on adding sodium humate solutions. However, this mechanism was unlikely to be important in Theng and Scharpenseel's systems because the humate solution was added to clay samples which had previously been dried. Banin and Lahav (1968), for example, have shown that resuspension in water of dry montmorillonite, promotes tactoid formation or the association of unit layers. This is true even for Na^+-montmorillonite whose layers would otherwise be completely dissociated. The data of Narkis et al. (1970) further indicated that the reduction in the size of Ca^{2+}-montmorillonite tactoids was not proportional to the amount of sodium humate added. This is clearly inconsistent with the requirement for a linear increase in adsorption with rising humate concentration.

Since the interlayer or intracrystalline surface was apparently inaccessible to the solute, the upper limit to adsorption would be set by the extent of the intercrystalline or intradomain area of the clay sample used. Calculations using a value of 1.52 $m^2\ g^{-1}$ for the average area occupied by humic acid (Cameron et al., 1972) showed that, with the possible exception of the iron clay system, adsorption was not limited by the surface requirement of the polyanion. The absence of a flat plateau was therefore ascribable to the fact that the solutions were insufficiently concentrated to complete the "monolayer." It is interesting to note that, on this basis, the maximum in the curves for Ca^{2+}- and Fe^{3+}-montmorillonite of Valla et al. (1972a), referred to earlier, corresponded closely to the monolayer capacity of their clay sample. The observed decline in uptake beyond this point was ascribed to "aggregation" of the clay crystals (induced by the relatively high humate concentration) by which the accessible surface was reduced. Bearing in mind that Valla et al. (1972a) added their humate solutions to aqueous suspensions of montmorillonite, such aggregation or, more correctly, flocculation might conceivably have occurred. It should be pointed out, however, that an increase in solute concentration over and above what is required for the attainment of a monolayer, would not necessarily lead to plateau adsorption if a second layer of solute molecules can form over the first one (Giles et al., 1974b). Thus, in the absence of interlayer uptake, the highest extent of adsorption in the linear isotherms of Evans and Russell (1959) for their H^{+}- and Ca^{2+}-montmorillonite samples, was considerably in excess of the available intercrystalline surface. This observation would indicate the formation of "multilayers" of humate molecules at the surface.

Another feature of C-type isotherms is that the logarithm of the slope, being proportional to the free energy change of the (constant) solute partition process, may be regarded as an index of solute—surface affinity (Greenland et al., 1965; Theng, 1972b). On the other side, the ionic potential of the exchangeable cation, defined as the ratio of valency (z) to ionic radius (r), is a measure of the polarising power of the cation, that is, the extent to which a cation exerts its influence on a neighbouring molecule. Fig. 12.4 shows that with the exception of the aluminium- and possibly of the iron-clay system as well, there is a good linear relationship between adsorption affinity and ionic potential which quantitatively expresses and confirms the overriding importance of the exchangeable cation in the clay—humic interaction.

By analogy with the behaviour of small organic molecules at hydrated clay mineral surfaces (Mortland, 1970; Theng, 1974), a water bridge type mechanism would appear to be the primary bonding mode between humic substances and clays in aqueous systems. The departure of the points for Al^{3+}, and possibly for Fe^{3+} also, from the straight line (Fig. 12.4) would suggest that other interactions were involved in these instances. Ligand-exchange reactions with polyhydroxy compounds of aluminium and iron come to mind here (Pohl and Schwiete, 1969; Parfitt et al., 1977). For steric reasons, the aluminium and iron in these compounds would be less "reactive" than if they were to exist as the corresponding free cations. The affinity of the solute for Al^{3+}- and Fe^{3+}-montmorillonite was therefore less than might be expected from the ionic potential of these ions. In addition, complexation or coordination of the humic (or fulvic) polyanion to such polyhydroxy compounds is possible, according to the following scheme (Van Dijk, 1971).

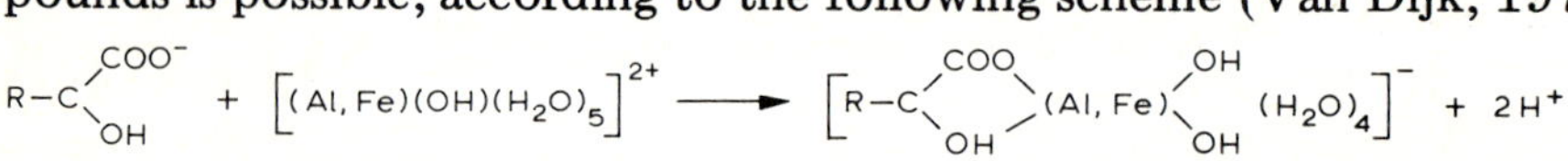

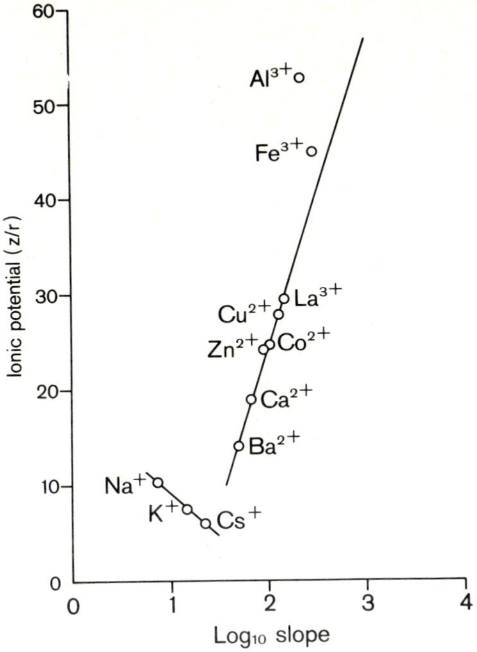

Fig. 12.4. Relationship between the logarithm of the isotherm slope and the ionic potential of the saturating cation (after Theng and Scharpenseel, 1975).

That this type of reaction probably occurred was indicated by the decrease in pH from 7 to ~4 of the supernatant solution in the presence of Al^{3+}- and Fe^{3+}-montmorillonite, a behaviour which was not observed for the other clay samples. Further, the amount of water-desorbable material was less (70% of the total) for the former than for the divalent ion systems (up to 80%). It was inferred that this difference corresponded to the portion held by ligand exchange to polyhydroxy compounds of aluminium and iron and one which could only be released by treating the respective humic (or fulvic) complexes with alkali (Schnitzer and Kodama, 1967; Rashid et al., 1972). Recent studies by Parfitt et al. (1977) have shown that for pure hydrous oxides of iron and aluminium, such as goethite and gibbsite, ligand exchange between the carboxylate groups of fulvic (and humic) acids and the surface hydroxyls of the minerals, is the principal mode of interaction.

The fact that much of the adsorbed humate could be released by elution with water — provided the complex had not been previously dried — was not inconsistent with a water bridge mode of bonding. Being polar, water molecules would compete effectively with the functional groups of the polyanion for hydrogen bonding sites around the primary hydration shell of the exchangeable cation. In aqueous systems Van der Waals interactions between solute and surface would not be of major importance. However, they must also be operative since some uncharged portions of the (extended) polyanion chain are likely to get sufficiently close to the surface to establish Van der Waals contact. Their relative contribution to the overall attraction would therefore depend on the number of uncharged segments in the chain and, all things being equal, on the molecular weight of the polymer. Thus, the line relating log slope to ionic potential (Fig. 12.4) for the fulvate fraction, consisting of relatively low molecular weight components, was displaced parallel to and slightly to the left of that for the corresponding (higher molecular weight) humate (Theng, 1976).

Although there were differences in the amount taken up among the samples saturated with monovalent cations, the extent of adsorption was appreciably lower than that for the corresponding polyvalent ion systems (Fig. 12.3). This reflects the limited ability of monovalent ions to act as a bridge between the polyanion and the clay surface. On the basis of their ionic potential, the affinity of the humate for the clay would be expected to decrease in the order $Na^+ > K^+ > Cs^+$. That the reverse sequence was obtained would indicate that the functional groups of the polyanion were directly coordinated to the exchangeable monovalent ions rather than through a water bridge. This is perhaps not altogether unexpected since K^+ and Cs^+ ions are at best only minimally hydrated (e.g. Nightingale, 1959) whilst the hydration energy of the Na^+ ion is but one-third of that for the least hydrated divalent ion (Ba^{2+}) in this series (Norrish, 1954; Mackenzie, 1964). If, as suggested, direct ion—dipole interactions were operative in monovalent ion systems, a decrease in hydration energy in the order $Na^+ > K^+ > Cs^+$ would favour adsorption. In other words, the extent of adsorption would show the reverse sequence and this was, in fact, observed (Fig. 12.3).

By analogy with negatively charged polysaccharides (section 10.3.2) and lignosulphonates (Chapter 11), the interactions of humic and fulvic acids with clays are sensitive to the ambient solution pH. Adsorption declines sharply as the solution pH increases from 3 to 6, beyond which it falls off more gently until at pH 9 it may become so low as to escape detection (Demolon and Barbier, 1929; Evans and Russell, 1959; Schnitzer and Kodama, 1966).

This observation may be explained in terms of the increased ionisation of the carboxyl groups on the polymer with rising pH. As the pH rises above 6 these groups will be largely ionised, their apparent p*K* value being between 5.5 and 4.5 (Posner, 1964, 1966; Schnitzer and Kodama, 1966) and hence, the polymer will tend to be repelled by the negatively charged clay surface. As remarked on earlier, the unfavourable effect of highly alkaline solutions (pH ~ 9) on uptake may be partly offset by the presence of polyvalent exchangeable cations at the surface. Indeed, Evans and Russell (1959) observed that the amount of humate and fulvate adsorbed by montmorillonite at pH 9 exceeded that at pH 7 for clay samples which had previously been treated with acid.

Similarly, the proportion of ionised carboxyls decreases as the pH falls below 5 until at pH < 3 the polymer would be more or less uncharged and may even acquire a small positive charge through protonation of amino groups belonging to amino acid and peptide residues. Acidic conditions also lead to the development of positive charges at crystal edges as well as the appearance at exchange sites of Al^{3+} ions released from the interior of the silicate layers (Coleman and Craig, 1961; Eeckman and Laudelout, 1961), all of which promote adsorption.

Besides influencing the charge characteristics of humic substances, variations in solution pH induce changes in polymer morphology (Flaig and Beutelspracher, 1951; Schnitzer and Kodama, 1975). Thus, in scanning electron micrographs, freeze-dried preparations of fulvic acid at pH 2—3 show up as fibres and bundles of fibres. At higher pH (4—7) the filaments tend to form a sponge-like structural network. Plastic-like sheets are seen at pH 8 whilst at pH 10 fulvic acid occurs as fine, homogeneous grains. Similar morphological changes to those observed for fulvic acid between pH 6 and 10 occur with humic acid (Chen et al., 1976). It is not certain, however, whether the shape and size of the particles as described would be maintained in aqueous solutions (Orlov et al., 1975).

As mentioned earlier, Theng and Scharpenseel (1975) and Theng (1976) failed to detect interlayer uptake of humic and fulvic acids by montmorillonite crystals at pH ~ 7, irrespective of the nature of the exchangeable cation. This observation is of general applicability to the interaction of humic substances with expanding 2 : 1 type layer silicates (Meyer, 1941; Jung, 1943; Evans and Russell, 1959; Gorbunov et al., 1971; Schnitzer and Khan, 1972; Tan and McCreery, 1975). Intercalation could only occur by taking the solution pH to below 4 when the polymers exist as more or less uncharged species. Under these conditions, however, humic acid is but sparingly soluble in water unless special precautions are taken (Martin Martinez and Perez Rodriguez, 1969), and this restricts its use as a solute. Most of the studies on interlayer adsorption have therefore been carried out with fulvic acid.

Although the ability of humic substances to penetrate the montmorillonite interlayers is primarly determined by the charge characteristics of the polymer, the morphology and shape of the molecule must also play a part in the intercalation process. Schnitzer and Kodama have paid a great deal of attention to the factors influencing the intercalation of fulvic acid into montmorillonite, laying particular emphasis on the properties of the resulting complexes. Their work up to 1971, published in a series of papers, has been summarised by Schnitzer and Khan (1972).

The effect of solution pH and the amount adsorbed (mg C $\times$ 2) on the intercalation of fulvic acid by Na^{+}-montmorillonite, is shown in Fig. 12.5 (Schnitzer and Kodama, 1966). The fulvic acid sample, extracted from the B_h horizon of a podzol, had a relatively low number average molecular weight (M_n 670) and was water-soluble over the entire pH range. The inter-

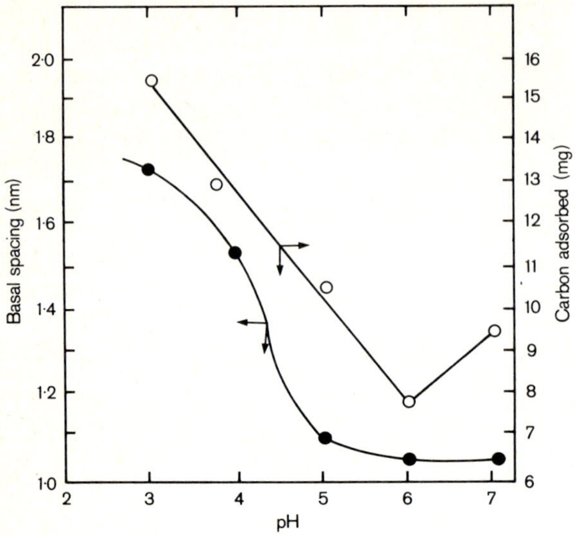

Fig. 12.5. The effect of solution pH on the adsorption and intercalation of fulvic acid by Na^+-montmorillonite (after Schnitzer and Kodama, 1966). Copyright 1966 by the American Association for the Advancement of Science.

calation process (at pH 2.5) was fairly rapid, the basal spacing increasing from 0.98 to 1.55 nm after 60 s, reaching its maximum (1.76 nm) after 5 h of bringing the reactants together, and was little influenced by temperature (Schnitzer and Kodama, 1967, 1968). As indicated in Fig. 12.5, interlayer expansion reached its highest extent at pH 2.5 when slightly more than half of the adsorbed amount was present in the interlayer space, the remainder being attached to external crystal surfaces. At this point most of the carboxyl groups on the polymer would have been unionised since their average p*K* value was close to 4.5, a conclusion which was further supported by differential thermogravimetric and infrared spectroscopic examination of the resulting complex. This meant that the molecule behaved and was adsorbed as an uncharged species rather than as a polyanion as was the case in Theng's (1976) work. Fig. 12.6 further illustrates the strong dependence of interlayer uptake on solute concentration at pH 2.5 as compared to that at neutral pH.

These observations may be explained in terms of competition between the (uncharged) polymer and water for adsorption sites around the exchangeable Na^+ ion. Although such a mechanism was also suggested by Kodama and Schnitzer (1968), they referred to the process as "ligand exchange" on the ground that the adsorbed material was retained against washing with 1 M NaCl whereas it was effectively desorbed by shaking the complex with 0.1 M

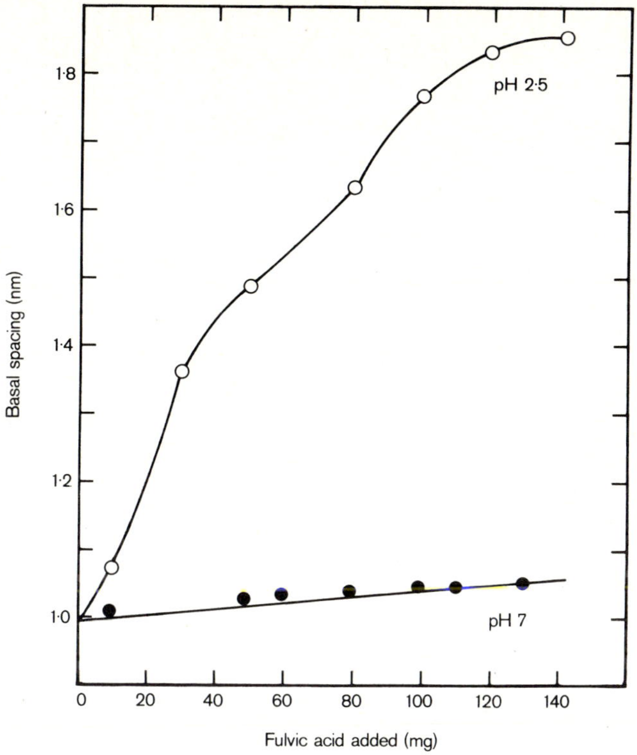

Fig. 12.6. Interlayer expansion of Na^+-montmorillonite in function of fulvic acid concentration and solution pH (after Schnitzer and Kodama, 1967).

NaOH. However, from what was said in the preceding section and with reference to Table 12.3, the bonding mechanism involved seems more compatible with ion—dipole interactions. The observed resistance of the complex to NaCl treatment may be ascribed to the fact that the exchange sites of the parent clay have already been predominantly occupied by Na^+ ions, whilst raising the solution pH by addition of dilute NaOH to the system would lead to ionisation of the carboxyl groups and hence, to the repulsion of the polyanion from the clay surface. A certain amount of ligand exchange could conceivably occur because some Al^{3+} ions, released from the silicate interior at low pH, would migrate to exchange sites and here form a polyhydroxy compound. A reaction of this type has recently been suggested by Moinereau (1977) during the intercalation of fulvic acid, separated from an andosol, by H^+/Al^{3+}-montmorillonite over the pH range 1.5—5.

The increase in basal spacing with the amount of fulvic acid added or present in solution, as indicated in Fig. 12.6, is reminiscent of the behaviour of lysozyme towards montmorillonite (cf. Fig. 7.3). By analogy, we may infer that layer segregation occurs during the intercalation of fulvic acid. The adsorption data further indicate that, on the basis of the spatial requirement of the interlayer polymer, a basal spacing of 1.78 nm (Δ-value ~0.8 nm) is compatible with the presence of a double layer of fulvic acid in the interlayer space (Schnitzer and Kodama, 1967). If so, the hint of an inflection near a basal spacing of 1.4 and 1.8 nm (Fig. 12.6) may correspond to the completion of a single layer and a double layer of intercalated molecules, at which stages about 10 and 20 mg of fulvic acid per 40 mg clay was adsorbed, respectively.

A monolayer thickness of the order of 0.4 nm implies that the polymer is adsorbed in an extended, more or less two-dimensional conformation in the interlayer space. In this respect fulvic acid behaves quite differently from lysozyme which appears to retain its gross globular shape on intercalation (cf. Fig. 7.4). Assuming that the solution conformation of fulvic acid approximates that of its counterpart in the freeze-dry condition in which the polymers show up as fibres under the electron microscope (Chen et al., 1976), the long axis of the fibre must lie nearly parallel to the silicate layers. For such a surface conformation secondary interactions (Table 12.3) would become important because of the possible establishment of numerous segment—surface contacts. It is not certain, however, whether a similar conformation would obtain on external crystal surfaces. This uncertainty should be considered in relating the extent of interlayer expansion to the amount adsorbed, a topic which will be taken up later.

The reactions of humic acid with Na^+-montmorillonite, in function of pH and solute concentration, gave entirely analogous results (Martin Martinez and Perez Rodriguez, 1969; Perez Rodriguez and Martin Martinez, 1970). However, because of its relatively large molecular weight and bulky dimension, the extent of interlayer expansion here was greater than for fulvic acid. Thus, the corresponding inflections in the curve relating basal spacing to the amount present in solution (cf. Fig. 12.6) occurred near 2 and 3 nm, respectively. These observations together with the adsorption data indicate that unlike fulvic acid, the humic acid coil or chain does not collapse extensively to assume a flat conformation on interlayer surfaces.

Fig. 12.7 shows that the dependence of interlayer expansion on the amount of fulvic acid adsorbed (at pH 2.3) varies with the nature of the exchangeable cation. The variation or sequence, however, does not relate in a systematic way to the corresponding metal-fulvic acid stability constants (Schnitzer and Hansen, 1970). Nor is there a straightforward relationship between either adsorption or interlayer expansion and the ionic potential of the cations as Theng (1976) has shown for the uptake of his fulvic acid sample (on external crystal surfaces of montmorillonite) and one which would be expected if intercalation were primarily determined by ion—dipole interactions (cf. Fig. 12.4). Thus, the Na^+- and Pb^{2+}-montmorillonites of Kodama

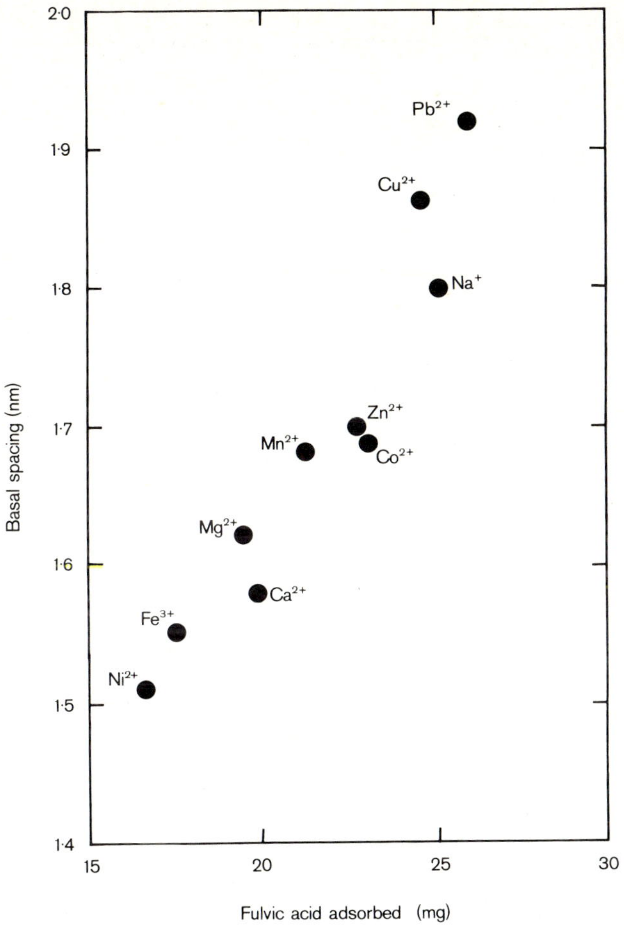

Fig. 12.7. Intercalation of fulvic acid by montmorillonite saturated with different cations (after Kodama and Schnitzer, 1968).

and Schnitzer (1968) adsorbed more, and the magnesium- and nickel-saturated samples less fulvic acid than would be predicted from the ionic potential of these cations. Of special note is that the iron clay system which, on the basis of the high Fe^{3+}-fulvic acid stability constant and ionic potential of Fe^{3+}, should be the most reactive, actually took up next to the least amount in the series (Fig. 12.7).

No satisfactory explanation for these observations can as yet be advanced. However, as already alluded to, other variables enter and complicate the picture. One of these is the probable participation in the process of Al^{3+} ions which occupy exchange positions on being released from the crystal interior by proton attack. Another relates to the contribution of attractive forces other than ion—dipole interactions, to the overall adsorption energy. Because of chain collapse and the resulting intimate contact between solute and (interlayer) surface, such secondary interactions may become significant in these systems. Indeed, for the Cu^{2+}-montmorillonite complex at least, the adsorbed fulvic acid was so tenaciously held as to resist heat decomposition at 1273 K (Schnitzer and Kodama, 1972). In this instance, the infrared spectra indicated that β-diketone groups of the polymer reacted with the Cu^{2+} ions on the clay to form acetylacetonate-type chelates. This would partly account for the high adsorption capacity of the copper-saturated clay and the heat stability of the adsorbed polymer. Incidentally, the occurrence of β-diketone groups and related ketonic carbonyl structures in soil humic acid, has previously been postulated by Theng and Posner (1967) on the basis of chemical and infrared spectroscopic studies. In addition, it seems probable that, with the possible exception of the sodium clay system, the fulvic acid polymer would meet with varying degrees of physical hindrance as it penetrates the montmorillonite interlayers since the interlayer separation in crystals of the parent (untreated) clay samples is dependent on the nature of the saturating cation (Norrish, 1954).

Besides electron microscopy, infrared spectroscopy and X-ray diffractometry to which we have already referred, thermal methods of analysis can be usefully applied to the characterisation of the complexes formed between clays and humic substances. In the case of interlayer montmorillonite—fulvic acid complexes, differential thermal analysis (DTA), in combination with differential thermogravimetry (DTG) and isothermal decomposition, is further capable of distinguishing between the intercalated material and that adsorbed on external crystal surfaces (Kodama and Schnitzer, 1969). This is illustrated in Fig. 12.8 which shows the DTA patterns of fulvic acid (FA), Na^{+}-montmorillonite (Mt), a mixture of FA with Mt, a complex of FA with unheated Mt in which part of the adsorbed material is present in the interlayer space, and a complex of FA with heat-collapsed Mt containing no interlayer polymer.

The broad weak endotherms between 80 and 110°C in all the curves arise from dehydration of the samples. For FA alone (curve A), the exotherms at about 330 (shoulder) and 450°C are ascribed to decarboxylation and oxidation of the "nucleus", respectively. The endotherm at 675°C and the exotherm near 920°C in curve (B) for Mt alone are due to dehydroxylation of the silicate structure and a phase change or recrystallisation, respectively. As would be expected, the pattern for the mixture of FA with Mt (curve C) is essentially a composite of the individual components. In the pattern of the FA-unheated Mt complex (D), the broad exotherm between 400 and 450°C indicates that the adsorbed material decomposes over a wide temperature range. Of particular interest and significance here is that instead of an endotherm (cf. curve B), a broad exotherm appears centering at 670°C.

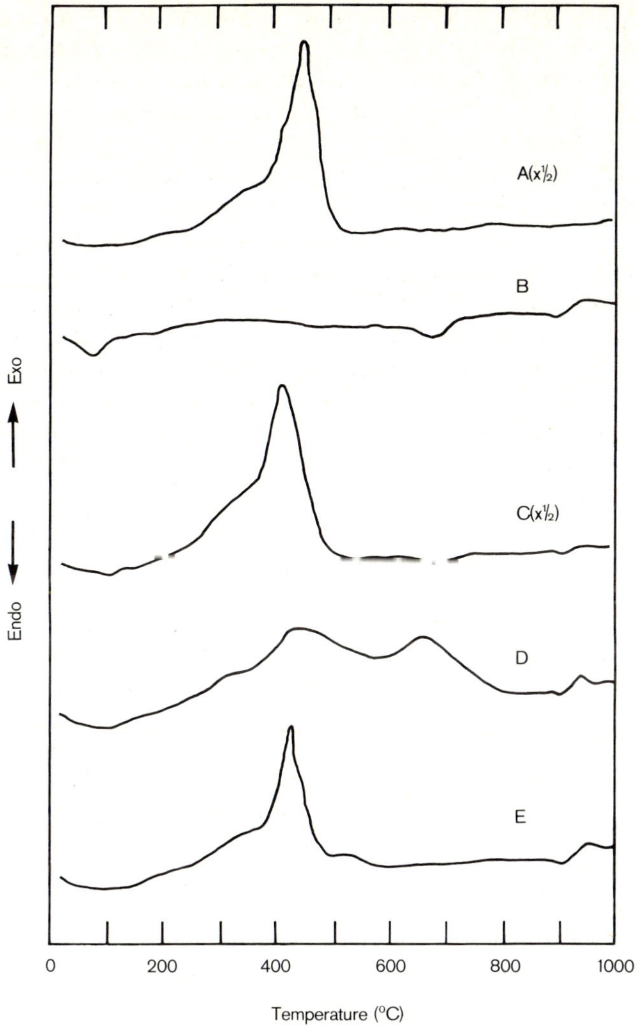

Fig. 12.8. Differential thermal analysis curves of (A) parent fulvic acid; (B) parent montmorillonite; (C) a physical mixture of (A) and (B); (D) fulvic acid complex with unheated montmorillonite; (E) complex with heat-collapsed montmorillonite (after Kodama and Schnitzer, 1969).

Since no corresponding exotherm occurs in the curve for the FA-heated Mt sample (E), this high temperature exotherm may be ascribed to the combustion of the interlayer material. If so, it would be diagnostic of interlayer complex formation. By the same token, it seems reasonable to suppose that the low temperature exotherm (350—550°C) arises from the oxidation of fulvic acid adsorbed on external crystal surfaces.

The apparent ability of DTA to differentiate between the externally and internally adsorbed material has been used by Kodama and Schnitzer (1971) to examine naturally occurring clay—organic matter complexes. In this connection, it is relevant to recall the attempts by many to isolate interlayer clay—organic complexes from soil. Such efforts, however, have at best met with limited success. This is so because interlayer complexes, as we have seen, only form at pH < 4 and fairly high solute concentrations. This combination of factors is seldom found under natural field conditions. In addition, considerable practical difficulties attach to the isolation and identification of such complexes. Although Khan (1969) has stated that newly formed humic substances in soil could penetrate the interlayer space of expanding 2 : 1 layer silicates, other workers (Arshad and Lowe, 1966; Dudas and Pawluk, 1969) have failed to substantiate this claim. Using an ultrasonic dispersion method, Lowe and Parasher (1971) separated the <2 μm fraction from different horizons of some Canadian soils which were rich in organic carbon. Basal spacings between 1.4 and 1.8 nm were recorded which they attributed to interlayer complex formation between clays and organic matter but their evidence was not entirely unequivocal.

Using a similar soil (orthic humo-ferric podzol), Kodama and Schnitzer (1971) isolated the fine (<0.2 μm) clay fraction and subjected it to DTA, before and after treatment with hydrogen peroxide (Fig. 12.9). The curve of the untreated sample (A) shows a strong exotherm near 340°C and a weak one at about 540°C but does not give the endotherm between 500 and 700°C due to dehydroxylation of the silicate structure. On treatment with H_2O_2 which destroys the organic matter, an endotherm appears near

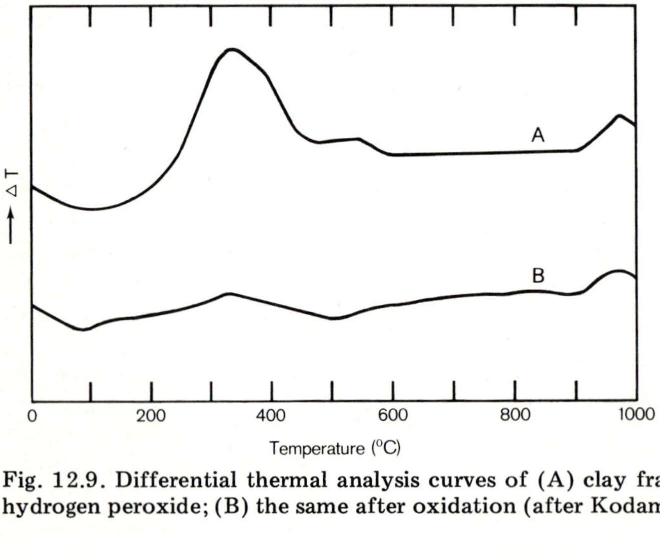

Fig. 12.9. Differential thermal analysis curves of (A) clay fraction before oxidation with hydrogen peroxide; (B) the same after oxidation (after Kodama and Schnitzer, 1971).

510°C (curve B). From these observations and by analogy with the data of Fig. 12.8, they inferred that the 540°C exotherm in curve A indicated the presence of interlayer organic matter, after whose destruction the dehydroxylation endotherm appeared near this temperature. They also suggested that the organic matter was likely to be a fulvic acid which was complexed with Fe^{3+} ions. Using similar techniques, Satoh and Yamane (1971) have also obtained evidence for the presence of interlayer organic matter in montmorillonite complexes, separated from the A horizon of an acid volcanic soil. More recently, Moinereau (1977) has argued for the existence in andosols of a humic substance, either as such or combined with polyhydroxy aluminium complexes, in the interlayer space of montmorillonite-type minerals.

Clearly, thermal methods of analysis, in conjunction with chemical and infrared spectroscopic measurements, provide a promising approach to solving the problem of characterising and identifying naturally occurring organo-mineral complexes. In conclusion, we can say that the existence of such complexes in soil is beyond any reasonable doubt. However, the question as to whether organic substances can be intercalated into expanding 2 : 1 type layer silicates under conditions prevailing in normal soils, must remain somewhat open.

So far we have been concerned almost exclusively with the reactions between humic substances and montmorillonite. Although the term "montmorillonite" has been used to denote the mineral species of that name, it may reasonably be assumed that the other species of the smectite or montmorillonite group of minerals would behave similarly. There are relatively few data on the interactions of humic substances with other minerals belonging to the 1 : 1 and 2 : 1 type phyllosilicates (cf. Table 1.1). What meagre information is available indicates that the factors affecting, and the mechanisms underlying, the adsorption process, are similar to those observed and postulated for montmorillonite. There is, however, one notable difference in that, with the possible exception of vermiculite and (hydrated) halloysite, none of these minerals offers scope for intercalation to occur irrespective of pH and solute concentration.

The reactions of humic substances with the 1 : 1 type layer silicates, kaolinite and halloysite, have been examined by several workers (Evans and Russell, 1959; Sen, 1960; Scharpenseel, 1966; Inoue and Wada, 1968; Gorbunov et al., 1971; Orlov et al., 1973; Kodama and Schnitzer, 1974). As is the case with montmorillonite, adsorption markedly increases with falling pH, a trend which is more pronounced when polyvalent cations, notably iron and aluminium, are present at the exchange complex of the clay. However, the adsorptive capacity is appreciably (30—50%) less than that of montmorillonite and this is a reflection of the relatively low accessible surface and exchange capacity of kaolinite (and halloysite). Using a low molecular weight (M_n 945) fulvic acid, Kodama and Schnitzer (1974), for example, found that uptake was proportional to the crystal area of their kaolinite size fractions.

Similarly, the relatively coarse grain micaceous minerals adsorb less humic acid than does montmorillonite at comparable pH and solution concentration, the extent of adsorption decreasing in the order montmorillonite > vermiculite > kaolinite > chlorite > biotite > muscovite (Orlov et al., 1973). It is also interesting to note that L-type isotherms obtain for these minerals in situations and under conditions where montmorillonite gives rise to linear curves (Evans and Russell, 1959; Orlov et al., 1973; Theng and Scharpenseel, 1975). L-curves indicate that the number of accessible sites at, and the affinity of the solute for, the surface diminish as adsorption progresses (Giles et al., 1960, 1974a). By analogy with the montmorillonite system, discussed previously, the humic acid polyanion is apparently incapable of forcing its way into the intercrystalline (intradomain) pores of micaceous minerals, rendering a proportion of their external crystal surface inaccessible to the solute. This interpretation is further supported by the observation that in the pseudo-layer silicates where entry into the pore structure of the minerals is less restricted, the extent of adsorption is at least comparable to that obtained for montmorillonite. Thus, Kodama and Schnitzer (1974) noted that sepiolite took up nearly three times as much fulvic acid as did muscovite at a given pH and solute concentration. This was ascribed to the ability of fulvic acid to penetrate and be accommodated in the channels of sepiolite from which the polymer displaced part of the structural (zeolitic) water initially present. Infrared spectroscopic examination of the complex indicated that uptake was further favoured by association of the carboxylated groups of fulvic acid with Mg^{2+} ions at the edges of the long tubular talc-like layers (cf. Fig. 1.9).

A similar situation is encountered with soil minerals which lack long-range order, such as allophane and imogolite. The analogy with sepiolite is particularly apt for imogolite since this mineral also consists of thread-like units enclosing narrow (~1 nm inner diameter) channels. Using a water-soluble humified clover extract resembling fulvic acid, and different samples of imogolite, Inoue and Wada (1971a, b), have been able to estimate the portion of the adsorbed material which was accommodated in the channel structure ("inter-unit" pores) by displacement of water. Their results are summarised in Table 12.4. That pore entry did occur was further supported by the observation that adsorption became vanishingly small for imogolite samples which had previously been heated to 378 K, a treatment known to cause the collapse or coalescence of inter-unit pores.

The mode of bonding is primarily one of ligand exchange by which the carboxylate groups of the polymer become incorporated into the coordination shell of surface aluminium, as Parfitt et al. (1977) have subsequently shown. Entropy effects due to water displacement and secondary (hydrogen bonding and Van der Waals) interactions must also be operative. The contri-

TABLE 12.4

Accommodation of humified clover extract into the channel and pore structure of imogolite and allophane together with the resulting displacement of water from the clay system (from Inoue and Wada, 1971b)

Clay [a]	Amount of material accommodated		Amount of water displaced	
	mg per g clay	% of total adsorbed	mg per g clay	% of total adsorbed
Ki-G	75	39	54	12
	90	33	64	14
Ka-G	63	67	45	14
	113	61	81	25
	134	53	96	30
	157	51	112	35
"905"	39	45	28	11
	68	42	47	19
	90	45	64	26
	106	44	76	30

[a] Ki-G and Ka-G refer to the Kitakami and Kanuma gels, respectively, which consist largely of imogolite (Miyauchi and Aomine, 1966; Wada and Matsubara, 1968). The "905" sample contains both allophane and imogolite in a concentration ratio of 10 : 4 (Yoshinaga and Aomine, 1962a,b).

bution of the latter to the overall adsorption energy would be important for those solute molecules which adsorb with their chains parallel to the long axis of the imogolite threads. The steep initial rise in the isotherm and the apparent irreversibility of the adsorption process at low solute concentration suggest that the molecules, adsorbed under these conditions, may adopt this type of orientation. As adsorption progresses, pore entry becomes increasingly more difficult due to physical and steric hindrance so that later arrivals would tend to occupy external sites in such an orientation that only relatively few polymer segments are in contact with the surface. The affinity of the adsorbed material for the surface at high solute concentration (>0.5 mg ml^{-1}) is therefore reduced and this is reflected by the gradual decrease in the slope of the isotherm. A pictorial representation of complex formation between humic substances and imogolite is depicted in Fig. 12.10. A similar process presumably obtains with allophane as well, except that the channel or pore structure here is less regular than in imogolite, being formed by the more or less random aggregation of loosely packed, silico-alumina spherules (Kitagawa, 1971; Wada and Wada, 1977; cf. Fig. 1.11).

At this point, it seems appropriate to recall that soils derived from volcanic ash tend to accumulate and retain large amounts of organic matter

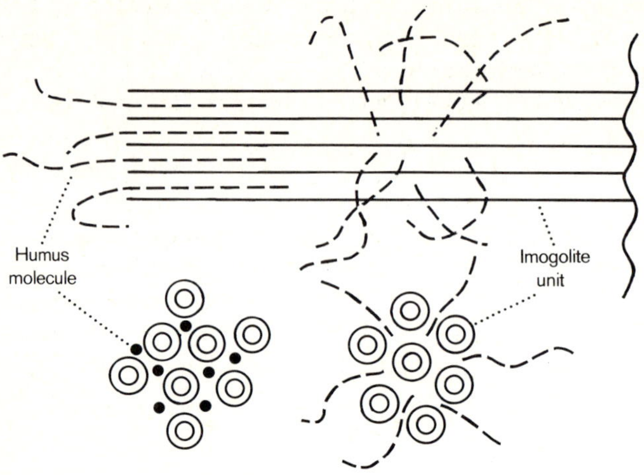

Fig. 12.10. Diagram showing the interaction of humic substances with imogolite (after Inoue and Wada, 1971b).

(Aomine and Kodama, 1956; Jackman, 1964; Tokudome and Kanno, 1965, 1968; Siefferman, 1969). Such soils, referred to as andosols or andepts, contain allophane and, in most cases, also imogolite as their dominant clay minerals (e.g. Aomine and Mizota, 1972; cf. Table 12.4). On the basis of these field observations it is inferred that the resistance to microbial decomposition of organic matter in andosols is due to the formation of a stable allophane—humus complex. The importance of (amorphous) alumina in this process and the part played by organic matter in the genesis of andosols, have been discussed by several workers (Tan, 1966; Inoue, 1973; Hétier, 1975; Duchaufour, 1976). It would appear that only fulvic and humic acids are stabilised by interacting with allophane since fresh organic residues are generally rapidly decomposed in andosols. Duchaufour (1976) has pointed out that this interaction is "reciprocal" in that on the one hand, an abundance of humic substances tends to inhibit the formation of crystalline clay minerals and, on the other, the presence of allophane retards the decomposition of these substances. More details on the properties of allophane and of andosols, in general, have been given by Quantin (1972) and Fieldes and Claridge (1975).

By comparing the rate of carbon and nitrogen mineralisation between soils rich in either allophane or layer silicates (halloysite, montmorillonite), Broadbent et al. (1964) have been able to confirm that the high stability of organic matter in volcanic ash soils is due to complex formation with allophane rather than to any nutrient deficiency or the composition of the microbial population. Following up this work, Wada and Inoue (1967) have measured mineralisation rates in relation to uptake characteristics and to changes in the nature of the organic material which took place on adsorption. As before, they used a water extract of rotting clover leaves as the solute and two soils with comparable clay contents, one containing allophane and the other montmorillonite as the dominant mineral species. The results which are summarised in Table 12.5, clearly show that both the rate and extent of uptake (at near neutral pH) are strongly influenced by the mineralogical composition of the clay fraction. The rate of subsequent material loss through either migration down the soil column or mineralisation was appreciably lower for the allophanic Choyo soil than for the montmorillonite-rich Nakajyo soil. This is consistent with the reaction and bonding mechanisms which we have previously described for allophane systems (cf. Fig. 12.10). Of interest is the change which occurs in the adsorbed material during the early stages (first few weeks) of interaction. Thus, the humic substance extracted from the Nakajyo soil was darker in colour per unit carbon and apparently richer in humic acid-type polymers as compared with the corresponding extract from the Choyo soil. This would suggest that the former material had a relatively high concentration of condensed aromatic structures as well as a high degree of polymerisation.

Earlier, Sen (1961) estimated rates of humic acid decomposition in the presence of either montmorillonite or illite from the amount of carbon dioxide evolved with time. In both instances, this rate was decreased relative to the clay-free system, the extent of reduction being slightly greater when montmorillonite was present. A much greater decrease in mineralisation rate

TABLE 12.5

Differences in the early reactions of humic substances from rotting clover leaves and soils which largely contain either montmorillonite or allophane in their clay fraction (from Wada and Inoue, 1967)

Reaction characteristics	Soil	
	Nakajyo (montmorillonitic)	Choyo (allophanic)
Attainment of adsorption equilibrium	Rapid	Slow
Adsorption capacity (mg C per g soil)	6.5	28—44
Stability of the adsorbed material [a]	Relatively small	Relatively great
Degree of humification of the adsorbed material [b]	Relatively high	Relatively low

[a] Based on the loss of material through migration down the soil column and on the rate of mineralisation of the adsorbed material.

[b] Based on the changes in the visible spectrum of the material extracted from the clay complex and on the content of humic acid-type constituents in the extract.

was observed on addition of Al^{3+} ions to the clay—humic acid system. In the light of the preceding discussion on bonding mechanisms, these observations may be interpreted in terms of cation bridge formation involving aluminium, rather than of any adverse effect of aluminium on microbiological activity as Sen has suggested. Cation binding has also been invoked by Juste et al. (1975) to account for the biological stability of different metal humates.

It should be pointed out in this context that humic substances are inherently stable as compared with such biopolymers as proteins and polysaccharides. As indicated in section 12.1, this stability is ascribable to the random manner in which the constituent units in the molecule of humic and fulvic acids are joined together whatever the underlying polymerisation mechanisms may be. A large variety of enzymes would be required at any given time and place before such a "heteropolycondensate" could be extensively hydrolysed (Swaby and Ladd, 1962; Kleinhempel, 1970). The presence in the system of polyvalent cations would impart greater stability to the structure by rendering the functional groups of the polymer inactive through complexation or chelation with the metal ions. If, in addition, such cations form part of the exchange complex of clays, other strong attractive interactions become operative between solute and surface (cf. Table 12.3). These together with such factors as physical and steric inaccessibility of the adsorbed material to microorganisms or extracellular enzymes (e.g. Craswell and Waring, 1972) would enhance even further the biostability of humic substances in the soil environment.

12.4. MISCELLANEOUS REACTIONS

We have already touched on the ability of humic substances to form stable coordination or chelation complexes with metal ions. Because these ions in soil normally associate with or are incorporated into the structural framework of minerals, the formation of organo-metal complexes in this way may lead to mineral decomposition and dissolution. In this regard, fulvic acid is especially important due to its water-solubility under both acid and alkaline pH conditions and its relatively high concentration of oxygen-containing, ionisable functional groups (cf. Table 12.1).

Another situation in which the chelating properties of humic substances have important consequences is encountered in natural water systems. There is good evidence to indicate that the bulk of the dissolved organic matter in natural waters consists of polymeric compounds resembling soil humic substances (Christman, 1970; Christman and Minear, 1971; Beck et al., 1974). They may, in fact, be derived from soil organic matter through fractional elution by meteoric waters (Reuter and Perdue, 1977). Besides being actively involved in complex formation with various metal ions (Gamble and Schnitzer, 1973; Reuter and Perdue, 1977), the presence of dissolved humic compounds in these systems interferes with the flocculation of clays and other suspended minerals. Narkis et al. (1968), for example, found that higher doses of polycationic flocculants were required to remove the turbidity of waters, due to suspended montmorillonite particles, than would otherwise have been.

Equally relevant in this connection are the organo-metallic interactions which contribute to or are involved in soil formation (e.g. Duchaufour, 1976). Podzolisation is perhaps the classic and outstanding example which comes to mind here. Although differing views have been expressed as to the intricacies of the chemistry of podzol formation (Stobbe and Wright, 1959; Martin and Reeve, 1960; Duchaufour, 1968; Dudal, 1969; Russell, 1973), it is generally agreed that the process essentially involves the degradation of primary and secondary minerals by humic or humic-like substances in the upper layers (A, eluvial, or albic horizon) of the soil, probably through the formation of organo-metal complexes. By this means and in this form, various metals, notably iron and aluminium, are mobilised and translocated down the soil profile where they accumulate in the B (illuvial or spodic) horizon. Other examples of the effect of organic substances on mineral decomposition in relation to pedogenesis, have been given inter alia by Wright and Schnitzer (1963), Kononova et al. (1964), Tan (1966), and Ponomareva and Ragim-Zade (1969). We might add here that simple, polyfunctional organic compounds such as amino acids, polycarboxylic acids, and polyphenols are equally capable of complexing metals and so assist in mineral decomposition (Duff et al., 1963; Schalscha et al., 1967; Baker, 1973; Malcolm et al., 1975). Since these compounds may occur in soil or enter the soil system as leaf leachates and waste products, they have been used as model substances in experiments on chemical weathering and pedogenesis (Schatz et al., 1954; Bloomfield, 1957; Davies et al., 1960; Huang and Keller, 1971, 1972). However, the overall contribution of non-polymeric organics to soil-forming processes may not be very significant. This is so because these compounds, as such, would be either rapidly broken down by soil microorganisms or incorporated into humic substances (cf. Fig. 12.1) under conditions which prevail in the environment of most soils. On the other hand, humic and fulvic acids are stable and persistent soil organic constituents besides being powerful complexing and solubilising agents for a wide variety of metals and minerals (Rosell and Babcock, 1968; Ponomareva and Ragim-Zade, 1969; Schnitzer and Hansen, 1970; Baker, 1973).

The question arises as to whether metal—organic matter complexes can be isolated from natural soils and if so, whether their properties are in any way similar to their synthetic counterparts. Of the few examples mentioned in the literature, that of Schnitzer and Skinner (1964) is illuminating. Using a dilute (0.1 M) solution of hydrochloric acid, these workers succeeded in extracting a metal—fulvic acid complex from the B_h horizon of a podzol. The material showed analytical (chemical, infrared spectroscopic, and thermogravimetric) properties resembling those given by the "low" (molar ratio) metal—fulvic complexes, prepared in the laboratory. The data could be rationalised in terms of the complexation of iron and aluminium, probably in the form of $Fe(OH)^{2+}$ and $Al(OH)^{2+}$, respectively.

As remarked on above, the reactivity of soil organic matter preparations towards metals is closely related to and lies at the root of, the mineral-degrading ability of humic substances. Thus, Kodama and Schnitzer (1972) were able to show that an aqueous solution (0.2% w/v) of fulvic acid at pH 2.5 was capable of decomposing chlorite minerals. Indeed, fulvic acid was more effective in this regard than a dilute hydrochloric acid solution at the same pH. Likewise, fulvic acid was effective in attacking different micas (Schnitzer and Kodama, 1976). From the proportions of the various metals (Al, Fe, K, Mg, and Si) brought into solution (in 710 h at room temperature), the following sequence of stability was deduced: biotite > phlogopite > muscovite. Examination of the fulvic acid, which had been in contact with the mineral, by infrared and electron spin resonance spectroscopy indicated that the operative mechanism was one of metal—fulvic complex formation. As with the chlorites, the mica species containing the highest amount of iron was the least stable towards fulvic acid. This behaviour is apparently related to the relatively low crystal field stability energy of Fe^{2+} and Fe^{3+} (in the high spin states) existing in octahedral coordination within the silicate structure. Earlier, Baker (1973) observed that more Ca, Fe, K, Na, and Mg were extracted than Al and Si when various silicate minerals were treated with a 0.1% w/v aqueous solution of humic acid from a podzol. This would again suggest that mineral dissolution was initiated by metal ion removal from the silicate framework through complex formation with the polyanion. Tan (1975) has inferred similarly from changes in the infrared spectra of humic (and fulvic) acids which had previously been in contact with either kaolinite or montmorillonite.

Interestingly, humic substances may also induce the formation of clay minerals from their constituent parts at ambient temperature and pressure. Linares and Huertas (1971), for example, noted that kaolinite was formed when fulvic acid was added to a system containing silica and alumina with a molar ratio greater than unity, in the pH range 3—9. Similarly, La Iglesia Fernandez and Martin Vivaldi (1971) were able to synthesise kaolinite by bringing together solutions of monomeric silica, aluminium chloride, and fulvic acid for 40—60 days at room temperature, no crystals being formed in the absence of fulvic acid. Crystallisation was promoted under acid conditions (pH 4) when the solubility of both the gibbsite phase and kaolinite itself was at a minimum. In all instances, an amorphous phase was formed (besides kaolinite). This material could be transformed into kaolinite by autoclaving at 405 K and 2 bar pressure, a process which also required the presence of fulvic acid.

The underlying mechanism is as yet little understood but a coprecipitation of silica and Al^{3+} ions seems to be involved. By analogy with the action of simple organic acids, such as citric, oxalic, and tartaric acids, in similar situations (Wey and Siffert, 1961), the explanation should be sought in the ability of fulvic acid to complex or chelate Al^{3+} ions, maintaining them in octahedral coordination. Similarly, Gallardo Lancho et al. (1976) have inferred from thermodynamic data that the gibbsite found in some Spanish soils could not have been formed through inorganic processes alone. Rather, it seems probable that complexation of aluminium by fulvic acid was involved in its genesis. The gibbsite so formed could then condense with silica to form kaolinite.

In conclusion we wish to mention briefly the effect of clays on the synthesis of humic substances by microorganisms. Using casein as a source of carbon and nitrogen, Filip (1968) noted that the addition of montmorillonite to sand and soil cultures increased the amount of organic matter formed in the system. Subsequently, Filip et al. (1972a) extended this work to solution cultures containing *Epicoccum nigrum* and *Stachybotrys chartarum*. These fungal species are known to synthesise a variety of phenolic compounds which convert to humic acid type substances by oxidative polymerisation (Martin and Haider, 1971).

Fig. 12.11 shows that the presence of montmorillonite greatly acceler-

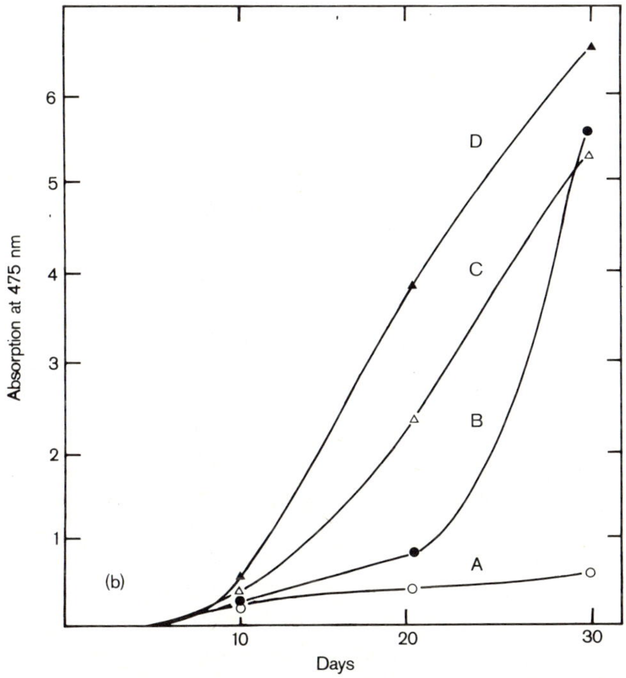

Fig. 12.11 Influence of montmorillonite on the synthesis of humic substances in deep stationary cultures of *Epicoccum nigrum*. Curve A, control; curve B, 0.25% clay; curve C, 0.5% clay; curve D, 1.0% clay (after Filip et al., 1972a).

TABLE 12.6

Ratio of total amount of humic acid-type polymers formed to biomass of *E. nigrum* and to glucose consumed, in stationary cultures (from Filip et al., 1972a)

Montmorillonite added (% w/v)	Incubation time			
	20 days		30 days	
	HA/biomass [a]	HA/glucose [b]	HA/biomass [a]	HA/glucose [b]
0	0	0	4.1	1.5
0.25	18.2	4.7	17.5	4.2
0.5	15.7	5.8	18.0	4.5
1.0	15.8	4.1	29.6	5.4

[a] g humic acid × 100 per g cells.
[b] g humic acid × 100 per g glucose.

ated the synthesis of these substances and increased the amount formed both in solution and within the cells. Indeed, in cultures at an initial pH of 6, the formation of intracellular humic-like polymers preceded their release into the solution phase. Kaolinite and finely ground quartz produced similar changes but their effect was less pronounced. The events described above occurred concomitantly with an increase in the rate and amount of biomass production, glucose consumption, and nitrogen incorporation into the cells (Filip et al., 1972b). On the basis of these observations it was suggested that clays indirectly influenced phenolic polymer formation by modifying the growth and metabolism of the organisms involved.

It seems probable, however, that the observed increase in organic matter synthesis is also promoted by the adsorption of the polymers onto the clay surface. This view is supported by the observation that the measured amount represents a net gain since the ratios of humic acid to total cell mass and to glucose consumed were also increased by the presence of montmorillonite (Table 12.6). Further, montmorillonite with its relatively larger extent of accessible surface was more effective in this respect than either kaolinite or quartz, and the humic acid-type polymers produced contained appreciable amounts of ash. However, the presence of clays did not appear to alter the nature and chemical properties of the polymers formed.

REFERENCES

Alexandrova, L.N., 1966. Possible ways of fulvic acid formation in soil. Transactions of the Meeting of Commissions II and IV of the International Society of Soil Science, Aberdeen, pp. 73—77.

Allison, F.E., 1973. Soil Organic Matter and Its Role in Crop Production. Elsevier, Amsterdam, 637 pp.

Aomine, S. and Kodama, I., 1956. Clay minerals in some arable soils in Miyazaki prefecture. Journal, Faculty of Agriculture, Kyushu University, 10: 325—344.

Aomine, S. and Mizota, C., 1972. Distribution and genesis of imogolite in volcanic ash soils of Northern Kanto, Japan. Proceedings of the International Clay Conference, Madrid, pp. 207—213.

Arshad, M.A. and Lowe, L.E., 1966. Fractionation and characterization of naturally occurring organo-clay complexes. Soil Science Society of America Proceedings, 30: 731—725.

Aylmore, L.A.G. and Quirk, J.P., 1960. Domain or turbostratic structure of clays. Nature (London), 187: 1046—1048.

Aylmore, L.A.G. and Quirk, J.P., 1962. The structural status of clay systems. Clays and Clay Minerals, 9: 104—130.

Baker, W.E., 1973. Role of humic acids from Tasmanian podzolic soils in mineral degradation and metal mobilization. Geochimica et Cosmochimica Acta, 37: 269—281.

Banin, A. and Lahav, N., 1968. Effect of various treatments on the optical properties of montmorillonite suspensions. Israel Journal of Chemistry, 6: 285—294.

Beck, K.C., Reuter, J.H. and Perdue, E.M., 1974. Organic and inorganic geochemistry of some coastal plain rivers of the Southeastern United States. Geochimica et Cosmochimica Acta, 38: 341—364.

Beutelspracher, H., 1955. Wechselwirkung zwischen anorganischen und organischen Kolloiden des Bodens. Zeitschrift für Pflanzenernährung, Düngung, Bodenkunde, 69: 108—115.

Bloomfield, C., 1957. The possible significance of polyphenols in soil formation. Journal of the Science of Food and Agriculture, 8: 389—392.

Bohn, H.L., 1976. Estimate of organic carbon in world soils. Soil Science Society of America Journal, 40: 468—470.

Broadbent, F.E., Jackman, R.H. and McNicoll, J., 1964. Mineralization of carbon and nitrogen in some New Zealand allophanic soils. Soil Science, 98: 118—128.

Cameron, R.S., Thornton, B.K., Swift, R.S. and Posner, A.M., 1972. Molecular weight and shape of humic acid from sedimentation and diffusion measurements on fractionated extracts. The Journal of Soil Science, 23: 394—408.

Chen, Y., Banin, A. and Schnitzer, M., 1976. Use of scanning electron microscope for structural studies on soils and soil components. Proceedings of the 9th Annual Scanning Electron Microscope Symposium, Chicago, Part III, pp. 425—432.

Chichester, F.W., 1969. Nitrogen in soil organo-mineral sedimentation fractions. Soil Science, 107: 356—363.

Christman, R.F., 1970. Chemical structures of colour producing organic substances in water. In: D.W. Hood (Editor), Organic Matter in Natural Waters. Institute of Marine Science, University of Alaska, College, Alaska, pp. 181—198.

Christman, R.F. and Minear, R.A., 1971. Organics in lakes. In: S.J. Faust and J.V. Hunter (Editors), Organic Compounds in Aquatic Environments. Marcel Dekker, New York, N.Y., pp. 119—143.

Coleman, N.T. and Craig, D., 1961. The spontaneous alteration of hydrogen clay. Soil Science, 91: 14—18.

Craswell, E.T. and Waring, S.A., 1972. Effect of grinding on the decomposition of soil organic matter. I. The mineralization of organic nitrogen in relation to soil type. Soil Biology and Biochemistry, 4: 427—433.

Davies, R.I., Coulson, C.B. and Lewis, D.A., 1960. Polyphenols in soil profile development. Scientific Proceedings of the Royal Dublin Society, Series A, 1: 183—190.

Demolon, A. and Barbier, G., 1929. Conditions de formation et constitution du complexe argilo-humique des sols. Comptes Rendus Hebdomadaires des Séances de l'Académie des Sciences, 188: 654—656.

Dubach, P. and Mehta, N.C., 1963. The chemistry of soil humic substances. Soils and Fertilizers, 26: 293—300.

Duchaufour, P., 1968. L'Évolution des Sols, Essai sur la Dynamique des Profils. Masson et Cie, Paris, 94 pp.

Duchaufour, P., 1976. Dynamics of organic matter in soils of temperate regions: its action on pedogenesis. Geoderma, 15: 31—40.

Dudal, R., 1969. Ninety years of "podzolic" soils. Paper presented at the 1st Congress of the Bulgarian Society of Soil Science, Sofia.

Dudas, M.J. and Pawluk, S., 1969. Naturally occurring organo-clay complexes of orthic black chernozems. Geoderma, 3: 5—17.

Duff, R.B., Webley, D.M. and Scott, R.L., 1963. Solubilization of minerals and related materials by 2-ketogluconic acid-producing bacteria. Soil Science, 95: 104—114.

Edwards, A.P. and Bremner, J.M., 1967. Microaggregates in soils. The Journal of Soil Science, 18: 64—73.

Eeckman, J.P. and Laudelout, H., 1961. Chemical stability of hydrogen montmorillonite suspensions. Kolloid Zeitschrift, 178: 99—107.

Evans, L.T. and Russell, E.W., 1959. The adsorption of humic and fulvic acids by clays. The Journal of Soil Science, 10: 119—132.

Felbeck, G.T., Jr, 1965. Structural chemistry of soil humic substances. Advances in Agronomy, 17: 327—368.

Felbeck, G.T., Jr, 1971. Chemical and biological characterization of humic matter. In: A.D. McLaren and J. Skujinš (Editors), Soil Biochemistry, Vol. 2. Marcel Dekker, New York, N.Y., pp. 36—59.

Fieldes, M. and Claridge, G.G.C., 1975. Allophane. In: J.E. Gieseking (Editor), Soil Components, Vol. 2. Inorganic Components. Springer Verlag, Berlin, pp. 351—393.

Filip, Z., 1968. Growth of microorganisms and formation of humic substances in media containing varying amounts of bentonite. Pochvovedenie, No. 9: 55—61.

Filip, Z., Haider, K. and Martin, J.P., 1972a. Influence of clay minerals on the formation of humic substances by *Epicoccum nigrum* and *Stachybotrys chartarum*. Soil Biology and Biochemistry, 4: 147—154.

Filip, Z., Haider, K. and Martin, J.P., 1972b. Influence of clay minerals on growth and metabolic activity of *Epicoccum nigrum* and *Stachybotrys chartarum*. Soil Biology and Biochemistry, 4: 135—145.

Flaig, W. and Beutelspracher, H., 1951. Electronenmikroskopische Untersuchungen an natürlichen und synthetischen Huminsäuren. Zeitschrift für Pflanzenernährung, Düngung, Bodenkunde, 52: 1—21.

Flaig, W., Beutelspracher, H. and Rietz, E., 1975. Chemical composition and physical properties of humic substances. In: J.E. Gieseking (Editor), Soil Components, Vol. 1. Organic Components. Springer Verlag, Berlin, pp. 1—211.

Gallardo Lancho, J., Sanchez Camazano, M., Saavedra Alonso, J. and Garcia Sanchez, A., 1976. Influence of organic matter on the formation of gibbsite and kaolinite in granitic soils of west-central Spain. Clay Minerals, 11: 241—249.

Gamble, D.S. and Schnitzer, M., 1973. The chemistry of fulvic acid and its reactions with metal ions. In: P.C. Singer (Editor), Trace Metals and Metal—Organic Interactions in Natural Waters. Ann Arbor Science Publishers, Ann Arbor, Michigan, pp. 265—302.

Giles, C.H., MacEwan, T.H., Nakhwa, S.N. and Smith, D., 1960. Studies in adsorption. Part XI. A system of classification of solution adsorption isotherms, and its use in diagnosis of adsorption mechanisms and in measurement of specific surface areas of solids. Journal of the Chemical Society, 3973—3993.

Giles, C.H., Smith, D. and Huitson, A., 1974a. A general treatment and classification of the solute adsorption isotherm. I. Theoretical. Journal of Colloid and Interface Science, 47: 755—765.

Giles, C.H., D'Silva, A.P. and Easton, I.A., 1974b. A general treatment and classification of the solute adsorption isotherm. II. Experimental interpretation. Journal of Colloid and Interface Science, 47: 766—777.

Gorbunov, N.I., Yerokhina, G.L. and Shchurina, G.N., 1971. Relationship between soil minerals and humic substances. Pochvovedenie, No. 7: 117—128.

Greenland, D.J., 1965a. Interaction between clays and organic compounds in soils. I. Mechanism of interaction between clays and defined organic compounds. Soils and Fertilizers, 28: 415—425.

Greenland, D.J., 1965b. Interaction between clays and organic compounds in soils. II. Adsorption of soil organic compounds and its effects on soil properties. Soils and Fertilizers, 28: 521—532.

Greenland, D.J., 1971. Interactions between humic and fulvic acids and clays. Soil Science, 111: 34—41.

Greenland, D.J. and Ford, G.W., 1964. Separation of partially humified organic materials from soils by ultrasonic dispersion. Transactions of the 8th International Congress of Soil Science, Bucharest, 3: 137—148.

Greenland, D.J., Laby, R.H. and Quirk, J.P., 1962. Adsorption of glycine and its di-, tri-, and tetra-peptides by montmorillonite. Transactions of the Faraday Society, 58: 829—841.

Greenland, D.J., Laby, R.H. and Quirk, J.P., 1965. Adsorption of amino acids and peptides by montmorillonite and illite. Part 2. Physical adsorption. Transactions of the Faraday Society, 61: 2024—2035.

Haider, K., Martin, J.P. and Filip, Z., 1975. Humus biochemistry. In: E.A. Paul and A.D. McLaren (Editors), Soil Biochemistry, Vol. 3. Marcel Dekker, New York, N.Y., pp. 195—244.

Hénin, S. and Turc, L., 1950. Essai de fractionnement des matières organiques du sol. Transactions of the 4th International Congress of Soil Science, Amsterdam, 1: 152—154.

Hétier, J.M., 1975. Formation et Évolution des Andosols en Climat Tempéré. Thesis, University of Nancy, 195 pp.

Hingston, F.J., Atkinson, R.J., Posner, A.M. and Quirk, J.P., 1967. Specific adsorption of anions. Nature (London), 215: 1459—1461.

Hingston, F.J., Atkinson, R.J., Posner, A.M. and Quirk, J.P., 1968. Specific adsorption of anions on goethite. Transactions of the 9th International Congress of Soil Science, Adelaide, 1: 669—678.

Huang, W.H. and Keller, W.D., 1971. Dissolution of clay minerals in dilute organic acids at room temperature. The American Mineralogist, 56: 1082—1095.

Huang, W.H. and Keller, W.D., 1972. Kinetics and mechanisms of dissolution of Fithian illite in two complexing organic acids. Proceedings of the International Clay Conference, Madrid, pp. 321—331.

Inoue, T., 1973. A preliminary study of organo-mineral complexes in a glassy volcanic ash soil containing imogolite and allophane as principal clay minerals. Pedorojisuto, 17: 26—36.

Inoue, T. and Wada, K., 1968. Adsorption of humified clover extracts by various clays. Transactions of the 9th International Congress of Soil Science, Adelaide, 3: 289—298.

Inoue, T. and Wada, K., 1971a. Reactions between humified clover extract and imogolite as a model of humus—clay interaction: Part I. Clay Science, 4: 61—70.

Inoue, T. and Wada, K., 1971b. Reactions between humified clover extract and imogolite as a model of humus—clay interaction: Part II. Clay Science, 4: 71—80.

Jackman, R.H., 1964. Accumulation of organic matter in some New Zealand soils under permanent pasture. New Zealand Journal of Agricultural Research, 1: 445—479.

Jung, E., 1943. Zur Kenntnis der Ton-Humusbildung. Bodenkunde und Pflanzenernährung, 32: 325—336.

Juste, C., Delas, J. and Langon, M., 1975. Comparison de la stabilité biologique de différents humates métalliques. Comptes Rendus Hebdomadaires des Séances de l'Académie des Sciences, D, 281: 1685—1688.

Khan, D.V., 1959. The composition of humic substances and their combination with the mineral fraction of the soil. Pochvovedenie, No. 1: 10—18.

Khan, D.V., 1969. Organo-mineral Compounds and Soil Structure. Nauka Publishing House, Moscow, 142 pp.

Kitagawa, Y., 1971. The "unit particle" of allophane. The American Mineralogist, 56: 465—475.

Kleinhempel, D., 1970. Ein Beitrag zur Theorie des Huminstoffzustandes. Albrecht-Thaer-Archiv, 14: 3—14.

Kobo, K. and Fujisawa, T., 1963. Studies on the clay—humus complex. 3. Adsorption of humic acid by clay. Soil Science and Plant Nutrition (Tokyo), 9: 36—37.

Kodama, H. and Schnitzer, M., 1968. Effects of interlayer cations on the adsorption of a soil humic compound by montmorillonite. Soil Science, 106: 73—74.

Kodama, H. and Schnitzer, M., 1969. Thermal analysis of a fulvic acid—montmorillonite complex. Proceedings of the International Clay Conference, Tokyo, 1: 765—774.

Kodama, H. and Schnitzer, M., 1971. Evidence for interlamellar adsorption of organic matter by clay in a podzol soil. Canadian Journal of Soil Science, 51: 509—512.

Kodama, H. and Schnitzer, M., 1972. Dissolution of chlorite minerals by fulvic acid. Canadian Journal of Soil Science, 53: 240—243.

Kodama, H. and Schnitzer, M., 1974. Adsorption of fulvic acid by non-expanding clay minerals. Transactions of the 10th International Congress of Soil Science, Moscow, 2: 51—56.

Kononova, M.M., 1966. Soil Organic Matter, 2nd Edition. Pergamon Press, Oxford, 544 pp.

Kononova, M.M., Alexandrova, I.V. and Titova, N.A., 1964. Decomposition of silicates by soil organic matter. Pochvovedenie, No. 10: 1—12.

La Iglesia Fernandez, A. and Martin Vivaldi, J.L., 1972. A contribution to the synthesis of kaolinite. Proceedings of the International Clay Conference, Madrid, pp. 173—185.

Levy, R. and Francis, C.W., 1976. Adsorption and desorption of cadmium by synthetic and natural organo-clay complexes. Geoderma, 15: 361—370.

Linares, J. and Huertas, F., 1971. Kaolinite synthesis at room temperature. Science, 171: 896—897.

Lowe, L.E. and Parasher, C.D., 1971. Observations on clay-size organo-mineral complexes isolated from soil by ultrasonic dispersion. Canadian Journal of Soil Science, 51: 136—137.

Mackenzie, R.C., 1964. Hydratationseigenschaften von Montmorillonit. Berichte der Deutschen Keramischen Gesellschaft, 41: 696—703.

Malcolm, R.L., Leenheer, J.A. and Weed, S.B., 1975. Dissolution of aquafer clay minerals during deep-well disposal of industrial organic wastes. Proceedings of the International Clay Conference, Mexico City, pp. 477—493.

Martin, J.P. and Haider, K., 1971. Microbial activity in relation to soil humus formation. Soil Science, 111: 54—63.

Martin, A.E. and Reeve, R., 1960. Chemical studies of podzolic illuvial horizons. IV. The flocculation of humus by aluminium. The Journal of Soil Science, 11: 369—381.

Martin Martinez, F. and Perez Rodriguez, J.L., 1969. Interlamellar adsorption of a black-earth humic acid on Na-montmorillonite. Zeitschrift für Pflanzenernährung und Bodenkunde, 124: 52—57.

Mattson, S., 1935. The pH and base saturation of the podzol profile. Transactions of the 3rd International Congress of Soil Science, Oxford, 1: 67—70.

Meyer, L., 1941. Über Entstehung und Bildung der Ton-Humus-Komplexe. Forschungdienst Sonderhefte, 17: 38—41.

Miyauchi, N. and Aomine, S., 1966. Mineralogy of gel-like substance in the pumice bed in Kanuma and Kitakami districts. Soil Science and Plant Nutrition (Tokyo), 12: 187—190.

Moinereau, J., 1977. Absorption de composes humiques par une montmorillonite H^+, Al^{3+}. Présence de smectites à couches interfoliaires organo-minérales dans les andosols. Clay Minerals, 12: 75—82.

Monnier, G., Turc, L. and Jeanson-Luusinang, C., 1962. Une méthode de fractionnement densimétrique par centrifugation des matières organiques du sol. Annales Agronomiques, 13: 55—63.

Mortland, M.M., 1970. Clay—organic complexes and interactions. Advances in Agronomy, 22: 75—117.

Myers, H.E., 1937. Physiochemical reactions between organic and inorganic soil colloids, related to aggregate formation. Soil Science, 44: 331—359.

Narkis, N., Rebhun, M. and Sperber, H., 1968. Flocculation of clay suspensions in the presence of humic and fulvic acids. Israel Journal of Chemistry, 6: 295—305.

Narkis, N., Rebhun, M., Lahav, N. and Banin, A., 1970. An optical-transmission study of interaction between montmorillonite and humic acids. Israel Journal of Chemistry, 8: 383—389.

Nightingale, E.R., Jr, 1959. Phenomenological theory of ion solvation. Effective radii of hydrated ions. The Journal of Physical Chemistry, 63: 1381—1387.

Norrish, K., 1954. The swelling of montmorillonite. Discussions of the Faraday Society, No. 18: 120—134.

Orlov, D.S., Ammosova, Ya.M. and Glebova, G.I., 1975. Molecular parameters of humic acids. Geoderma, 13: 211—229.

Orlov, D.S., Pivovarova, I.A. and Gorbunov, N.I., 1973. Interaction of humic substances with minerals and the nature of their bond. A review. Agrokhimiya, 9: 140—153.

Parasher, C.D. and Lowe, L.E., 1970. Isolation of clay-size organo-mineral complexes from soils of the lower Fraser valley. Canadian Journal of Soil Science, 50: 403—407.

Parfitt, R.L., Fraser, A.R. and Farmer, V.C., 1977. Adsorption on hydrous oxides. III. Fulvic acid and humic acid on goethite, gibbsite and imogolite. The Journal of Soil Science, 28: 289—296.

Perez Rodriguez, J.L. and Martin Martinez, F., 1970. Interaccion entre materia humica y montmorillonita sodica. Reunion Hispano-Belga de Minerales de la Arcilla, Madrid, pp. 98—102.

Pohl, K.D. and Schwiete, H.E., 1969. Untersuchungen zum Problem der Bindung organischer Bestandteile an Tonminerale. Berichte der Deutschen Keramischen Gesellschaft, 46: 587—593.

Ponomareva, V.V. and Ragim-Zade, A.I., 1969. Comparative study of fulvic and humic acids as the agents of silicate mineral decomposition. Pochvovedenie, No. 3: 26—36.

Posner, A.M., 1964. Titration curves of humic acids. Transactions of the 8th International Congress of Soil Science, Bucharest, 2: 161—173.

Posner, A.M., 1966. The humic acids extracted by various reagents from a soil. I. Yield, inorganic components and titration curves. The Journal of Soil Science, 17: 65—78.

Quantin, P., 1972. Les andosols. Revue bibliographique des connaissances actuelles. Cahiers de l'office de la Recherche Scientifique et Technique Outre-Mer (O.R.S.T.O.M.), Séries Pédologie, 10: 273—301.

Rashid, M.A., Buckley, D.E. and Robertson, K.R., 1972. Interactions of a marine humic acid with clay minerals and a natural sediment. Geoderma, 8: 11—27.

Reuter, J.H. and Perdue, E.M., 1977. Importance of heavy metal—organic matter interactions in natural waters. Geochimica et Cosmochimica Acta, 41: 325—334.

Rosell, R.A. and Babcock, K.L., 1968. Precipitated manganese isotopically exchanged with ^{54}Mn and chelated by soil organic matter. In: Isotopes and Radiation in Soil Organic Matter Studies. I.A.E.A., Vienna, pp. 453—469.

Roulet, N., Dubach, P., Mehta, N.C., Müller-Vonmoos, L. and Deuel, H., 1963. Verteilung der organischen Substanz und der Kohlenhydrate bei der Gewinnung "wurzelfreien" Bodenmaterials durch Schlämmsiebung. Zeitschrift für Pflanzenernährung, Düngung, Bodenkunde, 101: 210—214.

Russell, E.W., 1973. Soil Conditions and Plant Growth, 10th Edition. Longman Group Ltd, London, 849 pp.

Satoh, T., 1976. Isolation and characterization of naturally occurring organo-mineral complexes in some volcanic ash soils. Soil Science and Plant Nutrition (Tokyo), 22: 125—136.

Satoh, T. and Yamane, I., 1971. On the interlamellar complex between montmorillonite and organic substance in certain soil. Soil Science and Plant Nutrition (Tokyo), 17: 181—185.

Satoh, T. and Yamane, I., 1972a. Studies on the separation of the naturally occurring organo-mineral complexes and their characteristics. 1. Studies by means of the particle size fractionation. Journal of the Science of Soil and Manure, Japan, 43: 41—45.

Satoh, T. and Yamane, I., 1972b. Studies on the separation of the naturally occurring organo-mineral complexes and their characteristics. 2. Studies by means of the densimetric fractionation. Journal of the Science of Soil and Manure, Japan, 43: 61—65.

Schalscha, E.B., Appelt, H. and Schatz, A., 1967. Chelation as a weathering mechanism. I. Effect of complexing agents on the solubilization of iron from minerals and granodiorite. Geochimica et Cosmochimica Acta, 31: 587—596.

Scharpenseel, H.W., 1966. Tracer investigations on synthesis and radiometric combination of soil organo-mineral complexes. Transactions of the Meeting of Commissions II and IV of the International Society of Soil Science, Aberdeen, pp. 41—52.

Schatz, A., Cheronis, N.D., Schatz. V. and Trewlawney, G.S., 1954. Chelation (sequestration) as a biological weathering factor in pedogenesis. Proceedings of the Pennsylvania Academy of Science, 28: 44—51.

Schloesing, Th., 1874. Détermination de l'argile dans la terre arable. Comptes Rendus Hebdomadaires des Séances de l'Académie des Sciences, 78: 1276—1279.

Schnitzer, M. and Hansen, E.H., 1970. Organo-metalic interactions in soils. 8. An evaluation of methods for the determination of stability constants for metal—fulvic acid complexes. Soil Science, 109: 333—340.

Schnitzer, M. and Khan, S.U., 1972. Humic Substances in the Environment. Marcel Dekker, New York, N.Y., 327 pp.

Schnitzer, M. and Kodama, H., 1966. Montmorillonite: Effect of pH on its adsorption of a soil humic compound. Science, 153: 70—71.

Schnitzer, M. and Kodama, H., 1967. Reactions between a podzol fulvic acid and Na-montmorillonite. Soil Science Society of America Proceedings, 31: 632—636.

Schnitzer, M. and Kodama, H., 1972. Reactions between fulvic acid and Cu^{2+}-montmorillonite. Clays and Clay Minerals, 20: 359—367.

Schnitzer, M. and Kodama, H., 1975. An electron microscopic examination of fulvic acid. Geoderma, 13: 279—287.

Schnitzer, M. and Kodama, H., 1976. The dissolution of micas by fulvic acid. Geoderma, 15: 381—391.

Schnitzer, M. and Skinner, S.I.M., 1964. Organo-metallic interactions in soils. 3. Properties of iron— and aluminum—organic matter complexes, prepared in the laboratory and extracted from a soil. Soil Science, 98: 197—203.

Sen, B.C., 1960. Studies on the adsorption of humic acid on H-clays and the role of metal cations in humus adsorption. Journal of the Indian Chemical Society, 37: 793—797.

Sen, B.C., 1961. Studies on the bacterial decomposition of humic acid in the clay—humus mixtures. Journal of the Indian Chemical Society, 38: 737—740.

Sen, B.C., 1964. Studies on the effect of different constituents of soil humic acid in the formation of clay—humus complexes. Journal of the Indian Chemical Society, 41: 145—149.

Sideri, D.I., 1938. On the formation of structure in soil. IV. The structure of mixed clay—sand and clay—humus formations. Soil Science, 46: 127—137.

Siefferman, G., 1969. Les Sols de Quelques Régions Volcaniques du Cameroun. Thesis, University of Strasbourg, 285 pp.

Sommerauer, A., Sussman, D.L. and Stumm, W., 1968. The role of complex formation in the flocculation of negatively charged sols with anionic polyelectrolytes. Kolloid Zeitschrift und Zeitschrift für Polymere, 225: 147—154.

Spycher, G. and Young, J.L., 1977. Density fractionation of water-dispersible soil organic-mineral particles. Communications in Soil Science and Plant Analysis, 8: 37—48.

Stevenson, F.J. and Butler, J.H.A., 1969. Chemistry of humic acids and related pigments. In: G. Eglinton and M.T.J. Murphy (Editors), Organic Geochemistry. Springer Verlag, Berlin, pp. 534—557.

Stobbe, P.C. and Wright, J.R., 1959. Modern concepts of the genesis of podzols. Soil Science Society of America Proceedings, 23: 161—164.

Swaby, R.J. and Ladd, J.N., 1962. Chemical nature, microbial resistance, and origin of soil humus. Transactions of Commissions IV and V of the International Society of Soil Science, New Zealand, pp. 3—8.

Swift, R.S., Thornton, B.K. and Posner, A.M., 1970. Spectral characteristics of a humic acid fractionated with respect to molecular weight using an agar gel. Soil Science, 110: 93—99.

Szalay, A., 1964. Cation exchange properties of humic acids and their importance in the geochemical enrichment of UO_2^{2+} and other cations. Geochimica et Cosmochimica Acta, 28: 1605—1614.

Tan, K.H., 1966. On the pedogenetic role of organic matter in volcanic ash soils under tropical conditions. Soil Science and Plant Nutrition (Tokyo), 12: 80—84.

Tan, K.H., 1975. The catalytic decomposition of clay minerals by complex reaction with humic and fulvic acid. Soil Science, 120: 188—194.

Tan, K.H., 1976. Complex formation between humic acid and clays as revealed by gel filtration and infrared spectroscopy. Soil Biology and Biochemistry, 8: 235—239.

Tan, K.H. and McCreery, R.A., 1975. Humic acid complex formation and intermicellar adsorption by bentonite. Proceedings of the International Clay Conference, Mexico City, pp. 629—641.

Theng, B.K.G., 1972a. Formation, properties, and practical applications of clay—organic complexes. Journal of the Royal Society of New Zealand, 2: 437—457.

Theng, B.K.G., 1972b. Adsorption of ammonium and some primary n-alkylammonium cations by soil allophane. Nature (London), 238: 150—151.

Theng, B.K.G., 1974. The Chemistry of Clay—Organic Reactions. Adam Hilger Ltd, London, 343 pp.

Theng, B.K.G., 1976. Interactions between montmorillonite and fulvic acid. Geoderma, 15: 243—251.

Theng, B.K.G. and Posner, A.M., 1967. Nature of the carbonyl groups in soil humic acid. Soil Science, 104: 191—201.

Theng, B.K.G. and Scharpenseel, H.W., 1975. The adsorption of ^{14}C-labelled humic acid by montmorillonite. Proceedings of the International Clay Conference, Mexico City, pp. 643—653.

Theng, B.K.G., Wake, J.R.H. and Posner, A.M., 1967. The humic acids extracted by various reagents from a soil. II. Infra-red, visible, and ultra-violet absorption spectra. The Journal of Soil Science, 18: 349—363.

Theng, B.K.G., Wake, J.R.H. and Posner, A.M., 1968. The fractional precipitation of soil humic acid by ammonium sulphate. Plant and Soil, 29: 305—316.

Tokudome, S. and Kanno, I., 1965. Nature of the humus of humic allophane soils in Japan. Soil Science and Plant Nutrition (Tokyo), 11: 185—199.

Tokudome, S. and Kanno, I., 1968. Nature of the humus of some Japanese soils. Transactions of the 9th International Congress of Soil Science, Adelaide, 3: 161—173.

Turchenek, L.W. and Oades, J.M., 1974. Size and density fractionation of naturally occurring organo-mineral complexes. Transactions of the 10th International Congress of Soil Science, Moscow, 2: 65—72.

Tyulin, A.Th., 1938. The composition and structure of soil organo-mineral gels and soil fertility. Soil Science, 45: 343—357.

Valla, M., Guckert, A. and Jacquin, F., 1972a. Formation de complexes entre acides humiques et montmorillonite. I. Étude de mécanismes d'adsorption d'acides humiques-^{14}C en fonction des cations. Bulletin de l'École Nationale Superieure d'Agronomie et des Industries Alimentaires, 14: 227—238.

Valla, M., Guckert, A. and Jacquin, F., 1972b. Formation de complexes entre acides humiques et montmorillonite. II. Étude de la stabilité des complexes acides humiques-^{14}C montmorillonite a l'égard des réactifs d'extraction chimiques. Bulletin de l'École Nationale Superieure d'Agronomie et des Industries Alimentaires, 14: 239—244.

Van Dijk, H., 1971. Cation binding of humic acids. Geoderma, 5: 53—67.

Wada, K. and Inoue, T., 1967. Retention of humic substances derived from rotted clover leaves on soils containing montmorillonite and allophane. Soil Science and Plant Nutrition (Tokyo), 13: 9—16.

Wada, K. and Matsubara, I., 1968. Differential formation of allophane, imogolite and gibbsite in the Kitakami pumice bed. Transactions of the 9th International Congress of Soil Science, Adelaide, 3: 123—131.

Wada, S. and Wada, K., 1977. Density and structure of allophane. Clay Minerals, 12: 289—298.

Watson, J.R. and Parsons, J.W., 1974. Studies of soil organo-mineral fractions. I. Isolation by ultrasonic dispersion. The Journal of Soil Science, 25: 1—8.

Wey, R. and Siffert, B., 1961. Reactions de la silice monomoléculaire en solution avec les ions Al et Mg. In: Genèse et Synthèse des Argiles. Colloques Internationaux du Centre National de la Recherche Scientifique (C.N.R.S.), 105: 11—24.

Wright, J.R. and Schnitzer, M., 1963. Metallo-organic interaction associated with podzolization. Soil Science Society of America Proceedings, 27: 171—176.

Yoshinaga, N. and Aomine, S., 1962a. Allophane in some Ando soils. Soil Science and Plant Nutrition (Tokyo), 8: 6—13.

Yoshinaga, N. and Aomine, S., 1962b. Imogolite in some Ando soils. Soil Science and Plant Nutrition (Tokyo), 8: 114—121.

APPENDIX

The structural formulas of some of the synthetic and naturally occurring polymers which have been used as complex formers with clay minerals.

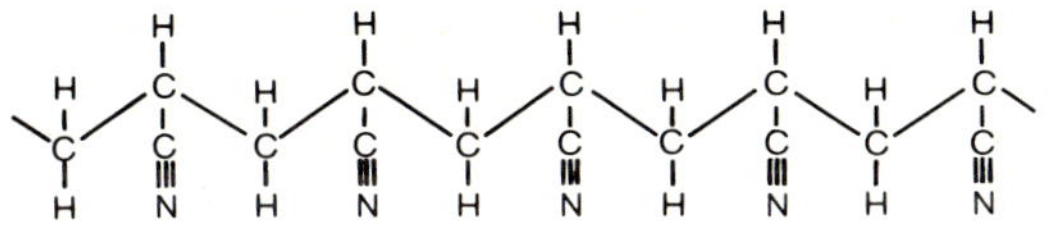

Polyacrylonitrile

Hydrolysed polyacrylonitrile (HPAN)

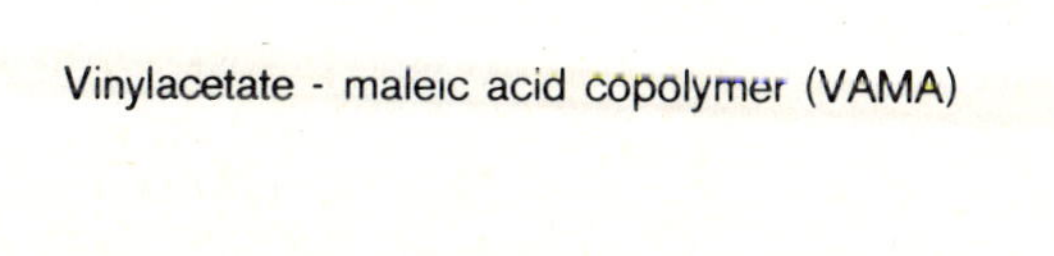

Vinylacetate - maleic acid copolymer (VAMA)

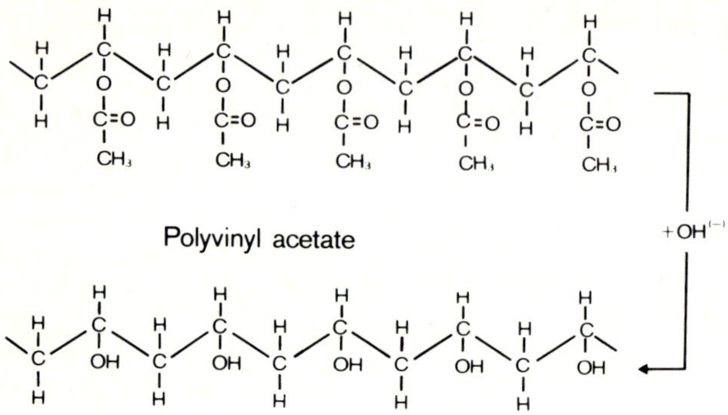

Polyvinyl acetate

Polyvinyl alcohol

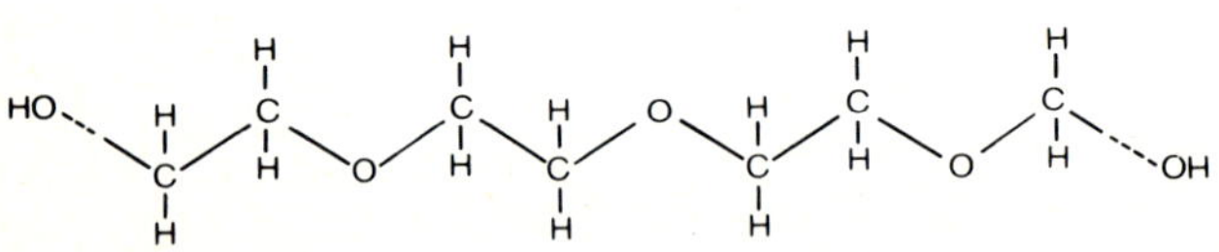

Polyethylene oxide or Polyethylene glycol

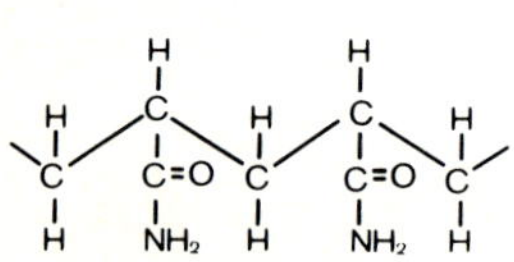

Polyacrylamide

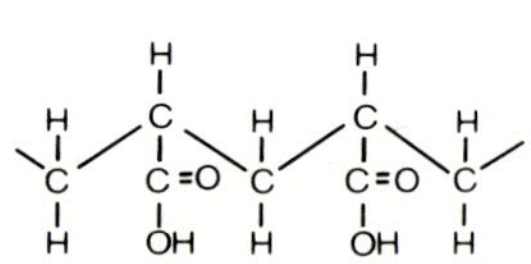

Polyacrylic acid

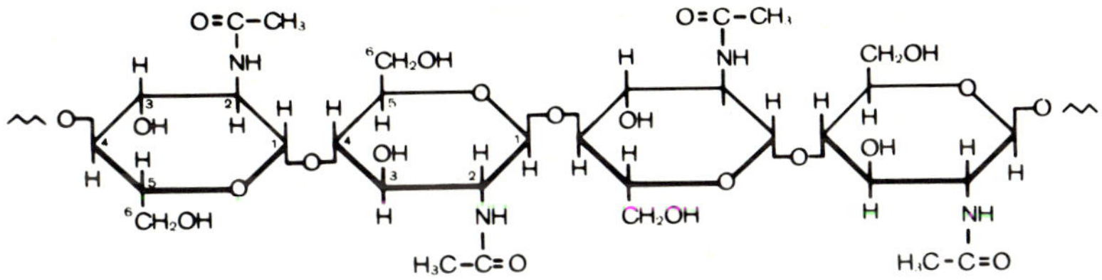

Chitin (β–(1→4)– linked N-acetyl-glucoseamine units)

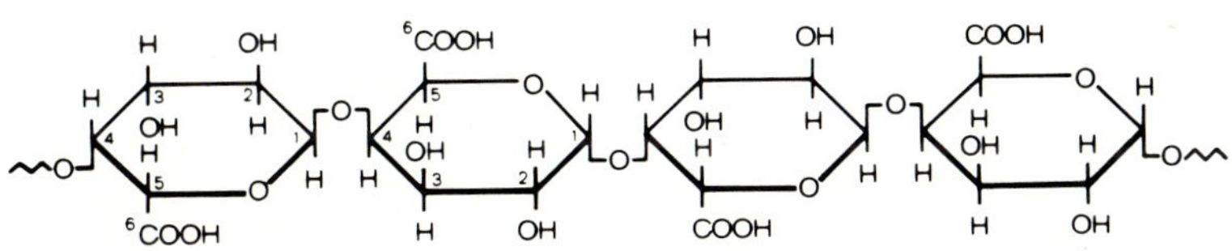

Pectinic or Polygalacturonic acid (α-(1→4)-linked galacturonic acid units)
Pectin is the fully or partially methylated derivative

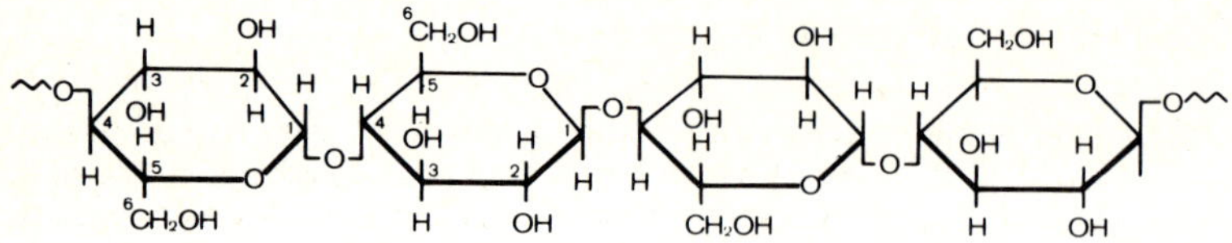

Cellulose (β - (1→4) - linked glucose units)
In Amylose , the linkage is α - (1→4)

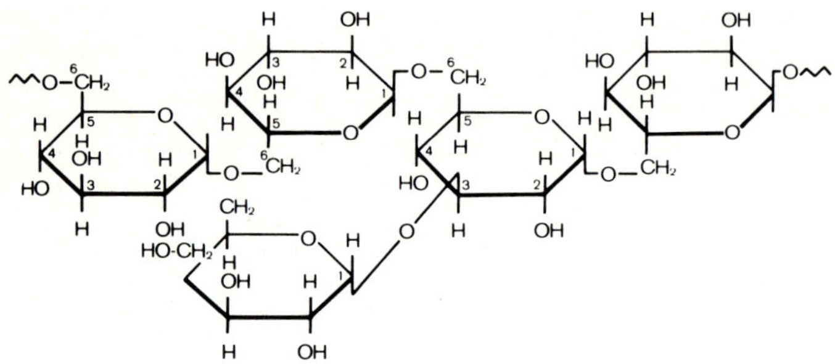

Dextran (α - (1→6) - and α - (1→3) -linked glucose units)

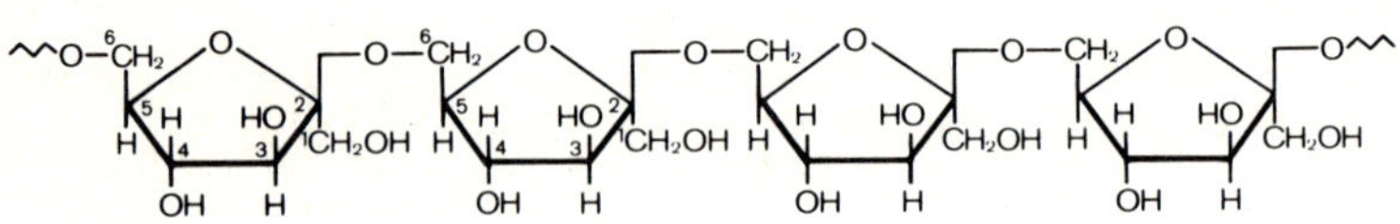

Levan (β - (2→6) - linked fructose units)

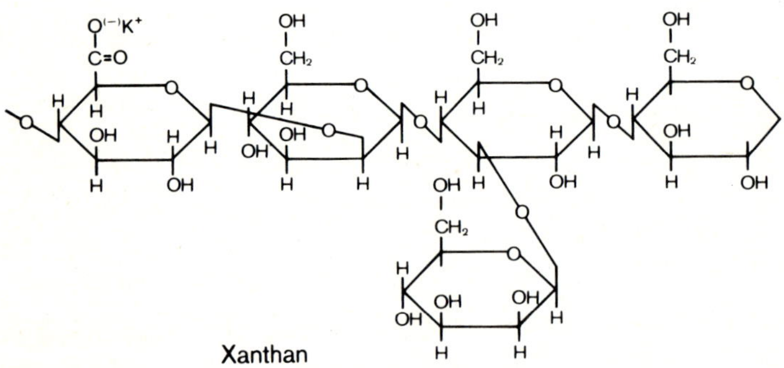

Xanthan

AUTHOR INDEX

The number(s) directly after the initial(s) of the author is (are) the page number(s) on which the author (or his work) is mentioned in the text. Numbers in italics give the page on which the complete reference is listed. Entries followed only by italic numbers refer to editors of multi-author works whose names do not appear in the text.

SUBJECT INDEX